Darwin's Walk & The Last Wave

Disappearing Landscapes, Declining Species

Richard Krooth

Hamilton Books

Lanham • Boulder • New York • Toronto • Plymouth, UK

4501 Forbes Boulevard, Suite 200, Lanham, Maryland 20706
Hamilton Books Acquisitions Department (301) 459-3366

Unit A, Whitacre Mews, 26-34 Stannary Street,
London SE11 4AB, United Kingdom

Printed in the United States of America
British Library Cataloguing in Publication Information Available

Library of Congress Control Number: 2017938901
ISBN: 978-0-7618-6922-1 (cloth : alk. paper)—ISBN: 978-0-7618-6923-8 (electronic)

∞™ The paper used in this publication meets the minimum requirements of American National Standard for Information Sciences Permanence of Paper for Printed Library Materials, ANSI/NISO Z39.48-1992.

“Why can’t we preserve and cherish the only home we’ve ever known?”
—Carl Sagan

“Greenhouse gases are accumulating in Earth’s atmosphere as a result of human activities, causing surface air temperatures and subsurface ocean temperatures to rise.”
—Ralph Cicerone, Chairman, Academy Panel,
commissioned by President George W. Bush

“We have assessed impacts as they are happening in natural and human systems on all continents and oceans. No one on this planet will be untouched by climate change.”
—Rajendra Pachauri, chair of the Intergovernmental Panel on Fourth
major assessment on global warming, March 31, 2014,
Climate Change (IPCC), jointly established by
the U.N. and the World Meteorological Organization.

The hourglass of Earth has tipped
The sands of time are running out

Contents

List of Figures

List of Tables

Preface

Can human technology swamp biogenesis, ending its reproduction as renewal or even its existence—and ours?

Many are not asking this or related questions. Instead many go on peddling greed, racism, sectarian hatred and other distractions in a society dulled by consumerism and preoccupied by the proliferation of empty choices and disinformation. Yet, no matter what, conscious or not, if we go on poisoning the Earth, it may take humankind down with it, as this study will point out.

Out in front of the Horsemen of the Apocalypse rides a paler horse and paler rider, Ecocide, in its train that fearful four of pestilence, war, starvation, and death. The leader is scarcely to be comprehended, a system of electro-mechanical parts connected and moved by also-dangerous sources of power, trampling down the vintage of a ravished and outraged Nature preparing for the last harvest. The dust thrown up by the riders is lethal heat and corrosive toxins, effusive gases and biospheric poison. Land, first, then sea and air have been soiled and overawed, seeded with effluents, whose result our planet does not need to harvest: they will harvest Earth's remains.

Inventors and investors, technicians and tinkerers blindly consider the devastation as progress, any problem denied or expected to be cured by more putrefactions of the same machine. Some say they must take part because in no other way can they retain self-respect, career pride, and earnings.

Naturalist thinkers like myself challenge those wedded to or at least embraced by technology, arguing that the discharges and excretions of the collection of machines endanger the environment and unravel the fabric of the biological domain. Indeed, the universe is an interlocked whole, working out a time cycle of thermodynamics involving Sun, Moon, Planet and Galaxy, together shaping the conditions essential for species evolution and development on Earth.

Under the reign of natural development, these gyres are expressed as processes and concepts, which this study will try to elaborate—including the Fibonacci sequence of species linkages, evolving, developing and expiring; the second law of thermodynamics, affecting energy inputs that are diminished by heat on a global scale, reducing work-output and bringing on entropy; and the advance of technologies using replaceable parts, driven by inanimate energy, that by friction and transmission directly or indirectly heat the planet and is leading to the ultimate destruction of species.

In the torrid zone, Planet Earth probably will not suddenly explode or expire, with extinctions of most species. But wrenching Nature may well annihilate millions of planetary forms of life and lay the foundation for the march of the surviving few to further evolve and develop as new forms most stunning, fit to reproduce.

Today entropic heat is building, enclosing Earth's sea, land and atmosphere in a nearly sealed sphere. It has set loose the specter of a world enflamed, pressing on species and their varieties lacking genetically inherited capability to resist, or even to build their ability to adapt—compromising their fitness to survive and reproduce.

Hope is fading. There is seemingly no stopping the advance of global heat under present relations of human societies that are driven to produce their own needs as well as accumulate wealth by powering machines using fossil fuels, replacing the power of the Sun and natural forces, as well as human labor or domesticated animals.

Behind this build-up of entropic heat is the back-story. For the forces driving thermodynamic technologies are built with replaceable parts sustaining machine production; they are fueled by inanimate energies that directly or indirectly compromise or destroy the biosphere and undermine the essential conditions supporting species life.

This study thus attempts to investigate these sources searing Earth and planetary life. We hope to illustrate how, through its own operation over time, the input of human, animal or inanimate energy for productive work leads to diminishing output, caused by the spread of heat to the coldest portions of a machine, to a vessel, to Earth itself—the loss of energy appearing as heat, exhaust, and waste. To elaborate the theories and methods used in the account you are about to read, I have relied on a number of great scholars and teachers.

- Astronomer Milutin Milankovich discovered cycles of 5,000 to 55,000 years, during which Hot House Earth ends and a new Ice Age begins sometime between these estimated points, transforming planetary conditions for species life.

- Twenty-three centuries ago, Aristotle taught us that species evolve, are interlocked, and use geographic corridors to spread their progeny.
- Fibonacci's time-based mathematical progression, applied to species evolution, reveals that they increase their biological complexity and pace of successful reproduction.
- Following Linnaeus's rules for classifying species, we learn that the most basic biological category is a group of living things that look alike and normally mate only with each other.
- Alexander von Humboldt's holistic concept of Cosmos, synthesizes the unity of Nature's forces.
- Charles Darwin's and Alfred Russell Wallace's catalogue of the struggle for existence within and among species, shows how this leads to their offensive and defensive transformations—evolving favorable variations, and modifying structural and organic specializations that tend to be preserved, unfavorable ones to be destroyed, forming new species through adaptation to niche environments improving their reproductive fitness.
- Explanations by Stephen Jay Gould, Richard Leakey, Roger Lewin, Richard Dawkins, Edward O. Wilson, Carl Zimmer, Richard Attenborough, and Australia's Richard Smith document evolution of sister species, each turning on genetic or other adaptations to environmental changes, laying the foundations for, and formation of, new species.
- Hugh Falconer's metaphoric predictions portray how future evolutionary speciation will be reconstructed with a different superstructural architecture.
- Lynn Margulis proves that a single species' endogenous interdependence on other species, creates a multi-structural web of mutual existence under the roof of a single, encapsulated species form.
- Fritjof Capra reveals exogenous species' interdependence, creating a web of life; and
- Sean Carroll's investigations of species adaptations to sudden environmental change by precise forces, if still partly unknown, turn on the biological functions of the structure of DNA's nucleotide bases, connecting the two backbones of the DNA's helical structure.

A detailed account of the trajectory of species evolution and development has been further explored and elaborated by hundreds of other colleagues, some mentioned in the acknowledgements.

Our goal is to reveal later evidence of how species survived or not under changes in the natural flows of energy and human schemes and inventions, that together alter environments, jeopardizing life diversities, and place the entire planet at risk. In other words, we seek a general theory explaining entropy and a possible disclimax on a global scale.

These are just a beginning. Entropy is the underlying danger to Planet Earth and its species today.

Richard Krooth
Berkeley, California
November, 2016

Acknowledgments

This book began with research in Australia, expanded its scope in the United States, then it went global.

I owe special thanks for the willingness of Australian and Western hardscrabble explorers and scholars to share their knowledge; for maintaining their commitment to secure wilderness areas, species diversity and botanical and zoological wonders; for their grand nature centers and libraries; especially for their timely writings that I have drawn upon; and for the energy some of them generously offered to me. For my brief study, the following people, information centers, and resources were essential.

- Staff, Melbourne Botanical Gardens
- Rangers, Blackhurst Nature Center, Blue Mountains, Australia
- Librarians, State Library of New South Wales
- Guides, Sydney Historical Museum
- Curators and Mentors, Sydney and Melbourne Zoos
- Personnel and guides, International & Australian Youth Hostels, Sydney, Blue Mountains, Melbourne and Phillips Island
- John Merson, head of the Institute of Environmental Studies at the University of New South Wales
- Microbiologist Ben Cebo, Melbourne, Australia
- Professors Joy and Denys Walter, St. Helens, Tasmania
- Film Director, Peter Weir, "The Last Wave" (Criterion Collection, 2001)
- David Karoly, Ph.D., ARC Federation Fellow and Professor, University of Melbourne, Australia
- Brandy Kuhl, Librarian, San Francisco Botanical Garden/Strybing Arboretum Society

- John Bruno, associate professor of marine sciences at the University of North Carolina, Chapel Hill
- Librarians, Harvard University, Museum of Comparative Zoology, Cambridge, Massachusetts
- Ray Bradley, Ph.D., Director of the Climate System Research Center, University of Massachusetts
- Michael Mann, Ph.D., Director, Earth System Science Center, Pennsylvania State University
- Jonathan Overpeck, Ph.D., Professor of Geosciences and Atmospheric Sciences, University of Arizona
- Ben Santer, Ph.D., Research Scientist, Lawrence Livermore National Laboratory
- Gavin Schmidt, Ph.D., Climate Scientist, NASA Goddard Institute for Space Studies
- Kevin Trenberth, Sc.D., Distinguished Senior Scientist, Climate Analysis Section, National Center for Atmospheric Research
- Scientists, researchers, professors and environmental writers: Charles Darwin, A. P. Elkin, A. W. Reed, Margaret Baker, Robin Corringham, Peter Singer, Tim Flannery, Fritjof Capra, E. O. Wilson, Lynn Margulis, Sean B. Carroll, Richard Leaky, Richard Dawkins, John Pernetta, Edward E. Berg, Hiroshi Fukurai, Morris Edelson, Ove Hoegh-Guldberg
- Special thanks for guidance in editing to Holly Buchanan, Assistant Acquisitions Editor, and Beverly Shellem, Senior Production Editor at Hamilton Books; to researcher, editor and photographer Ann Baxandall Krooth; to botanist Lana Martartella for manuscript layout and technical help; to Hideki Ito for constructing tables; and to Alex Chavez for computer oversight.

A mere list of individuals who rose on the tide of the environmental movement is only a partial picture of what they accomplished. Many pushed to expose the ecological and public health consequences of industrialization, focusing on pollution, exposure to toxic chemicals, and clear cutting of forests causing the loss of native species. Public spokesmen like Judge Russell E. Train helped build and transform the World Wildlife Fund into a global force for conservation. Such efforts led not only to the reorganization of U.S. national environmental priorities, but also to the 1961 founding of the African Wildlife Leadership Foundation, an international organization to train African communities in wildlife habitat and national park management.

As well, in the U.S. under Russell Train's conservation leadership, a new approach to economic growth turned on environmental values that needed to be incorporated into public and private decision-making about what to build and where to build it. With his prompting, professor Ian McHarg, of landscape architecture at the University of Pennsylvania, wrote the seminal 1969 book, "Design With Nature," incorporating the principles of ecology into landscape construction.

Even earlier, overlapping that moment, the environmental movement was off and running. Rachael Carson's 1962, *Silent Spring* documented jeopardy to species, including ourselves, from a toxic environment. Corroborated by the selfless efforts of political science professor Lynton Keith Caldwell, at Indiana University, the pathway led to the setting of federal standards for environmental impact statements before technological projects set their heel. And Ralph J. Cicerone, 2005-2016 president of the National Academy of Sciences was instrumental in promoting the global warming pacts. starting with the Kyoto Protocol.

There also remain many hundreds of accomplished ecologists and crusading environmentalists in Australia, New Zealand, Germany, Denmark and the Netherlands, to name a few. Following Petr Alekseevich Kropotkin's 19th century writing on evolution and ecology in Russia and Eastern Europe, as early as 1960, Vladimir Nobokov and George Shaw Wheeler led a movement to protect species diversity and make both nation-states and industry responsible for securing and maintaining environmental integrity and safety.

Beyond countries in our moment, the world has become a corridor for thousands of individuals to seek answers to environmental questions that gnaw at and crowd their thoughts. Among them, like all authors hurriedly or otherwise sending their insights into print and the ethernet, I need to release responsibility for this book's contents from the intellectual shoulders of my colleagues, my environmental aides, and my willing captives in the wilds.

Despite all those who graciously aided me in this fast-forward expedition, I alone am accountable for the information gathered and the conclusions drawn.

Richard Krooth
Berkeley, California
November 2016

Introduction

This study describes the reasons humankind may be facing its last moments on Planet Earth. A few among our species undoubtedly will hang on for decades, even centuries, living in dismal conditions of a pre-industrial world.

Evolution that brought our species onto the global stage may well bring on an iconic sea wave encompassing continents; then, at a far distant future time, gradually freeze and destroy most forms of now existing earthly life, to be replaced by evolved forms of surviving species. To detail the passages to such an end, this account elaborates on three stages of Earth's existence: its 4.8 billion-years relationship to the sun; a near-incomprehensible stretch of time making evolution possible; and, in our time, the many ways in which our species has overstepped the bounds of existence by burning fossil fuels, creating a toxic and gaseous blanket covering the globe—absorbing solar light and heat re-radiated from Earth's surface—ultimately pushing the thermometer above the maximum thermal limits that most species can long endure.

1. ANTHROPOCENTRIC ERA

By contrast with the millennial struggle of plants and animals to survive and evolve on a globe with limited means to sustain them, the present Anthropocentric Era produces vast surpluses, though at a heavy environmental cost. With the technological output of sustenance, resources and commodities has come lethal heat and perilous outpours of gases, moving towards overwhelming Earth's biosphere, climate, and the ability of the vast majority of life forms to modify and evolve their organic structures to survive and reproduce.

These moments have embodied at least three thresholds of Earth's existence. Periodic planetary temperatures, moving from an Age of Ice to an Age of Thermal Heat every 5,000-to-55,000 years, initially set the stages for evolution. These extended time cycles turned on Earth's relationship to the sun, either as the planet's circular orbit around the sun created an equable climate for evolution; or, as Earth's orbit became elliptical (oval-shaped) and brought on ice covering land and sea, altering the climate in which a relatively few species were able to geographically move, evolve, modify their organic structures, and reproduce. The repeat of these cycles encompassed the Holocene that already has lasted 200,000 years. Today Earth's circular orbit around the sun offers optimal temperatures and conditions for evolution, especially of humankind.

During the first half of these 200,000 years, humans developed languages, physical skills and hand-made tools to shape and craft essential natural resources. In the next 100,000 years, humankind developed written symbols and script recording events and ideas, as well as developed sophisticated tools using the physical power of animals and directing the forces of water and wind, and burning fossil fuels.

These developments were elaborated during the third, current moment, known as the Anthropocene, so far lasting a mere 260 years, overlapping industrial production. During this time, our species has invented complex technical means to produce, powered by burning fossil fuels, aerating their residues building up Earth's translucent, atmospheric cover of carbon dioxide, methane, and other greenhouse gases, as well as releasing acids and particle debris.

2. EARTH'S TEMPERATURE

Today the sun's direct energy that strikes Earth's surface has been re-radiated as light and heat to its atmospheric cover—sometimes stored as long as a century—with its potential for warming Earth generally calculated as carbon-dioxide parts per million of the atmosphere's air molecules.

As this store of heat accumulates, so too does Earth's temperature rise, already exceeding the maximum thermal tolerance of many evolved species.

This heat also is rapidly melting the remains of the last Ice Age glaciers, frozen polar seas, and permafrost frontiers. The melt is raising sea levels; washing over island and continental shorelines; invading both wild and agricultural areas; and reducing food supplies for the planet's ever-rising populations.

In these ways, our species' technological sphere invades the biosphere, jeopardizing environmental conditions essential for species reproductive fitness.

Humankind is bringing to a close the once equable climate of the Holocene, blocking unnumbered forms of evolution on Earth, possibly leading towards a distant end-point of mass species annihilation.

3. GAIA AS METAPHOR

Gaia as metaphor represents the force of a living world. She cycles Earth's molten rocks and lava, tectonic plates, land mass, seas, and atmosphere as if alive as a composite force—sustaining the *mélange* of species that once evolved, went extinct, or have continued to develop through natural selection. Today those alive appear as species communities, including the great number of the most viable members, flourishing best and rearing the greatest number of offspring.

Species evolution and development would have been, and continues to be, impossible without Gaia's natural cyclical flows of energy and an equitable climate on Earth. But now our species' greed-driven, anthropocentric power may one day set off violent storms and a last wave of sea that will snuff out the remaining lights of planetary existence.

If ever there was, or remains, a Gaian empathetic regard for the fate of those within a given species or among symbiotic species, it is steadily and unrelentingly being destroyed by the advance of technological relationships imposed on populations and nature during the Anthropocentric Age. This is largely because, quite blindly, humans have imprisoned themselves by developing production forces, using the technological creations of their own hands that they then worship as if an outside force, imposing on themselves a fetish to accumulate wealth. Within this technological sphere, class structures have evolved. Slavery, feudalism, marketplace relations, and colonialism have taken surplus-value from some to award others, allowing vast accumulations of wealth that, in our time, are placed in production remitting further financial wealth in a timed cycle that ignores the vast majority, Nature and the condition of other species. The door has been pried open for multiplying mechanical and technological methods unsettling the biosphere and destroying species diversity. Step-by-terrible-step the essential heartbeat of Nature has slipped from daily human concerns, leading to a deathly cascade of species so vast, so destructive, that the web of life itself is being torn.

4. PRODUCTION AND SOCIAL ORDER

As the forces of production have overwhelmed the biological sphere, more and more production on land and raiding the seas turns on burning fossil and nuclear fuels.

Added to the release of carbon dioxide, methane and fissionable radiation to the atmosphere are plant and animal transpiration as well as human-generated gases, heat, water vapor and particle matter.

Overloading the Storage Vaults for Effluents

Overloading the atmospheric storage vault of aerated carbons and other greenhouse gases absorbing heat re-radiated from the sun striking Earth's surface, the warmth of living plants and animals and more than seven billion people has been released to the air enlarging the discharges of perspiration and fluids.

With at least half of world population working at means of production and farming, their extended needs and invasive destruction of forests and savannah, inland waterways and seas, eliminate the storage sinks absorbing surplus carbon dioxide and other greenhouse gases—raising surface temperatures and undermining the maximum thermal temperatures human and other species are able to long endure.

5. THERMAL POWER AND DISORDER

If past millennial cycles are any guide, thermal disasters that boiled the oceans deep and cracked the surface of the planet 55 million years ago might recur during the present Anthroprocentic Era.

Evidence from the Palaeocene-Eocene geologic time-boundary is revealed in the Norwegian Sea. A vast natural, gas accumulation of fossil fuel—hydrocarbons consisting of methane gas—was buried in deep sea sediments. These icy methane concentrations, called clathrates, were ignited in a massive submarine explosion by super-heated volcanic rock magna that squeezed through the Earth's crust.

The remaining methane then combined with and eliminated seawater oxygen, forming carbon dioxide breaking the ocean surface and aerating, with any remaining carbon dioxide turning sea depths so acidic that most sea creatures went extinct. Possibly only new blooms of surface plankton reabsorbed the excess carbon over the following 20,000 years or so.

Details reveal that some 1,500 to 3,000 gigatons of carbon were released into the atmosphere, pushing up carbon concentrations from around 500 to 2,000 parts per million of air molecules over an eyelash of decades, possibly lasting even centuries.

Oceans turned acidic, leading to massive extinctions, threatening tiny marine foraminifera at the bottom of the food chain. Thereafter, it took millions of years for the evolution of new forms of foraminifera along coasts and in open seas.

On land, torrid heat incinerated plant and animal life, fueling an evapotranspiration crisis that then led to less rainfall, abrupt warming, and periodic migrations of any remaining Asian fauna over rough land-bridges into Europe and North America. Some flora also escaped arid conditions through birds carrying their pollens and defecating the remains of seeds in their gut. Seedlings and burs were possibly carried by furry animals and by air currents—throwing all newly-arrived species into a struggle with one another and existing species for survival to secure soils and nutrients. This encounter drove many ancient life forms and creatures to extinction.

6. INDUSTRY AND BURNING FOSSIL FUEL

Anthropogenic dangers to contemporary species might seem miniscule. But considering that during the surge of the English Industrial Revolution about 1775, and year 2010, atmospheric carbon dioxide alone rose from about 270 to 387 parts per million. By the late 20th century, the burning of fossil fuels yearly had already set free about 10 billion tons of carbon dioxide. Some 5 billion tons were recycled in soils, vegetation and ocean sinks—but the other 5 billion aerated each year already had cumulatively heated the planet for over 270 years impacting species survival.

Between the atmospheric layer surrounding Earth and terrain and the deep oceans below, plants and animals and sea life remained dependent on the cycling of water and oxygen, nitrogen, carbon dioxide, nitrous oxide, and phosphorous. All these needs came into the crossfire of significant environmental changes.

7. DECLINE: ATMOSPHERE, DESERTS, DELUGE

Humankind has not patiently, or most otherwise, accepted Nature's wonders. During the past 250 years or so, fossil fuels have been mobilized to drive machinery, aerating gases, debris and poisons, many becoming long-lasting components of cloud formation and atmosphere.

Geographic expansion of the industrial foundations for production also has impacted environmental conditions, altering flows and circulation of air, water and soils. On practically all continents, organized engines of war have brought further degradations of Nature. And most recently power plants have released nuclides and toxins globally, challenging species adaptation and ability to survive.

Some cite as causes the wealth accumulating drives of great enterprises, avariciously digging mines and extracting resources from land and sea. Others view the profit-quest as secondary to using the bounty of Nature to secure human needs. Still others blame population expansion and human spatial designs as infringing on the natural domain. And considering these forces imposing on the natural world, there is little doubt our species has developed a colossal blind spot that leaves little hope for securing unmolested future natural cycles.

Humankind also seems intent on falling over its own feet, walking over a proverbial cliff when it comes to calibrating the directions in which its own actions and technological advances may take them. As a species we either produce means of destruction with consequences our societies are unable to control, or our social drives overwhelm the existing means to sustain us. Either way entropy sets its heel, leading to local disorders and disasters, punctuated by social dystopias that rapidly spread to other regions and merge or otherwise consolidate into an overarching disclimax, jeopardizing species reproductive fitness.

The disclimax itself being the end of a biological community of plants, animals and fungi living in a sustainable state of being, the question remains: Can we restructure the pathway of the means of production and life when they overreach the natural order?

8. NATURAL CYCLES AND ENTROPY

Our species' rapid transformation of the climate is taking place at a rate 1,000 percent more quickly than any known previous changes, save the moments of the five great extinctions of the past.

Species on land and in the sea cannot keep pace with such a tempo by reproducing, evolving, adapting, or developing. Their survivability also turns on our species' myopia and purposeful ruination of wetlands and surrounding territories, the release of invasive species, and outright destruction of those life varieties considered expendable for our safety and welfare. Here we offer a brief Anthropocentric recount of one of the most egregious impositions on and invasions of the natural remains of the day.

Essence of Natural Interlocks of Atmosphere, Terrain and Sea

Most academic studies ignore the Gaian flows of energy that surge through the biosphere—the self-regulating and self-organizing interlock between the sun and moon and Earth's elemental surface and subsurface, atmosphere and oceans, together supporting the planet's species.

In essence, the Gaian loop links global hydrological systems to meteorological winds and gas releases and territorial conditions. Today these forces impact and are impacted by human activities disrupting the evolution of species diversities on Earth, threatening and destroying the conditions supporting and sustaining life.

The future looks increasingly grim as human anthropocentric releases and rising concentrations of greenhouse gases in the atmosphere are projected to increase temperatures, particularly in the Indian Ocean, and to accentuate the differential between temperatures in the North and South Atlantic. Seasonal rains also matter enormously to human food production, from the Indian subcontinent monsoon to the American Southwest. And evaporation from the seven seas determine precipitation intensity and its geographic reach.

Yet, a better understanding of the forces that will shape these features of the climate system in coming decades is a very difficult challenge given the many factors in play. And from the atmospheric, many start with looking to the condition of the seas.

9. DANCE OF PELAGIC DEATH

Sea life itself is precariously balanced, and today is being undermined by human overfishing, disposal of sewage and runoff of agricultural fertilizers reducing nutrient supplies, leading to bursts of algal blooms toxifying water in a dance of death.

Without phytoplankton, the entire aquatic life system is being brought to the verge of collapse, as phytoplankton is consumed by zooplankton that feeds the microcosm of sea life—up to the top predator white shark that has swum the oceans for over eleven million years, adults growing to 21.5 feet, weighing up to seven thousand pounds. The white shark's diet is vast, including unnumbered fish varieties, other sharks, seals, sea lions, porpoises, dolphins, skates, rays, sea turtles, mollusks, crustaceans, and seabirds.

So why worry if these apex predators are eliminated and no longer control the size, abundance and behavior of innumerable other species in the sea? Because much like other apex predators, the loss of white sharks reducing biodiversity might disrupt the entire ecosystem, switching to a completely different feeding chain among a lesser set of competitive species. Fishing fleets ensnarl these gigantic creatures, nonetheless; and their death stroke is the fate of whales of all description, up the feeding chain to krill, phytoplankton and all that swim.

10. BIOGEOGRAPHICAL DEEP

Human ignorance of places in the farthest seas and deepest oceans may be compared with myopic ancient Greek attachment to the fatherland as one's only world that a denizen could see from the Acropolis of his native city—the Greeks largely viewing their city state as the center of a ring of coasts around a sea of islands, not knowing a continental expanse as its motherland. Much later came the European discoveries of Columbus and Vasco da Gama that extended the geographical horizon without limit, bringing the world-sea into the same relation with continents as the universe of Copernican space held with Earth—making Western Europe a slender province in a gigantic whole of the planet.

The sea as center today similarly holds Earth's landmass as islands, with relatively little known of its many creatures and their life cycles. Still less is known about the ocean's geographic moats and mountains, ridges, vents and circulatory gyres supporting webs of existence and feeding chains, distributing its foods, chemistries and gases—and, with blind human intervention, releasing possible enzootic microbes and unknown toxins or venoms.

There are sea floors never viewed by our species, as well as ancient species that have thrived and evolved unseen by humankind.

11. BIOSPHERIC COLLAPSE

Human ignorance of global land connections to, and impact on and from the seas, also has blocked an understanding of the convergence of regional environmental disturbances and broader geographical disruptions, leading to overlapping entropies and a possible world-wide disclimax.

Some say the disruption of Earth will take hundreds of years, with periodic signs of impending death appearing for certain landscapes and among the largest species and their phylum. But it may happen more quickly in still unknown ways from forces directly causing global warming; or from advancing technologies endangering the biosphere; or possibly from changed environmental conditions undermining species survival; and from a cascade of species set in motion by still unforeseen forces blocking evolution and development of existing species life.

12. FORCES CAUSING GLOBAL WARMING

Though scientific and other opinions vary widely, the time clock for global warming is running out. By most calculations there is more than sufficient aeration of carbon dioxide, other greenhouse gases and particle matter to lead to a surge in global warming, the massive melt of glacial ice, the deepening of oceans, the inundation of land, and the reduction in species diversities.

Before our species took charge of other life varieties on the planet, over eons past extremes of bitter cold and lethal heat periodically had endangered life diversities. These extremes were relayed by global hydrological systems linked to atmospheric and territorial conditions in a great Gaian loop that affected, and still affects, the evolution and development of species life.

Accompanying overall global warming of some five degrees centigrade, Earth conditions sustaining surviving species lasted some 200,000 years.

Our goal is to reveal evidence of how species survived or not under changes in the natural flows of energy and human schemes and inventions, that together altered environments, jeopardizing life diversities, and placed the entire planet at risk. In other words, we seek a general theory explaining entropy and a possible disclimax on a global scale.

13. HOTHOUSE EARTH & EMERGENT DYSTOPIAS

Earth's lung is nearly out of breath. The exchange of energy and gases between living species and the biosphere is being blocked by anthropocentric forces, bringing on deathly downward cascades of species and potential extinction of whole phylum of life forms.

Scientists viewing the brave new world of peril meanwhile are in dispute over the precise causes of entropy, although many are innovating ways to save the world's more than seven billion humans, the hundreds of billions of micro-species and six-or-seven million larger species. It is a treacherous crossroads that Earth faces, leading to possible extinction.

The sun's radiant energy combined with anthropocentric misdeeds are well on the way to creating Hothouse Earth—consuming the planet's life configurations. To stop a massive break down, a disclimax threatening the natural world, humankind has rationalized, yet barely searched for possible ways to delay what seems possible.

Nature meanwhile is balanced on a razor's edge: a deadly cut in species diversity is already nearly unavoidable. To cope with and adapt to the rapid extension of torrid global climates, a possible reprieve might come from reflecting sunlight into space; or a renewed Ice Age, with the Earth cooling more rapidly than warming; and, in still unknown ways, restoring Gaia's natural cycles and patterns of species evolution and development.

14. DARWIN'S WALK AND THE LAST WAVE

The human learning curve is slow and often skewed. Today we may know Earth is facing a test of survival. But we continually put off reversing the course, leading to an eventual day of reckoning, watching one catastrophe after another, again resolving something must be done to stop global warming, species loss, and an ever-lethal atmosphere. Clearly our species has lost its bearings. Nation states and their diverse social classes are unable to respond to negative turning points towards entropy, criticality and possible disclimax.

The concept of Darwinian organic, evolving life is in conflict with the development of technology and the second law of thermodynamics, which reveals the relative decline of energy usefulness through the spread of wasted heat to the coolest parts of vessel Earth. This process also encapsulates the buildup of scorching heat, creating friction and waste, diminishing output returns from energy inputs produced by burning fossil fuels or deploying fissionable nuclear isotopes.

These parallel forces have clashed during the Anthropocentric Age, as organic life weakens while human technological output is diminished as escaping heat. When these two antipodes come into conflict, human inventions of new techno-thermal arsenals of production continually emit still more heat to Earth, interrupting or destroying the foundations for continued species' evolution and development—anticipating a possible future moment when humankind will destroy the biosphere and itself.

It is now fairly clear that humans have yet to learn these lessons from either Darwinian or thermodynamics theory, driving towards dystopia and a disastrous end to species diversity on Earth.

15. GAIA'S NATURAL CYCLES AS METAPHOR

Nature's cycles of terrain and subterrain, oceans and atmosphere, are being interrupted by humans severing the many strands of the web of life. Species are being threatened in many ways, as lethal anthropocentric temperatures melt glaciers causing the rise of oceans crashing over island and continental shorelines—their combined forces merging with the possibility of shifting tectonic plates, tsunamis, ARK storms sweeping oceans onto land, and acid rain poured on soil and seas, spreading acidity, destroying species. There are no limits to possible future dangers.

The gyre of the planet around its axis could become more than a wobble. As the weight and depth of deepening seas, moving around the Equator as a center, shift the centrifugal force thrusting outward towards Earth's perimeters, they could suddenly alter the axis—loosening world-girdling violent winds, uprooting terrain, changing land and ocean feeding chains, and bringing on a new Ice Age—replacing both moderate and Hot House conditions supporting existing species. This has happened before.

A deadly combination of ARK storms carrying huge volumes of sea water saturating continental coastlines and mountains might deluge entire regions, invading and drowning cities and farmland—to be quickly abandoned by a quarter of the world's population. At the same time, farmland bathed in brine, would produce little food and support no cattle. Even drying soil might temporarily draw sea water uphill through funnel-like capillary action.

Once glaciers melt and tundra dries at the North and South poles, as well in the northern hemisphere, the release of great volumes of methane could reduce the proportion of atmospheric oxygen below 17 percent, jeopardizing the lives of breathing animals. And aerated methane, combining with oxygen, also could enlarge the carbon dioxide blanket circling Earth, capturing heat re-radiated from the sun's rays reaching the planet.

If these forces converge on a torrid Earth, civil society would collapse. Surviving populations might move inland or upland, only to find rising temperatures, insufficient food and fresh water, and a desperate struggle for necessities. Malthusian predictions of spreading starvation, disease, pandemics, even warfare would become everyday realities.

The future remains unknown, of course; but there are questions that must be answered in constructing a reasonable view of global peril, bringing on a Last Wave covering most of Earth, undermining the biosphere, and causing the death perhaps of millions of species shattering diversity, a rare few surviving and evolving along Darwin's path.

I

House of Cards

Chapter One

Advancing Technosphere; Endangered Biosphere

> Arguing about whether or not today's extreme events are early indicators of climate change does nothing to advance the priority of global warming and other pressing environmental issues on our national policy agenda. The real significance of extreme events is as harbingers, not just of a changing climate but also of a changing world in human society and the infrastructure that supports it . . . becoming increasingly vulnerable to natural disasters. The mounting disruptions in their wake reveal the progressive deterioration of our planetary life-support systems. Extreme events are, indeed, teachable moments: "wake-up calls" that an environmental crisis of global proportions is imminent—much more so than the subtle and sometimes ambiguous early warning signs of global warming might lead us to believe.
> Prof. John Mitchell Wallace, 2012[1]

Joe Romm, one of the most significant environmentalists in the U.S., says,"If Earth were an engine ceaselessly burning fossil and isotope fuels, firing its boilers and nuclear power plants to produce energy and release lethal heat to its outermost, cooler peripheries, it eventually would reach a critical state extinguishing almost all earthly forms of life."[2]

In control of fire-breathing engines, humans certainly could turn them off before such a point was reached. But what if they did not, or refused to? What if they lacked foresight during the process of burning energy and drawing on resources in order to produce food and goods, accumulate wealth, and provide for expanding populations?

Experiencing the greatest migrations ever seen on the planet, what if those in control still could not provide for the billions of migrants without an Earth two-and-a-half times larger? And to answer their needs, what would happen if populations emigrated to resource-rich nations creating an imbalance between people, usable materials, food and exploitable energy, thereby generating massive discontent and intense competition to secure the means of life by degrading one another and appropriating parts of the natural environment essentials for their own survival?

It is conceivable that the expanded technological sphere would be pushed to the limits where it was no longer able to secure the needs of populations, unraveling the biosphere, accompanied by massive environmental disorders. And what would follow if existing states and institutions lacked the means to control enlarging populations on the many fronts of an ongoing civil war for survival . . . with still unknown consequences?

LONG-TERM CYCLES RULE WORLD TEMPERATURES AND CLIMATES

In the long stretch of 5,000 to 50,000 years from now, Hot House Earth will end and a new Ice Age will begin. According to the now-proven Milankovich cycle, the next glaciation will begin sometime between these estimated points.[3] This theory has been verified by sedimentary studies. It is based on the interaction between the Earth and Sun, known as *orbital variation*, that changes and has changed throughout the 4.8 billion year history of our planet. This sets the stage for periodic planetary temperatures moving from an Age of Ice to an Age of Thermal Heat every 5,000-to-55,000 years or so.

The dynamic gyre of Earth's orbit around the Sun appears either as circular, creating an equable climate for evolution; or, as elliptical (oval-shaped), bringing on ice covering land and sea, altering the climate in which only a few species are able to evolve and reproduce. This gyre has absolutely nothing to do with the amount of carbon dioxide, methane, water vapor, or any other greenhouse gas in the atmosphere.

Rather the switch of an Earth covered by ice or thermal heat turns on ecentricity, that refers to the change in the shape of the Earth's orbit around the Sun. Currently, our planet's orbit is *almost a perfect circle*. There is only about a 3% difference in distance between the time when we are closest to the Sun (perihelion) and the time when we are farthest from the Sun (aphelion). Over a 95,000 year cycle, however, the Earth's orbit around the Sun changes several times from a thin ellipse (oval) to a circle and back again. So within the 95,000 year cycle, there could be many 5,000-to-55,000 years of pathways covering the Age of Ice to an Age of Thermal Heat.

When the orbit around the Sun is at the most elliptical point, there is larger difference in the distance between the Earth and Sun at aphelion, farthest from the Sun, than at perihelion closest to the Sun. Though the current three million mile difference in distance doesn't much change the amount of solar energy Earth receives, a larger difference would modify the amount of solar energy received and would make perihelion, closest to the Sun, a much warmer time of the year than aphelion, furthest from the Sun.

CHANGING ANNUAL ANGLE OF THE SUN STRIKING EARTH

Earth not only changes its orbit around the sun from circular to elliptical. It also rotates on an unseen axis that changes its face presented to the sun. As the ancient Grecian mathematicians knew, klinein, or "slope", described how the angle of the sun's rays annually hitting Earth varied in different regions at different times of the year.

Aristotle had proved the world was round, with a southern and a northern sphere, rotating on a presumed axis presenting different faces of Earth towards the Sun, varying by season and geography. Perihelion would occur on January 3rd, when Earth is 91.4 million miles away from the Sun; and at that point the Sun strikes Earth at an acute angle in the northern hemisphere and is directly overhead in the southern hemisphere. At aphelion, July 4th, our planet is 94.5 million miles from the Sun, when the Sun is at its highest moment overhead as seen from Earth in the northern hemisphere, and at an acute slope in the southern hemisphere. This *annual* change in the slope of the sun was different, though, than the *longterm* change of Earth's orbit around the sun from circular to elliptical.

EARTH'S ORBIT CHANGES FROM CIRCULAR TO ELLIPTICAL ORBIT, AFFECTING CLIMATE

At the end of the present cycle of Earth's circular orbit, when it is replaced by an egg-shaped one, a new Ice Age will inevitably return, though to verify the consequences there may be no humans or most other now-known species on the planet. In the interim, scientists will heatedly debate their theories about the scales of greenhouse gases in the atmosphere bringing on premature global warming, causing seas to rise, landscapes to disappear, and species to disappear.

CLIMATE REVERSALS

The end of the last Ice Age, about 165,000 years ago, initially led to a warming era—then gradual cooling, passing the critical mark of 100,000 years ago, when recorded carbon dioxide levels plummeted—once again spiking upwards about 11,000 to 12,000 years ago, approximating the domestication of food production throughout not only Asia, but also the southern reaches of America. Here the first Clovis and other Asian peoples arrived between 15,000 and 20,000 years ago, carrying the mitochondrial genetic tracers of a common African mother, Eve.

During climate reversal, as the orbit of Earth *vis-à-vis* the Sun changed, the spread of ice appeared, as in Africa until about 72,000 years ago. As the ice froze water sources, many Africans already had fled the continent's arid interior, heading for the more benign southern African shore. Here access to fresh water, seafood and more plentiful plant and animal resources may have led to an increase in the population and encouraged technological advances. Escaping to the fringes of an ice covered Earth, similar population movements and survival techniques likely developed in the Americas and Northern Australia.[4]

EFFECTS OF CLIMATE CHANGE

Climate change was gradual, not a sudden burst from ice to warmth. Even today, the remains of the last Ice Age crown the world's melting polar caps and glaciers, signifying the beginning and possible end of an intermediate term, named the Holocene. So far this passage has lasted 200,000 years, sustaining our human forebears and present day civilization with an equable climate at livable temperatures, allowing initial human development using hand-tools and later sophisticated technologies.

During the first 100,000 years, humankind developed languages, had no script to record events, yet advanced their social relationships by developing totems representing family lineages, sacred emblems and rituals—finding in natural species and phenomena, as well as in symbolic objects of mythological import—sacramental bonds between themselves and nature, between past and present, and between the unseen and the seen. Over the next 100,000 years or so, humankind developed written script recording events, ideas, and the development of sophisticated tools, using the physical power of animals and the inanimate forces of water, wind, and fossil fuels.[5]

The question humankind then faced, and continues to ponder, was what actions to take at the final close of the present Ice Age, as we burn fossil fuels and raise temperatures beyond the thermal maximum limits that human and other species can bear without extreme stress, or ultimately expiring—sweeping the planet of most forms of life. Will there be no human changeover from burning wood and peat, coal, oil and gas? Will relief from Anthropoentric Hot House Earth only come when the reign of fossil fuel supply and flaming period have ended?

CONUNDRUMS AND CONTRADICTIONS PERSIST

In the last 500 million years, Earth has been through four Hot House periods, probably controlled by variations in cosmic ray flux, as the Solar System passes through the arms of the galaxy. Atmospheric CO_2 is estimated to have been much higher than during the present, for example, in the 4,000-8,000 parts per million range during the period 400-600 million years ago. And yet, these Hot House periods have virtually no direct connection with high CO_2 periods, as what counted was and remains Earth's orbital or eliptical shape and the flux of cosmic rays.

Sea level also varied greatly through geologic time, 120 meters lower only 18,000 years ago, but 250 meters higher at least going back to the Cretaceous Era. These depths, hothouse moments and melting glaciers, were shaped by the orbit of Earth vis-à-vis the Sun, not by the connection between sea depth and atmospheric carbon dioxode.

THREE PERIODS: ETHNOCENE, HOLOCENE, AND HUMAN-DRIVEN ANTHROPOCENE

At least three periods have punctuated the biogeography of Earth: the Ethnocene era that began with Earth's existence 4.8 billion years ago; the Holocene that began about 165 million years ago during the still-continuing transition from the Ice Age to Hothouse Earth; and the Anthropocene that began about 264 years ago build-up by the human technological sphere and the burning of fossil fuels. The final end of the Ice Age under the present Anthropocene may see the ultimate destruction of essential conditions for most forms of planetary species. Atmospheric heat and carbon dioxide have risen since about 1750, a bare moment of the human-driven technological sphere, variously made up of burning fossil and nuclear fuels, mobilizing machines and raw materials, producing unlimited varieties of manufactures, and releasing aerated gases, heat, putrefactions and waste.

The ever-enlarging and intensive range of such operations have had negative consequences, including environmental malaise and irradiation; extended human alienation of humankind from one another and Nature; and annihilation of species and their phylum. Overall these forces have brought on deep climate disorders.

Today the biosphere has been so disrupted that it not only faces regional crises, but such unrest is also so fundamental that its expansion from local and provincial spheres to continental and global ones have brought on overlapping, serial entropies threatening a distant, future planetary collapse.

ANTHROPOCENTRISM AND THE GEOGRAPHIC TRANSFER OF HEAT AND ENERGY

Humankind is not dealing with Einstein's or even Gaia's unbounded space, but rather with three-dimensional parameters that are closing down under pressure of near-blinded, anthropocentric human plans, actions and measures. The spread and threat of industrialization, burning fossil fuels. releasing lethal gases and acids have been aerated over more atmospheric cover. Temperatures are escalating, often warming winters even more than summers. Atmospheric fallout and blight are spreading as much at night as during day. Lethally heated air currents have brought on more warming over the land surface than over the seas. Oceans are being acidified, killing phytoplankton and animal life that live by the photonic chain of the Sun's energy.

More heat buildup over both the North and South polar caps than at the Equator has led to massive melting. As well, the heat has raised Atlantic wind temperatures and rain patterns; caused a fall in the Arctic cover of snow and ice; melted permafrost releasing methane into the atmosphere, combining it with oxygen that not only absorbs the Sun's radiation re-radiated from Earth, but also changes the air's oxygen balance that has been required for breathing creatures over many millions of years. In fact, additional CO_2 is produced by the ultimate oxidation of methane, though during its short lifetime as methane over a few years has a much greater (23-to-35-times) atmospheric warming impact than does carbon dioxide.

REACH OF GLOBAL WARMING AND DUST

Global warming is real, massive, and extends its reach over land and seas in unpredictable ways. Small air and dust swirls that begin off the African west coast, for example, packed with sand and minerals, suddenly produce lightening-like-friction and extreme energy; and drawing up warm ocean vapor, drive storms that sweep across vast fronts with unpredictable frequency, ferocity and magnitude, drowning sea vessels, pounding and eviscerating islands in their path.

Viewing a cross-section through a hurricane reveals its dynamics. A typical hurricane is some 1,000 miles (1,500 kilometers) across, consisting of bands of dense cumulus-type clouds arranged in spirals around the eye of a calm region of low-pressure. (The lower the air pressures at the center of the eye, the higher the wind speed.) Within these bands warmed sea air rises quickly as up-currents (cooler air sinking into the eye), drawing in winds that may gust at up to 220 m.p.h. (360 k.p.h.), carrying extremely dense, torrential rain and debris.[6]

Aerial super-storms on an ocean-basin-scale are based on the exchange of air molecules and ocean moisture, creating friction and heat, seen most clearly in the tropical regions of the Atlantic.

In these tropical areas, where the surface of the sea often exceeds 80 degrees Fahrenheit (27 degrees Centigrade), heat and water vapor are discharged into the atmosphere. As the air is warmed, it expands, becomes less dense, and rises rapidly, forming a spiral pattern; and as the warm, moisture-laden air rises and further expands, the heat from vapor condensation releases energy and the speed of the wind increases.[7]

As tropical cyclones produce extremely powerful winds and torrential rain, they also produce high waves, damaging storm surge, and tornadoes. But thought they develop over large bodies of warm water, they lose their strength once they move over land due to *increased surface friction* and *loss of the warm ocean as an energy source*. This is why coastal regions can receive significant damage from a tropical cyclone, while inland regions are relatively safe from receiving strong winds. Heavy rains, however, can produce significant flooding inland, and storm surges can produce extensive coastal flooding up to 40 kilometers (25 miles) from the coastline.

Although their effects on populations can be devastating, tropical cyclones can relieve drought conditions. They also *carry heat energy away from the tropics and transport it toward temperate latitudes*, which makes them an important part of global atmospheric circulation. As a result, tropical cyclones help to maintain equilibrium in the Earth's troposphere, as well as relatively stable and warm temperatures worldwide.[8]

UPSETTING BIOLOGICAL EQUALIBRIUM

Such equilibrium, however, has been upset during the present era of massive burning of fossil fuels, raising the planet's temperature of land, sea and atmosphere. As the human technological sphere dries terrain, that winds then loosen, for example, it intensifies the atmospheric cycling of both soils and heat. Especially in African and Asian deserts, dust swirls create small wind-like twisters, that grow into mini-cyclones, then powerful hurricanes carrying the agglomerated soil and dust mix across the Atlantic.

In fact, of Earth's discharge of an estimated two billion metric tons of dust each year, more than half comes from African deserts and dry-lands on a 6,400-kilometer ride on the Atlantic's western trade winds, much of it carried over the southern and eastern U.S. every summer, responsible for 75-to-80 percent of the dust that falls over Florida. Yearly an estimated 40 billion tons of dust loaded with life-sustaining minerals, including iron and phosphorous, also are dumped into the water-column of the sea, and on the carpet fertilizing Amazon's massive rain forest.[9]

HURRICANE FORCE, DESTRUCTION, NO RELIEF IN SIGHT

Moving from the western African coast heading towards the Americas, this trans-Atlantic transfer of soils also generates enormous friction, producing heat that then interlocks with and exacerbates the power and speed of hurricane winds. Merging these forces, as evaporation from heated ocean water comes in contact with hot atmospheric vapor and dust, it increases the thrust of winds and soils. The speeding, turbulent air quickly mixes with the heat-fed waters of the Gulf of Mexico and the Atlantic Ocean, generating atmospheric waves of electrical energy and water vapor. Hurricane winds then pressure tidal flows, creating giant surges of surf lashing coastlines, merging with on-land weather-systems of rain, snow and sleet—becoming a super-storm peeling off soil cover, trees and what passes as the fabric of a vaunted, civilized society.

As one seasonal storm exceeded another in 2012, Hurricane Sandy lashed the Bahamas, Cuba and parts of Florida, covered parts of New York City in 14-foot waves, moving north to devastate New Jersey, then charged inland into the Great Lakes region. The thrust in New Jersey, where the storm roared ashore with hurricane force, downed trees, tore off rooftops, flooded building and homes with backed-up sewage, and piled-up cars as if falling leaves. Police and fire officials, some with their own departments flooded, fanned out to rescue hundreds. Unknown numbers drowned or were killed by falling trees. "We are in the midst of urban search and rescue. Our teams are moving as fast as they can," Governor Chris Christie said. "The devastation on the Jersey Shore is some of the worst we've ever seen. The cost of the storm is incalculable at this point." Most homeless were saved, yet suffered—in tent-camp imprisonment.[10]

At least 7.4 million people across the East were without electricity. Airlines canceled more than 12,000 flights. Urban life came to a halt for days. Losses mounted, capped by widespread pollution of the Hudson River and New York Harbor by a variety of toxic chemicals, including petroleum and fluids from cars and boats; contaminants from flooded subways, roads, parking lots and tunnels; and effluents washed from shoreline industrial sites, as well as from commercial and residential buildings.

The nation's oldest nuclear plant declared an alert; and a second plant just 40 miles from New York City was forced to shut down power. Five different nuke plants in Hurricane Sandy's path faced disruptions during the storm. As the Hurricane closed in on the Northeast, it converged with tidal flows and a cold-weather system that turned it into a monstrous hybrid of rain, high wind, and even snow in West Virginia and other mountainous areas inland. As a measure of how big the storm was, high winds spinning off the edge of Sandy clobbered the Cleveland area, uprooting trees, cutting power to hundreds of thousands, closing schools and flooding major roads along Lake

Erie. Hundreds of miles from the storm's center, gusts topping 60 m.p.h. prompted officials to close the port of Portland, Maine, and rerouted several cruise ships.

Civilized populations and their infrastructures cannot withstand *the forces of Nature put in motion by ourselves*. Sandy's charge to Nature, in lives, anxiety, property and just places to live, work and bear their kin would remain unknown. Early estimates had Sandy costing New York City $60 billion in property damage. People in the City were terrified, especially in public housing, where nights without lighting spelled dangers on the street and in dark, high-rise hallways. An estimated 400,000 New Yorkers live in public housing and many of their institutional, brown brick buildings hug the waterfront, where 14-foot waves rushed into streets and foyers. Days after the super- storm, some 227 of the 2,600 buildings operated by the New York City Housing Authority remained without power, including many in low-lying neighborhoods like Coney Island in Brooklyn, Rockaway Beach in Queens, and Alphabet City in Manhattan, the areas most seriously affected by the storm.

There would be no immediate or complete relief for them, nor long-term cease to global warming, as the directors and scientists of gigantic corporations extended their plans to draw fossil fuels from Earth, surely knowing the high environmental costs of burning these loadstones. Into the future, they would continue to strip-mine coal, extract tar sands, shatter geologic shale for gas, and drill for oil.

CAUSAL FORCES; FETISHES ABOUND; A RISING THERMOMETER

The causal forces of Sandy and Earth's hot-house-blanket-cover reside in the deeply-enmeshed, technical sphere, where it is virtually impossible to separate the multiple, overlying layers of extraction of fossil and nuclear fuels from the delivery systems that bring them to machines, then burn some of them to produce energy in order to extract and then craft raw materials being processed. Not only is this energy burned as fuel exuding gases, but the productive output of materials and manufactures is accompanied by ever more friction, heat and waste poured into the natural domain.

From burning extracted fuels to the appearance of finished commodities, these collectively become the fetishes that, essential or not, seem to sustain our species pride of progress and source of advancement. Yet, as the material base putting the technological sphere in motion, such forces drive up the global thermometer, emasculate, compromise, jeopardize, disassemble, and often destroy parts of the biosphere making up the interlocked integrity of

Nature's natural circulation of energy supporting species on a planetary scale.

Nor is this an exaggeration of human activity driving up global warming. As early as1981, James Hansen of the National Aeronautics and Space Administration Goddard Institute for Space Studies and Columbia University Earth Institute, had *publicly* warned that:

> Political and economic forces affecting energy use and fuel choice make it unlikely that the CO_2 issue will have a major impact on energy policies until convincing observations of the global warming are in hand. In light of historical evidence that it takes several decades to complete a major change in fuel use, this makes large climate change almost inevitable. However, the degree of warming will depend strongly on the energy growth rate and choice of fuels for the next century. Thus, CO_2 effects on climate may make full exploitation of coal resources undesirable. An appropriate strategy may be to encourage energy conservation and develop alternative energy sources, while using fossil fuels as necessary during the next few decades.
>
> The climate change induced by anthropogenic release of CO_2 is likely to be the most fascinating global geophysical experiment that man will ever conduct. The scientific task is to help determine the nature of future climatic effects as early as possible. The required efforts in global observations and climate analysis are challenging, but the benefits from improved understanding of climate will surely warrant the work invested.

EARTH'S ENERGY IMBALANCE

Throughout a six-year period covering 2005-2010 at the scientific center at NASA Goddard Institute for Space Studies, James Hansen's elite team concluded that Earth had absorbed more than half (58 Watts) the solar energy per square meter than it let off. The imbalance of excess energy was more than twice as much as the reduction in the amount of Sun-supplied energy to the planet between maximum and minimum solar activity (0.25 Watts per square meter).

The imbalance strongly implied that atmospheric carbon dioxide levels, re-absorbing solar radiation striking Earth, needed to be reduced to about 350 parts per million to restore the energy balance to equilibrium. "The fact that we still see a positive imbalance despite the prolonged 2006-2010 solar minimum [of emitted radiation reaching Earth] isn't a surprise given what we've learned about the climate system, but it's worth noting because this provides unequivocal evidence that the Sun is not the dominant driver of global warming," Hansen concluded.[11]

Joe Romm of *NationofChange*, in 2012 reaffirmed:

> Pinpointing the magnitude of Earth's energy imbalance is fundamental to climate science because it offers a direct measure of the state of the climate.

> Energy imbalance calculations also serve as the foundation for projections of future climate change. If the imbalance is positive and more energy enters the system than exits, Earth grows warmer. If the imbalance is negative, the planet grows cooler. The most recent measurements show that carbon dioxide levels are currently 392 parts per million and scientists expect that concentration to continue to rise in the future.

And in the wake of Hurricane Sandy, on October 30, 2012, renowned U.S. linguistics professor George Lakoff publicly warned the worldwide web that:

> Global warming is real, and it is here. It is causing—yes, causing—death, destruction, and vast economic loss. And the causal effects are getting greater with time. We cannot merely adapt to it. The costs are incalculable. What we are facing is huge. Each day, the amount of extra energy accumulating via the heating of the Earth is the equivalent of 400,000 Hiroshima atomic bombs. Each day!

Yet to be scientifically proven are Lakoff's assertions that:

> Because the Earth itself is so huge, this energy is distributed over the Earth in a way that is not immediately perceptible by our bodies—only a fraction of a degree each day. But the accumulation of total heat energy over the Earth is increasing at an astronomical rate, even though the temperature numbers look small locally—0.8 degrees Celsius so far. If we hit 2.0 degrees Celsius, as we may before long, the Earth—and the living things on it—will not recover. Because of ice melt, the level of the oceans will rise 45 feet, while huge storms, fires, and droughts get worse each year. The international consensus is that by 2.0 degrees Celsius, all civilization would be threatened if not destroyed.[12]

Bloomberg's *Business Week* was more pointed, its November 6, 2012, masthead politicizing: "IT'S GLOBAL WARMING, STUPID."

ROLLING CLIMATE DICE

The fallout was immediately apparent. Already woodland, then the grass, soybeans, corn and cows were withering under rays of an unremitting Sun, dehydrating their essences in the spring and summer heat of 2012.

The season had melded into climate warming, metaphorically loading the planet's throw of dice. As the 2012 summer heat exceeded usual climate patterns and hot extremes, the Torrid Zone covered an enlarging portion of Earth's surface. Daily weather patterns had become inextricably linked to human-caused climate disorders, throwing two-thirds of the U.S. into extreme drought, spiking crop destruction, and sending heat storms into the lower 48 states. Breaking all-time heat and drought records, Earth's vessel then distributed lethal warmth to its diminishing cooler regions.

In his 2112 "Public Perception of Climate Change and the New Climate Dice," James Hansen again described this chance of unusually warm or cool seasons relative to climatology. The dice, he said, have become progressively "loaded" in the past 30 years, coincident with rapid global warming. The facts supported such a metaphor, he argued, as:

- The distribution of seasonal mean temperature anomalies had shifted toward higher temperatures; while the range of anomalies also had increased.
- As important was the emergence of a category of summer time extremely hot outlying areas (more than three standard deviations warmer than climatology would predict).
- And most importantly, this extreme which once covered much less than 1% of Earth's surface in traditional land areas, now typically covered about 10% of the planet.

So it followed that he could reaffirm George Lakoff, that extreme heat anomalies such as those in Texas and Oklahoma in 2011 and Moscow in 2010 were a consequence of global warming, and that their likelihood in the absence of global warming was exceedingly small.[13] This was no mere anecdotal evidence, but a sign of unremitting global warming. In March 2012, the U.N. special report of the Intergovernmental Panel on Climate Change corroborated that the world was in the midst of "unprecedented extreme weather and climate events"—undeniably linking everyday weather patterns with human-caused climate change.

During 2012, the forecast proved out almost immediately, experiencing the warmest winter and spring ever recorded in the U.S. history, as well as in the first 10 days of summer. Throughout June and July 113 million Americans were living under an extreme heat advisory.

THE VESSEL WE CALL EARTH

Extreme heat thereby had set a dominant trend line in 2012, linked to previous years' experience. Though a country the size of the U.S. is almost never, everywhere, hot all at once, the most significant thing about the monthly report of the National Climatic Data Center was that so many hot all-time records for heat were being broken. So how hot it was on a given day was not as important as the historical records consistently being broken, indicating a longer-term trend.

For the first six months of 2012, the average temperature across the contiguous United States was the warmest on record dating back to 1895. The average 2012 temperature January-June 30th was 57.4 degrees Fahrenheit—4.5 degrees higher than the long-term average for the same period. In fact, the percentage of the lower 48 states experiencing drought increased from 37 percent (representing the long-term average) to 56 percent in the first half of 2012—the largest area of drought seen so far in the 21st century.

So what had occurred in the various regions indicated daily heat records were part of a regional or global thermal trend. To be more accurate: the second law of thermodynamics was proving that the vessel we call Earth is distributing lethal heat to its cooler regions; and that a Sun-splashed planet was linking natural cycles of heat to the anthropocentric ones we humans have ignited.

Not everywhere was hot in the lower 48-states. According to Kevin Trenberth, a climate analyst at the National Center for Atmospheric Research in Boulder, Colorado (a federally-funded research and development center), said the first half of the year 2012 found Washington, Oregon and Minnesota some of the wettest and coolest on record. But don't take that as a saving grace, as:

(1) Most of the overall torrid heat occurred east of the Rocky Mountains, where it mixed with extremely dry weather; and the Central Plains, Ohio Valley, and Mid-Atlantic states were drier than average, making it their warmest year on record.

(2) The southeast, which some climate data has shown is warming more slowly than the rest of the country, had a very warm June as well. While the area started off cool, by the end of June, South Carolina and Georgia recorded what might be their hottest statewide temperatures to date, with daily highs respectively reaching 113 and 112 degrees. These kinds of records are usually broken in July and August and not in June—important indicators of overall climate trends.

(3) Colorado, where then-recent wildfires consumed forests and homes, extended this area of drought. Colorado's temperature in June was 6.4 degrees higher than its historical average. The heat in June followed an extremely dry winter in the state; and during 2012 winter, Colorado's mountains had the lowest snow pack in the last 40 years, snow that disappeared quickly in spring and an early summer. So by the beginning of June, there was no snow remaining in the Colorado Rocky Mountains.

THE TINDER BOX EXPLODES

A tinder box had been set in place in Cascade, Colorado, ready to burn 40 million surrounding acres of ghost forest—trees still standing, killed by an infestation of bark beetles that led to an epidemic as they metastasized in the warmer winters. Ecologist Edward Berg had already documented and warned of such infestation and burn potential in Alaska's Kenai Peninsula.[14]

A 2007 hot spell had set the stage for the first fire on the North Slope's tundra in 7,000 years, accelerating permafrost melting and its carbon emissions. With warmer 2011-winter temperatures in the Western United States and Canada, an epidemic of mountain pine beetles also killed hundred of thousands of hectares of trees, turning forests from carbon sinks absorbing CO_2 into carbon releases as dead trees decomposed and potentially would go up in flames from lightening and human match strikes. The danger was far more extreme in the West, where overbuilding in vulnerable habitat had situated 40 percent of the new homes over the last two decades. And burn the dead forests and new homes did, as a fire storm reeled down the mountain at Waldo Canyon, flaming debris shooting from the crown of forest canopy, tearing through half-million-dollar houses near the Garden of the Gods—forcing 32,000 people out of their homes—the most destructive wildfire in Colorado history.[15]

Siberia, too, is warming under direct solar hear, beginning to transform vast forests of dying larches into grounds for spruce and fir woodlands, raising the heat released to the atmosphere. The transformation is seen in winter when larches drop their leaves allowing the Sun's heat to reflect off the snow and return to space, but now potentially raising the planetary thermometer by 1.5 degrees Centigrade. Not so for spruces and fir that keep their winter needles, absorbing solar heat before it can reach the snow reflecting the heat to the outer heavens.[16]

The climate warming, bark beetles killing trees, forest fires threatening, continued in California during 2016.

JUMP INTO THE FUTURE: SPECULATION AND CONJECTURE

Turning Earth into a hothouse will undeniably spread firestorms, create desert-like conditions and, reaching their maximum thermal limits, spell the death of many forms of species life. It remains to be seen if this will take place gradually over the space of years or decades, even centuries—or things will worsen suddenly next summer.

And what will happen if our species cannot resist burning fossil fuel until atmospheric carbon dioxide levels exceed 450 p.p.m. and Earth's temperature rises another two degrees Centigrade? Would most earthly species and their phylum be destroyed as their respective maximum thermal limits are exceeded?

If the vast majority of people became aware of their destiny, facing extinction, and took charge of and stopped burning fossil fuel—or all exploitable fossil fuel was already burned and exhausted—would they be prepared for the return of an Ice Age in approximately 5,000 to 55,000 years? Would there be sufficient time and geographic space for say 10 billion people to retreat to ocean basins as the ice covered most land and froze seas—as indigenous people once did in Southern Africa 72,000 years ago and as did those who crossed the Timor Trough from New Guinea to northern Australia some 50,000 to 100,000 years ago and lived in caves among giant estuarine alligators?

Future speculation and conjecture also are based on working out the second law of thermodynamics. As a global heat engine, Earth has absorbed the Sun's heat rays, the most intense re-radiated into the atmosphere vapor blanket around the Equatorial belt. As nights come on, the rising air has cooled, condensing aerated water vapor to form falling rains. Plant life then explodes, feeding the regeneration of evolving omnivorous and carnivorous life. But this framework of the increasing torrid atmosphere was about to change, no longer confined to the tropics.

Under the onslaught of carbon dioxide, methane and other greenhouse gases invading the atmosphere, absorbing the Sun's heat re-radiation from Earth, deserts that once encircled Earth at 30 degrees latitudes North and South began to extend their reach towards the Polar Regions. Fossil fuel releases long had over-heated skies. And the cycle of supercharged hot air had pressed into the upper atmosphere, fostering a burst of pressure spreading dry hot air flows towards Earth's Polar Regions and releasing some irradiant heat into space.

As the air cooled it fell through the atmosphere and cut a gigantic hole without moisture-laden clouds to block the Sun's rays from again reaching Earth at ground-zero. This further dehydrated soils, evaporating water, pushing plant transpiration to the point where at one-inch soil depth temperatures reached 80 degrees Fahrenheit, two-inch depths 78 degrees, and the roots of both domesticated crops and wild grasses and certain trees dried and died. As each pulse of the rising torrent of ever-hotter air again reached towards the stratosphere, relasing some heat to space and spreading laterally in all directions, the meridian bands of desert-like conditions were extended in North America and Europe, with the maximum thermal limits being reached for some plant and animal species. Lacking rain, desertification and a fall in species diversity would follow.

The rest is guesswork. As a time machine traveling over thousands, perhaps millions, of years into the future, the cycles of global warming and glaciations will *not* be affected by the amount of CO_2 in the atmosphere. But *the present* Anthropocentric burning of fossil fuels will still exacerbate Hot House Earth, creating a moisture-laden, scalding atmosphere that will retain the human remains of the release of carbon dioxide and other greenhouse gases.

BLIND SPOTS

Perhaps we have left out some scientific facts, if not conjecture, in our zeal to make an ironclad connection between the Anthropogenic burning of fossil fuels, release of CO_2, rising atmospheric heat, melting glaciers and permafrost, warmer risings sea levels, increasingly violent hurricanes, and inundation of coastlines and islands. Our blind spot will remain until we can explain a number of anomalies.

Today as glaciers and sea ice melt away, the oceans possibly could rise by more than 90 or a hundred meters, vastly enlarging their geographic reach and weight, destroying the symbiotic balance of land plants and animal life. Could land life somehow be saved if, contrarily, oceans become the major sink of carbon dioxide, causing acidification dissolving carbonate minerals and the sea creatures dependent on them for shell formation?

Other unknowns threaten. Centrifugal force around the Equator might deepen its seas, changing the gyre and wobble around the world's axis, loosening hurricane wind storms destroying the web of life and destabilizing the planet's orbit—possibly modifying the cycles of the planet going through alternating stages of extreme heat and glaciation.

Could existing species survive such gyrations? Will Hothouse Earth become more than a metaphor, but rather a point of rare moment without return to habitation by any now known forms of life? No one yet knows.

UNKNOWNS

Likely two hundred thousand years from now, humankind will be extinct; continents will have drifted to again form one huge landmass, a sprawling super-continent named Pangaea-II. Meteorite strikes will have hit our planet, causing mass extinctions of plants, animals and micro-organisms. Another Ice Age will cover Earth; and once the ice melts, it will bring on a massive rise in sea levels—and possibly a last wave—though Darwinian evolution of *surviving species* undoubtedly will continue.[17] And yet, in the grand sweep of climate change, the beginnings often appear gradually in small places, revealing the onset of entropy, as Chapter 2 shows.

NOTES

1. John Mitchell Wallace, "Weather-and Climate-Related Extremes: Teachable Moments," *American Geophysical Union*, 2012.

2. Joe Romm, "NASA: Human Activity, Not Solar Activity, Drives Global Warming and Returning to 350 ppm is Needed to Stop It," NationofChange, http://www.nationofchange.org/nasa-human-activity-not-solar-activity-drives-global-warming-and-returning-350-ppm-needed-stop-it-13.

3. Astronomer Milutin Milankovitch developed the mathematical formulas upon which these orbital variations are based. He hypothesized that when some parts of the cyclic variations are combined and occur at the same time, they are responsible for major changes to the earth's climate (even Ice Ages). Milankovitch estimated climatic fluctuations over the last 450,000 years and described cold and warm periods. Though he did his work in the first half of the 20th century, Milankovich's results weren't proven until the 1970s. Some 12,000 years from now the Northern Hemisphere will experience summer in December and winter in June because the axis of the earth will be pointing at the star Vega instead of its current alignment with the North Star or Polaris. This seasonal reversal won't happen suddenly but the seasons will gradually shift over thousands of years. A 1976 study, published in the journal *Science* examined deep-sea sediment cores and found that Milankovich's theory corresponded to periods of climate change. Indeed, Ice Ages had occurred when the Earth was going through different stages of orbital variation. See Wikipedia, "Overview of Milankovitch cycles".

4. See 'Global Warming," *National Geographic*, September 2004. See also Steven Mithen, *After the Ice: A Global Human History, 20,000-5,000 BC* (Cambridge, Massachusetts; Harvard University Press, 2003), *passim*. Richard Krooth, *Gaia and the Fate of Midas*, 213, fn 513, citing James Shreeve, "Human Journey," *National Geographic*, Vol. 209 no, 3 (March 2006), 60-69, and chapters 11, 19. A.P. Elkin, *The Australian Aborigines* Fifth Edition, reprinted 1966 (Sydney, Australia: Angus and Robertson, 1938) (National Library of Australia, card number and ISBN 0 207 13009 4), Chapter 1. In 2012, Curtis W. Marean and his team concentrated their investigations in the caves and rock shelters overlooking the Indian Ocean. See also Kyle S. Brown, Curtis W. Marean, et al., "An early and enduring advanced technology originating 71,000 years ago in South Africa," *Nature online*, 07 November 2012. They described a previously unrecognized advanced stone tool technology from Pinnacle Point Site 5–6 on the south coast of South Africa, originating approximately 71,000 years ago. This technology was dominated by the production of small bladelets (microliths) primarily from heat-treated stone. There is agreement that microlithic technology was used to create composite tool components as part of advanced projectile weapons. Microliths were common worldwide by the mid-Holocene epoch, 100,000 years ago, but have a patchy pattern of first appearance that is rarely earlier than 40,000 years ago, and were thought to appear briefly between 65,000 and 60,000 years ago in South Africa and then disappeared. Marea's research extends this record to 71,000 years ago, showing that microlithic technology originated early in South Africa, evolved over a vast time span (11,000 years), and was typically coupled to complex heat treatment that persisted for nearly 100,000 years. Advanced technologies in Africa were early and enduring; a small sample of excavated sites in Africa is the best negation for any 'flickering' pattern.

5. See Elkin, *The Australian Aborigines passim*. See also Charles C. Mann, "State of the Species," *Orion*. November/December 2012, 16-27.

6. *DK Illustrated Oxford Dictionary* (New York/Oxford: Oxford University Press, 1998), 398. See also John Pernetta, *Guild to Oceans* (Buffalo, New York: Firefly Books, 2004), 55-56.

7. Pernetta, *Guild to Oceans*, 56. See also Jeffery Bartholet, "Swept from Africa to the Amazon," *ScientificAmerican.com*, Vol. 306 no. 2 (February 2012), 44-47.

8. See "Tropical cyclones," *Wikipedia*, http://en.wikipedia.org/wiki/Hurricane. See also the following, NASA Goddard's Scientific Visualization Studio, 2011. John Carey, "Global Warming: Faster than Expected? *ScientificAmerican.com*, Nov. 2012: 51-55. The U.S. National Oceanic and Atmospheric Administration Geophysical Fluid Dynamics Laboratory performed a simulation to determine if there is a statistical trend in the frequency or strength of tropical cyclones over time. The simulation concluded "the strongest hurricanes in the present climate may be upstaged by even more intense hurricanes over the next century as the earth's climate is warmed by increasing levels of greenhouse gases in the atmosphere."

9. Bartholet, *ScientificAmerican.com*, Vol. 306. no. 2, 44-47. See also Romm, "NASA: Human Activity, Not Solar Activity, Drives Global Warming, " *NationofChange*.

10. The aftermath of Hurricane Sandy witnessed personnel of the Federal Emergency Management Act (FEMA) disorganized, in need of adequate funding and setting up New Jersey tent cities that were organized like federal prisons. See Chuck Norris, "Horrors of FEMA Disaster Relief," WND commentary, Nov. 18. 2012, http://www.wnd.com/2012/11/horrors-of-fema-disaster-relief/.

11. James Hansen, et al., "Perception of climate change," March 29, 2012. See also Joe Romm, "NASA: Human Activity, Not Solar Activity, Drives Global Warming," *Nationof-Change*. http://www. nationofchange.org/nasa-human-activity-not-solar-activity-drives-global-warming-and-returning-350-ppm-needed-stop-it-13.

12. George Lakoff, "Global Warming Systematically Caused Hurricane Sandy," *Reader Supported News* October 30, 2012.

13. James Hansen, et al., "Perception of climate change," NASA Goddard Institute for Space Studies and Columbia University Earth Institute, March 29, 2012.

14. Timothy Eagan, "The Fires This Time," *The New York Times*, July 8, 2012. See also the following, Dr. Ed Berg, at Krooth, *Gaia and the Fate of Midas*, xxxi, fn. 18. Dr. James Hansen, Affiliation: National Aeronautics and Space Administration, NASA Goddard Institute for Space Studies, 2880 Broadway, New York, NY 10025. "Perception of climate change." Union of Concerned Scientists, *Heat in the Heartland: 60 Years of Warming in the Midwest* (UCS: Climate Change and Your Health, 2012).

15. Eagan, "The Fires This Time," *The New York Times*, July 8, 2012. See also Union of Concerned Scientists, *Heat in the Heartland,* 2012.

16. John Carey "Global Warming: Faster than Expected?" *ScientificAmerican.com*, Nov. 2012, 51-55.

17. See Richard Leaky and Roger Lewin, *The Sixth Extinction: Patterns of Life and the Future of Humankind* (New York: Doubleday/Anchor Books, 1995), *passim*. Also visit "The Future is Wild at the Sydney Aquarium" (Sydney, Australia), myfun.com.au/futureiswild.

Chapter Two

Fallout Bringing Down the House of Cards

> Political economy is a mere skeleton unless it has a little human covering and filling out. A little human bloom upon it, and a little human warmth in it.—Charles Dickens, *Household Words*, 1854

> How can a person discern long-term climate change, given the notorious variability of local weather and climate from day to day or year to year? This question assumes great practical importance because of the need for the public to appreciate the significance of human-made global warming. Actions to stem emissions of the gases that cause global warming are unlikely to approach what is needed until the public recognizes that human-made climate change is underway and perceives that it will have unacceptable consequences if effective actions are not taken to slow the climate change . . . . Early public recognition of climate change is critical. Stabilizing climate with conditions resembling those of the Holocene, the world in which civilization developed, can only be achieved if rapid reduction of fossil fuel emissions begins soon. —*Public Warning*, James Hansen, Makiko Sato, and Reto Ruedy, National Aeronautics and Space Administration Goddard Institute for Space Studies and Columbia University Earth Institute, March 4, 2012

Has the Holocene Era, so essential for civilized life as we know it, reached a final end during the onset of the present Anthropocene? The final close of the planet's 200,000 year-old Holocene may still be a short way off. But during the last two-and-a-half centuries, humans have driven towards this end, taking charge over the destiny of Earth and its species. The burning of fossil fuels has released great volumes of carbon dioxide, methane, other greenhouse gases and particulate matter into the atmosphere, where the Sun stores its heat—threatening to raise temperatures above the maximum thermal limits of earthly species.

SKETCHBOOK EARTH

World populations have more than doubled over little more than the last half-century. Forests have been mowed down to plant fields. Soil has been over-burdened without natural fertilization and fallow rest. Rangeland have been swept clear of wildlife, occupied by hoofed livestock with teeth and lips tearing up the roots of wild grasses. Water from lakes, rivers, wetlands and ponds have been overdrawn. Ceaseless irrigation has drawn on and eroded aquifers.

As one geography after another has been drawn into the vortex of the planet's torrid climate, desertification and ecocide threatens to diminish surviving species. It's still 5,000-to-55,000 years before renewal of the Ice Age. But unless the world slashes carbon dioxide levels during the ongoing century—keeping them under the 450-parts-per-million mark and returning to a more placid 350 ppm—"We will have started a process that is out of humanity's control," warns NASA's James Hansen, the most authoritative U.S. earth scientist.

In the present century, he adds, sea levels might climb as much as five meters (15+ feet), submerging coastal cities from Miami to Bangkok. Increased heat and drought also could bring massive famines. "We could be on the verge of a rapid, irreversible leap to a much warmer world."

Irreversible? So long as traditional industries, trade, urban populations and farming complexes all continue to burn fossil fuels, regional and global warming likely will overwhelm the remaining wilds, as well as diminish the supply of potable water, domesticated food supplies, and species diversity.

As three examples on the path approaching this end, this chapter briefly looks at the consequences for a planet headed towards its breaking point—on North America's western frontier, dehydrating its plains; at the northern polar cap, melting ice, re-radiating solar energy to the atmosphere and heating sea and air; and in the Midwest breadbasket, where high-pressure atmospheric domes block out cooler polar air flows, torrid heat searing crops and exceeding species' maximum thermal limits. In fact, all three of these examples are figures in the same continental sketchbook of an over-heated environment—its atmosphere filled with both gases and debris from burning fossil fuels, energy and heat re-radiated from Earth's surfaces under the gaze of the Sun.

BUFFALO COMMONS; LETHAL HEAT

North America's Buffalo Commons, the Great Plains once occupied by aboriginal, self-sustaining tribes and interlocking species, never can completely recover so long as the U.S. government's Department of Indian Affairs continues to control its territories. Even today, Native reservation self-rule exists nowhere.

Life on the Commons in part continues. Raptors and vultures still consume the prairie dogs that once pockmarked the Great Plains and broke the legs of horses of invading white pioneers, armies and ranchers. But with superior weapons and technologies, these wild conditions eventually were overwhelmed, destroying millions of buffalo that were the foundation for existence of the natives of the plains.

In the 1890s, frontier farmers driving horses pulling buckboard wooden plows could not bust the sod-cover held together by buffalo grass that rooted 20 inches below the surface. The foundation for the commercialization of frontier grain-farming only came with Jethro Wood's perfection of the cast-iron-tipped plough in 1819.

More than a century later during the late 1920s and 1930s, using fuel-oil-driven tractors tipped with metal blades, farmers cut furrows of buffalo grass to reach moisture to plant wheat, corn and sorghum, relying on rains that supposedly would follow the farmers. But rains fell irregularly. And winter snows and freeze often destroyed crop-seed and cattle, leading ranchers to ruin.

The mechanical force of farming with machines and oil thereafter was set against the cycles of natural grasses and species. When rains did not pour, the soil dried, roots of crops died, and 30-to-60 m.p.h. winds carried away the top-soil.

Millions of acres of prairie grasses for the cultivation of wheat were uprooting by these wind-storms, an aerial deluge of soil, sand and dust that eventually drove out 25 percent of the population of the southern Great Plains. And then kinetic energy produced by the friction of blowing soil and sands produced static electricity, locking up still more soil and carrying it away.

One calamity followed another. The lungs of cattle and people carried dust-pneumonia from fine particles lodged in their endocrine system, taking away their breath and very lives.

Continuing to follow earlier methods of domesticating soil and ranching also unbalanced species. Coyotes were the natural predators of rabbits. But the farmers killed coyotes to protect their cattle, cows and pigs. As millions of rabbits without predators then invaded gardens the farmers planted, to stop themselves from starving, the people killed the rabbits, . . . then faced other dangers partly of their own making.

"On Sundays," Timothy Egan writes, "a mob of people with clubs herded rabbits into a corral and smashed their skulls. The skies, for a time, were blotted with great clouds of grasshoppers. To kill the hoppers, the townsfolk mixed a blend of arsenic, molasses, and bran and spread it over the land, aided by the National Guard. . . . 'God didn't create this land around here to be plowed up, says [Oklahoma panhandle horseman Mell] White. 'He created it for Indians and buffalo. Folks raped this land. Raped it bad.'"[1]

As well, from cows grazing on remaining stubble, the desperate farmers killed their calves so their children could have milk to drink . . . and still watched their children shrivel, some falling 10 percent underweight, a grim reminder that the tillers of Earth and cattlemen respected neither the bounty of animal nor human fitness. Undermining their cultivation of wheat, loams of the Oklahoma panhandle, southern Kansas and Colorado were particularly vulnerable to wind storms in the wake of metal cross-plows uprooting millions of acres of prairie grass. From 1929 to 1936, some 80 percent of the Great Plains topsoil was swept away due to such wrong-headed plowing, dust storms and heat impacting 46 of the U.S. 48 states, while heat reached as high as 106 degrees Fahrenheit. And West-to-East nationwide a total of 100 million acres of top soil were blown away, showering black debris on far away cities like Chicago, St. Louis and New York, even covering New York harbor.[2]

There was a momentary happy ending: farming practices changed; a decade-long drought ended in the late 1930s, and life on the plains seemed to return to normal. The promise of the land boom and wheat fields that had led to the crisis was long forgotten; and the large-scale federal aid and intervention carried out under President Franklin D. Roosevelt's watch were a practical and moral imperative to pursue safe land practices. Forgetting this past, today the tablelands again have been re-occupied by dark forces. Farming corporations, well-heeled with petrol-guzzling, remote-controlled helicopters, chemical dusters and combines, operate edge-to-edge like an advancing tank core to produce human and animal feed in great volumes.

Their tenure will only last about 20 years, though, as long as the available fresh water lasts from the melt of the last Ice Age. The Ogallala aquifer in Nebraska, now half gone from its once 100 foot depth, has shrunk to 50 feet. With this water and sophisticated equipment, well-run corporations have driven out so-called inefficient small and middling farmers, whose cattle and pigs actually soaked up more water than corn or wheat. But the future of the corporate farms is also forbidding. Without water from the natural pressure of the once-gushing remains of the Ice-Age Ogallala, for example, the giant corn and wheat farms likely will turn their sights towards using high-pressure machines to pump the Ogallala aquifer completely dry. A new dust bowl may soon emerge, turning the plains from a bountiful savannah into a Sahara Desert. And should they fracture shale to extract gas to power machines, they will pollute the soil, water table, and all who remain on the plains. Given such a new beginning, stirring up loosened soil, old and new investors will continue to erect windmill-farms to supply remaining mega-ranches and cities with energy in their frenzied drive to accumulate wealth. So the plains will never again be returned to native nations, the buffalo, symbiotic life, or perhaps any life at all.

There is a lesson here that only overlapping fallibility can teach. The plains long have been swept of tribal nations by invading troops and white farmers wielding guns and ploughs. In turn, each modest farm family has been devoured by better-financed, larger farming families. But each larger unit, even those owned by corporations with thousands of acres, will succumb when the rains and aquifers fail. Without water, carbon dioxide alone cannot provide the macronutrients of carbon, hydrogen and oxygen required for plant life. The Ogallala aquifer, now half overdrawn, also will no longer provide natural pressure to irrigate farms, requiring expensive pumping that will drive many farming corporations out of business. Each to exit the fields of grain and raising cattle will leave behind desiccated soil to be blown away by the winds, extending the desert reach; and expanding deserts will circumscribe the foundations for species existence on the torrid lands. Reaching the limits of their maximum thermal tolerance, birds and large animals will take flight; the roots of grasses and straggling trace of trees will die; and only lifelessness will remain.[3] Under the watch of human ambitions this wondrous remaining parcel of Earth is becoming inhospitable to species evolution.

ARCTIC IMPACT: TEMPERATURES, CLIMATE, SPECIES DECLINE

As the Great Plains is turned into a desert, sooner still will the Arctic melt. Every year, the surface of the Arctic Ocean freezes during the long, dark winter, with the extent of sea ice usually peaking in March. Until recent decades, a high proportion of the ice was thick enough to survive the summer. But scientists now say the trend of planet warming has reduced the ice to just a shell in many places. This means it can easily melt in the round-the-clock sunshine that strikes the highest latitudes in summer. In the summer of 2012, the amount of sea ice in the Arctic had fallen to the lowest level on record, a confirmation of the drastic warming in the region and a likely harbinger of greater changes to come. Walt Meier, a top scientist at the Arctic Snow and Ice Center, observes that: "Parts of the Arctic have become like a giant Slushee this time of year."

As the sea ice melts, the average temperature of the region is rising more than twice as fast as that of the Earth as a whole, confirming a prediction first made in 1896: that increasing levels of carbon dioxide from the burning of fossil fuels would have an especially large impact in the Arctic. One reason is that the white surface of the ice once reflected a great deal of sunlight back to space. But now as the ice and glaciers melt, the exposed darker water and land lead to additional melting and more sunlight absorption—creating a feedback loop that scientists call Arctic amplification. Such amplification of heat that melts sea ice does not raise global sea levels, as the floating ice is already displacing its weight in seawater. But the sharp warming that is causing the sea ice to melt also threatens land ice, notably the Greenland ice sheet, which is melting at an increasing rate. As it pours into the ocean basin, melting land ice raises sea levels. And the heat or cold is transferred to other areas of the world, the loss of sea ice likely bringing harsh winter weather in the U.S. and Europe.[4]

To Midas thinkers bent on wealth accumulation, the melting does, however, offer some potential benefits, including new shipping routes through the Northwest Passage and easier access to oil and other mineral deposits. This has brought on a rush of stake claims and mineral exploration in the Arctic, partitioning its resources, landmass, and potential trade. An emerging epicenter of industry and trade is akin to the Mediterranean doubling of the breadth of the Suez Canal by Egypt seeking elevated profits at the expense of indigenous fish species facing invading predators. For the Arctic commercial opening promises access to a quarter of the world's estimated undiscovered oil and gas, massive deposits of valuable minerals, and a saving of thousands of miles during a journey between the Pacific and Atlantic oceans, making it a potential global maritime transport route.

Feeding and survival chains are being undermined. The price of invading the region's fragile ecosystem is death for the polar bears and tens of thousands of walruses robbed of melting ice floes inhabited by the seals they feast on. While these giant creatures are coming ashore to forage in northwest Alaska, sub-Arctic flora and fauna are migrating further northward. Frozen tundra are starting to revert to the swampland of 50 million years ago; and violent storms churning the newly opened sea waters are eroding shores and tumbling the homes of indigenous populations into the sea.

The reduction in sea ice also is altering weather patterns in the Arctic region, and perhaps beyond. It is putting stress on the ecology of the region and causing rapid erosion of shorelines that are now exposed to more vigorous waves. Jennifer A. Francis, a Rutgers University scientist who studies the effect of sea ice on weather patterns, is one of a small group of climate scientists arguing that the decline of sea ice is already having consequences for weather in the Northern Hemisphere's middle latitudes, including in the United States. She has published research suggesting that air circulation patterns are being altered in a way that favors more weather extremes, like heat waves and droughts. Though such ideas are not considered proven, they are getting more attention as the weather careens from one extreme to another—impacted by warming air currents flowing from the Arctic region southward to lower latitudes, that nonetheless often are unable to break through searing high-pressure heat domes.[5]

HIGH PRESSURE DOMES; STAGNANT JET STREAMS; BAD NEWS; SPREADING DESERTS

When an unmovable, massive dome of hot, high pressure embeds itself over a large landmass, spreading extreme drought conditions throughout the region, desertification is initiated. This combines with warmer polar regions due to a heated Earth and atmosphere, weakening the flow of cool air to surrounding regions—so that the longer the heat domes last, the more chance each species of life on land and in the sea will reach its thermal limits and expire.

Can humankind stop these consequences? Weather patterns and human responses are linked. The warmer that northern polar regions become, the greater is evaporation of moisture vapor falling as rain, the less snow falls, the weaker is the power of the cool northerly jet stream to penetrate the hot domes on its southward journey. Once people in the northern hemisphere can no longer resist the changing heated environmental conditions; and once governments run out of the ability to offset regional warming, people, nations and states alike begin to cut off the spigot of help for citizens elsewhere facing desperation. An example of such weakness speaks as much to the worldwide environmental crisis impacting small provincial locations. For these affect local food supplies sustaining populations, as well as the price of the basket of goods and services people must have to survive. Adding to the Malthusian dilemma of food production falling behind population growth, lethal environmental heat also promises to spread deserts, hunger, malnutrition, disease, pandemics and death.

Focusing on these issues in the single state of Wisconsin, the 2012 siege of Midwestern torrid heat wrinkled time, upsetting both the natural and domesticated domain. Among over 700,000 farms in the state, many small ones faced devastation, promising to be significantly cut as rains failed in the spring and summer months, killing the grass and grain crops necessary to feed cattle and especially cows. Failed crops, without income, farmers with monthly bills to pay faced hard times, even bankruptcy, though their debt levels were generally low by comparison to the previous 1980's last big drought and farm crisis that wiped out thousands of farms. The larger farms in 2013 had only a half-year buffer of grass and feed in storage for their cows; but the small dairy farms lived on the edge without silage for their cattle, many being forced to bring their livestock with calves to auction for slaughter, farmers and truckers lined up like stepping stones to an agrarian Dante's Hell.

It started with regional and global warming that few farmers had ever known much about or protested against. They did realize that without rain, they all would be competing for cattle feed. Corn would not grow as high as an elephant's eye, as from June 1st to July 13th the lack of rain had cast but

.31 inches. No drought-busting downpour was forecast for the following weeks, and some 2-to-3 inches was essential to salvage the state's southern corn crop. Without moisture the corn could not pollinate during the critical 10 days after it tasseled. "If it doesn't pollinate, there's no ear," advised Landmark Services Cooperative agronomist Joe Speech. "That's the reason it can become a total loss. You've got the 10-day window and that's it." As one-third of Wisconsin corn came from the 17% of the state's drought ridden zone, higher corn prices were expected. Cattle producers would run out of feed; and unable to feed livestock with reasonably-priced silage, forced to send their cows and hogs early to market—initially driving meat prices down, later following the cattle shortage, to escalate along with milk prices.[6] Cows that graze need grass; and the grass was browning, expiring, or literally on the verge of burning up, as your author documented traveling for days though southern Wisconsin all the way to the capitol in Madison and the Fox River Valley centered around Appleton.

When soil temperatures rise above 80 degrees Fahrenheit reaching down an inch deep, plant roots start to dehydrate, then die, killing the grass crowns. At two inches deep, 77 degrees is about the maximum of plant endurance; and soil temperatures nearing that level could kill grass, put trees in distress, causing them to curl and drop their leaves releasing carbon dioxide, and increasing the deadwood open to fires.[7] Bearers of bad news, by mid-June the media convinced most farmers that without rain, they would be competing for cattle feed. As Alfalfa had generally stopped growing in much of southern Wisconsin, farmers desperately searched for hay they could buy and store for winter, with the July 13th price reaching $250-to-$300 a ton, compared with a normal season of $150 for dry hay. A typical dairy cow eats an average of 35 pounds daily, 365 days each year consuming about two acres worth of hay and grain equal to 12,775 tons. At $250 per -ton per-cow that would be an annual outlay of about $3,193. If you've got 1,000 cows alive and kept well, that's about $3,193,750 for 2012.

Most cuts of golden hay for winter were already in the drying fields by early July. And it was dangerous for farmers to be out in the hayfields, as fires could spark by spontaneous combustion as the tightly compacted haybines dried out, building up bacterial heat. For those farmers whose pastures already had dried out, using corn silage and stored hay from 2011 to feed their hungry cattle would leave them short on feed for the 2012-13 winter, especially if rain didn't arrive soon and 2012 crops failed. Instead of risking their own saleable, high-priced grain for winter feed, early on some harvested stressed corn for silage. Chopping corn early for silage also was dangerous, as moisture trapped in the stalks caused silage to rot making it worthless as feed, and possibly leading to spontaneous combustion fires.

A HOBSON'S CHOICE

Global and regional warming also caused turmoil, internal stress and financial worry that farmers daily suffered, but few city dwellers barely understood. Cattlemen and ranchers faced a Hobson's choice at best: to pay transport costs to send their cows to verdant northern Wisconsin, where the grass was still green; or, without moving their critters, to pay for the rising price of silage to feed them; or to send their cows directly to the slaughter house. The high cost to transport and return each cow by truck to graze on rented turf had to be weighed against the escalating cost of delivered silage to the farm, as well as the varying market price per cow delivered to the slaughterhouse auctioneer. At an early July price of $17 per hundredweight, milk could hardly cover the rising cost of grazing up north or buying silage. So one had to calculate if immediate slaughter of a herd, or buying pricey hay for cows was more beneficial, risking the future rising price of feed. Would the long-run rising price of milk as fewer cows produced and more went to slaughter be advantageous, if the price received for a potential slaughtered cow actually fell once dehydrated and emaciated cows went to market and consumers refused to pay high prices for inferior cuts of beef? And for silage, would early cutting corn, and even mowing 25-year-old marsh grass now on hardened turf, not be the best choice to reduce costs if both rented grassland and the market price of corn for feed rose quickly?[8]

TIGHT MONEY, EXTREME DROUGHT

Money was tight, especially in the southern half of the state in desperate need of rain, where staggering farm losses cost $400 to $500 an acre to plant corn or soybean, covering the cost of seed, fertilizer, other chemicals, and running expensive equipment with petrol fuel. Plant 500 acres at $500 an acre and your crop is devastated by extreme drought—you're in deep financial trouble, if you're not insured for adverse soil conditions resulting in major crop and pasture losses, and/or precipitation deficit coverage for widespread water shortages or restrictions.

Some protest was voiced from 30% of farmers who had not insured their corn crop, faced non-renewal of the Federal drought relief provisions that had expired in 2011, and sought their own solutions at prayer meetings with boots and umbrellas. Wisconsin Extension center agricultural program faculty at University of Wisconsin-Madison, UW-Platteville and UW-River Falls tried to set up online brokering networks for stressed farmers to make deals on silage, corn and shipping cattle from southern Wisconsin to northern green pastures. But little was either said or done about global warming that the Anthropocene had set in motion, congealing with the Sun's radiation that reached Earth in eight minutes flat and was re-radiated to the atmospheric cover of greenhouse gases.[9]

ON THE DOLE; FEEDING THE NATION

What feeds the nation was at stake. There was no longer a commodity price-protective Ever-Normal Granary from President Delano Roosevelt days. Left were only four possible levels of awareness among Wisconsin cattle, soybean, corn and hay farmers: one's individual condition facing survival in the face of crop adversity; cooperative marketing efforts to deal with the heat blanket of destruction; collective questioning of the sources of rainlessness every eleven-to-twelve years or so; and political efforts to try and stop the human causes driving the lethal storm of heat. The last two possibilities were practically beyond most farmers' *active awareness* and/or ability to resist the rainless dance.

Five major corporations today control upwards of 90% of the production of meat, poultry, soybean and grain throughout the United States. Directly or through their subcontractors they use factory methods in the fields and in processing food. They lobby in Congress to secure the largest farm subsidies that today remain skewed. Favoring those cultivating the largest acreage, about $20 billion (in 2010 dollars) went to 2.9 million recipients, just about making up the top 1% of all farmers and agricultural corporations. From 1995 to 2010, the top 20% received 90% of all subsidies; the bottom 80% got 10%; and over half of those at the very bottom received no help at all.

Even for crop insurance, the main beneficiaries remained the big corporate farmers taking out the largest policies. Up to 2011 they received the lion's share, with the Federal crop insurance program paying them for losses amounting on average to 60% of all farmers' premiums ($2.4 billion in 2001, but some $8.7 billion in 2011, and an expected $9 billion in 2012 and coming years).

Is reform possible to secure the nation's future food requirements? America has needed and still needs an equitably allocated, Federally-funded farm insurance policy for *all* those who risk their fortunes and wellbeing against the variability of weather and intensified regional and global warming.

So far it has been up to the small-and middling-size farmers, ranking among the 62% who receive no government subsidies, to mobilize their voting power and turn out of office any in Congress who opposed funding them to adequately supply the rest of the nation with fair-priced food during crises in the era of global warming. Though they have been outmaneuvered by the lobbying pressure of large-scale corporate food processors, in future crises of want during severe regional warming, they may be able to secure fairness, equitably distributing all Federal funding—to each farmer, small or large—as the foundation for farm subsidies. In the long term, the U.S. can't have it both ways: Either the nation feeds itself by supporting its entire farming population during the present Anthropocentric era, or the majority of its people one day will go hungry.

OVERLAPPING DILEMMAS AND DANGERS[10]

The broader dilemma facing humankind is living well for the moment by burning fossil fuels to continue food and industrial production, and barely existing tomorrow in a lethal, suffocating atmosphere. Bookends of past geographic temperature and weather ranges sustaining the evolution of humankind and diverse species over the last 200,000 years or so, global plants and animals have been conditioned to survive at specific thermal limits. Breaching these conditions by raising global and regional temperatures above various species' maximum thermal limits of endurance already has led some to vanish from Earth, leaving survivors to cope with increasingly adverse environments that test their reproductive fitness.

ADVERSE REGIONAL CONDITIONS

Evidence of adverse regional conditions affecting species correspond to global warming patterns (using a based-line period 1951-1980, representing seasonal mean climate stability). These conditions include:

- changes in the sun's energy releases, affecting sea currents;
- melting ice in Greenland and Antarctica, leading to rising sea levels;
- widespread evaporation and transpiration of water suspended as vapor in the atmosphere, leading to irregular heavy downpours and floods in the northern hemisphere; and
- atmospheric summer heat sweeping the tropics and now northern latitudes, reducing species diversity.

Under atmospheric high-pressure, stagnant domes of torridity blocking cooler southward airflows from the Arctic, vegetation and many animal species can no longer survive heat above their thermal maximums. Ever warmer winters change the cycles of insect infestation, killing traditional pine forests that may then burn, and accelerating the dissemination of diseases jeopardizing the survival of other species.

Though some of these measures can be offset, enlarging food and grain output, producing momentary surges in carbon dioxide releases, a future moment will be reached when the sun's millennial store of energy in soil nutrients will be exhausted; when a heated atmosphere leads to rapid ground water evaporation and transpiration from plant and animal life; and when the atmosphere is loaded with vapor ready to unpredictably unleash rain, causing uncontrollable floods—leaving only human tears to harvest the grapes of wrath.

Clearly Earth's temperature is scaling upwards from what we humans are doing to the planet. The human-structured technological sphere already has loosened a cascade of dangerous forces, destroying one species after another, pushing the biosphere towards a state of disorder. And the end of the Holocene's species-tolerant temperature range well might bring us to a further moment of the destruction of most forms of life on Earth.

SUSPENDING BELIEFS

We have yet to see the future thrust of the global spread of heat and its outcome. If we can suspend our beliefs about the immutability of our planet's present surroundings, it is possible to imagine what's to come:

- that our species burns existing supplies of fossil fuel to the moment of an atmosphere loaded with 450-600 parts per million of carbon dioxide
- that the sun's energy reaching Earth is re-radiated to this atmosphere, raising its temperature and that of Earth's surface by 2 degrees Centigrade
- that the world's glaciers, sea ice and permafrost completely melt, raising sea levels between 5 and 6 meters (15 to 18 feet)
- that oceans submerge islands and cover large parts of continents with shallow seas where cities and wild and domesticated creatures and plants once survived
- that any remaining land areas are exposed to unbearable heat reaching desert conditions of 132 degrees Fahrenheit, killing all domesticated crops and leaving only alpha desert plants and small sun-rodents
- that aquifers—that had lasted from the melt of the last Ice Age some 165,000 years ago and were once drained for food production—are exhausted
- that starving and diseased world populations are reduced from 10 billion to less than 1 billion, returning them to a pre-industrial stage of survival; and
- that long before a renewal of the Ice Age 5,000 to 55,000 years from that moment, most of Earth's species will have expired.

HOPE AND DESPAIR

Imagine Earth as an overheated teapot, spinning off its unstable axis, steam pouring from beneath its fulminating lid, setting loose powerful winds and thrashing seas, a last wave of boiling water pouring forth, scalding most living land species. Such a Dante's hell we could not endure.

Before then hope still breeds strong that we the people of Earth can somehow save the planet, Bill McKibben, writing in the UK *Gurdian* November 1, 2012:

> It's experience that changes people: the summer's drought left more than half of American counties as federal disaster areas, and meteorologist Jeff Masters estimates [Hurricane] Sandy hit 100 million Americans with "extreme weather". Add in the largest forest fires in Colorado and New Mexico, the hottest month in US history, and the completely absurd summer-in-March heat wave that kicked off our year of living sweatily, and you can begin to understand why the percentage of Americans worrying about global warming has spiked sharply this year. Spiked high enough that even a few politicians are willing to speak out.

NOTES

1. Timothy Egan; *The Untold Story of Those who Survived the Great America Dust Bowl*, 8-9. See also Louis Thomas Jones, *Aboriginal American Oratory: The Tradition of Eloquence Among Indians of the United States* (Los Angeles, California, 1965), *passim*. Today under the Department of Interior's oversight, native reservations are flooded by dams to provide water, flood-control and hydroelectric services to cities. Their forests have been devastated by timber-tract companies. Reservation mines have been opened for burning coal to provide more urban electrical current. White ranchers raise food on the reservations to be packaged and sent to outside markets—while native people often go hungry, suffer obesity from diets of carbohydrates, salt, fats and sugar, with one in four at some stage of diabetes. Dental problems are rampant. And alcoholism and degenerative liver disease are common. Sexploitation is also rampant, native and outside courts turning a blinded eye.

2. See Egan; *The Untold Story, passim*. See also the following John Mohawk, "Steps to Sovereignty," *Harvest Quarterly*, No 1 (1976): 36-41. Richard Krooth, editor. *The Great Social Struggle*, Vol. I (Santa Barbara: Harvest Publishers, 1978), ix. Jones, *Aboriginal American Oratory, passim*. "The Dust Bowl," *PBS* November 18-19, 2012. 8:00 PM. Produced by Florentine Films and WETA. Directed by Ken Burns; written by Dayton Duncan; Mr. Duncan, Mr. Burns and Julie Dunfey, producers; Craig Mellish and Ryan Gifford, editors; Buddy Squires and Stephen McCarthy, cinematographers; Patricia Clarkson, Kevin Conway, Amy Madigan and Carolyn McCormick, voices; Peter Coyote, narrator.

3. Jeff Glaze, "Hot, dry weather takes a toll on trees, grass," *Wisconsin State Journal* (Madison, Wisconsin), July 13, 2012. See also Richard Krooth, "On the Road Again: Traveling the Plains and Texas," *Notebooks*, 2000-2012. Mohawk, "Steps to Sovereignty," 36-41.

4. See Charles H. Greene, "Climate Change: The Winters of Our Discontent," *ScientificAmerican.com*, December 2012: 51-55.

5. See Scott Borgerson, "The Coming Arctic Boom, As the Ice Melts, the Region Heats Up," *Foreign Affairs*, July/August 2003: 58-89. See also Justin Gillis, "Sea Ice in the Arctic has Fallen to the Lowest Level on Record," *The New York Times*, August 27, 2012.

6. Rob Schultz, "Area's corn crop may be total loss," *Wisconsin State Journal* (Madison, Wisconsin), July 13, 2012. Also Richard Krooth, "Talks with the Abe Eckstein Family: Cow and Corn Farmers," Larsen, Wisconsin, July-August 2012; July 2013.

7. Jeff Glaze, "Hot, dry weather takes a toll," *Wisconsin State Journal,* July 13, 2012. Also Krooth, "On the Road Again," Wisconsin *Notebooks*, July 2012.

8. See Karen Herzog, "Livestock feed runs short as crops fail in dry spell," *Journal Sentinel* (Milwaukee, Wisconsin), July 13, 2012. See also Mike Nichols, "Farmers' prospects starting to dry up," *The Post-Crescent*, (Appleton, Wisconsin), July 15, 2012.

9. So far there is no specific reference in physics, archeology, or paleontology to tree-ring dating records of 11-12 year droughts in Midwestern forests. Most Wisconsin old growth trees were cut for timber during the period 1890-1930. But see A.E. Douglass, *Climate, Cycles and Tree Growth*, Vols. I-III (Washington, D..C.: Carnegie Institution, 1936), *passim*. Radiocarbon dating is a method of estimating the age of an artifact or biological vestige based on the relative amounts of the different isotopes of carbon present in a sample, such as a series of concentric tree rings.

10. NASA Goddard's Scientific Visualization Studio, 2011. See also John Carey, "Global Warming: Faster than Expected?" *ScientificAmerican.com*, November 2012: 51-10.

Chapter Three

Technology and the Cascade of Species

> The process of modification [of species varieties] must be slow, and will generally affect only a few species at the same time; for the variability of each species is independent of that of all others. Whether such variations or individual differences as may arise will be accumulated through natural selection in a greater or lesser degree, thus causing a greater or lesser amount of permanent modification, will depend on many complex contingencies—on the variations being of a beneficial nature, on the freedom of intercrossing, on the slowly changing conditions of the country, on the immigration of a few colonists, and on the nature of the other inhabitants with which the varying species come into competition. Hence it is by no means surprising that one species should retain the same identical form much longer than others; or, if changing, should change in a less degree.—Charles Darwin [1]

> [*Vanity Fair*'s] image of New York City shows the effects of an 80-foot rise in sea levels. That's what would happen if not only the Greenland ice sheet but its counterpart in the Antarctic were to melt, says James Hansen, the director of NASA's Goddard Institute for Space Studies. . . . Hansen warns that, if global emissions continue on their current trajectory, the ice sheets [of Greenland and Antarctica] will not survive, because global temperatures will increase by 2 to 3 degrees Celsius by the end of this century. "The last time the earth was that warm, sea levels were 80 feet higher than today," he says. It will likely take hundreds of years for sea levels to rise the full 80 feet, but the process would be irreversible, and the rises would not be gradual. "You're going to be continually faced with a changing coastline, which will force coastal dwellers to constantly relocate," he says.—Mark Hertsgaard, 'Climate Change: While Washington Slept," *Vanity Fair*.[2]

Earth's energy sources are out of balance. The thermal forces of technical production are invading the biosphere and discharging heat and waste throughout the planet, threatening the biological sphere, as well as our species relationships, upsetting and endangering the social order.

In the previous two chapters we already have shown how the technological sphere has overwhelmed the biological domain in devastating ways. In this chapter we further detail how human-driven mechanical and electrical production on land and in the sea impact atmospheric and natural cycles by burning inanimate fossil and nuclear fuels, as well as by bioengineering undermining species natural selection.

Added to the discharges of human-generated heat, water-vapor and effluents into the atmosphere are those emitted from natural plant transpiration and animal secretions, releasing greenhouse gases.

As well, expansion of human populations working at means of production and on farms producing essentials have also extended their needs for resources and land, often leading to invasions and destruction of forests and savannahs, inland waterways and seas, jeopardizing the natural domain and species near and far. They, too, become extensions of the technological sphere jeopardizing the biosphere.

THE SECOND LAW OF THERMODYNAMICS & SPECIES MAXIMUM HEAT VARIABILITY

How exactly have the chasm and clash between the technical and biological spheres been exacerbated? We can explain this gap in two ways:

- by considering the fact that all species are confined by their environment for resources to survive and reproduce; and change that environment by disposal of their wastes
- and under the operation of the second law of thermodynamics, that posits in the course of production, output volume and quality will flag and be replaced in part by heat and waste that will advance and escape to cooler regions.

The evolution, modification and reproduction of species life, their reliance on and restructuring of their surrounding environment has limits. Especially for humankind, efforts producing and sustaining themselves, at some point, their efforts will flag and be replaced by heat and waste that then will increase and escape to cooler regions and lead to entropy—disorder in the flow of energy in the very gage they have created.

Along with friction, heat, and the inefficiencies of using inanimate fossil fuel heat energy in production, output from work is replaced in part by rising columns of aerated gases, waste, debris and chlorine-based dioxins spreading to cooler areas of the global environment. Together this transfer of heat and waste transforms and burdens natural environments, undercutting balanced global conditions essential for the fitness of species to survive and reproduce.

ESCAPING ENVIRONMENTAL CHANGE

Some species may seek new geographic environments that approximate their traditional, but now sweltering and drying, old ones. Some possessing genetic variability may adapt to a new environment and widen their range for survival and reproduction in the onrushing change in conditions. But all of them must adapt their organic and physical structures—or weaken, atrophy and expire.

In response to the shifting environment, for example, the tiny brown and orange argus butterfly (*Aricia agestis*) has spread about 49 miles (79 kilometers) northward on the British isle over 20 years. Although it's well known that climate change can prompt plants and animals to shift the ranges in which they live, the brown argus has widened its living quarters with unusual speed and focus, according to an analysis of decades of data collected by British volunteers. The butterfly's secret of adaptability was not internal, but external—a new host plant on which to feed its larvae.[3]

For those species with physical or organic abilities to adapt and reproduce, their progeny may gradually develop respiratory or other organic and structural appendages and metabolic processes to cope with *rapid* changes in the environment. Yet for those lacking genetic variability, or brain receptors to switch on or off to cope, they remain dependent on their earlier conditions for survival. And for those rooted in place, those who cannot move quickly enough to reach the needed geographic elevation and temperature range and resources needed for survival and reproduction—they sooner or later will shrivel in physical stature. Examples are many, *à la* the Wooley Mammoth's loss of food diminishing their size before extinction on Rangle Island during the the onset of Arctic freezing conditions. Or some 12,000-to-13,000 years ago, when human hunters took advantage of Antarctic's rapid warming spell and a fast-growing beech tree forests in Patagonia—cutting down the trees, reducing and scattering the original range, and causing the collapse of mega fauna mammoths and giant sloths. [Alan Cooper, et al., *Science Advances*, June 17, 2016.] Or, as we will see in Chapters 4 and 5, unable to scale Alexander von Humboldt's atmospheric, *aërial* mountain chains to reach the high plateaus of dry sun-swept soils of the *corrideros*, most species fell at death's door.

Thus are all species and biospheres vulnerable to natural and the human-designed technological sphere which one day may reach a moment of overlapping disorders, so violent and geographically extensive that an ecocidal impulse and disclimax will set in motion a massive downward cascade of species life—cutting the symbiotic relations of plants and respiratory animals, exchanging oxygen and carbon dioxide.

A poisoned, damning fog might then overlay land, sea and atmosphere, leading to the literal decay of Earth's biosphere, releasing methane, carbons and dioxins that capture the sun's radiant energy. This would drive up temperatures to the point at which atmospheric moisture literally boiled the planet's surface and eliminated terrestrial plants and animals as well as sea phytoplankton and krill sustaining the pelagic chain of life.

Only those chemosynthetic ecosystems, worms hosted by deep-sea thermal sulfur vents and heat-tolerant fishes, might remain.

ENGLISH SPRING, GLOBAL WARMING, CLIMATE CHANGE

Spring arrives in England three weeks earlier than it did 50 years ago. The Queen's own Sandringham country estate north of London overlooks Britain's premier bird-watching spot: the vast North Sea wetlands known as the Wash. Here V-shaped flocks of pink-footed geese descend every winter. But in recent years, says Mark Avery, conservation director of the Royal Society for the Protection of Birds, little egrets that are new to the area have arrived in the last 5 to 10 years. These shiny white birds native to Southern Europe have spread very rapidly to Northern Europe, perhaps because of rising temperatures.

Global warming melts glaciers and swells oceans as water expands when heated, threatening to raise sea levels as much as three feet by the end of the 21st century. This would leave much of eastern England, including areas near Sandringham, under water.

It is certain that temperatures are rising as greenhouse gases are trapping heat in the atmosphere. Carbon dioxide, the most prevalent of such gases, is released whenever fossil fuels are burned or forests catch fire. Global warming is expected to bring more heat waves like the one in the summer of 2003 that killed 31,000 people across Europe.

SCIENCE FICTION, SCIENCE FACT

In a reversal to cold, the greenhouse syndrome might be slowed, even shut down the Gulf Stream, the flow of warm water from the Gulf of Mexico that gives Europe its mild climate. If the Gulf Stream were to halt, and it has already slowed 30 percent between 1992-2012, Europe's temperatures would plunge, agriculture would collapse, London would no longer feel like New York but like Anchorage.[4]

A Malthusian fall of agricultural output might institute war on the helpless lacking food and warmth—first the desperately poor, sick and disabled would succumb to starvation, alternating extreme heat and cold; then the day-and-sweated laborers would be sundered; next would fall the varying degrees of middling classes; and, gradually, the once well-fed upper orders would go down, reaching the very apex of those with jet service trying to escape deluged cities, the lethal heat of lower geographic meridians, and the renewed frigidity of parts of Europe.

Life conditions might approximate the Little Ice Age, devastating food supplies, creating a mountain of decaying plant life releasing carbon dioxide, bringing on pandemics. The Holocene era would near an end. The treasury of human life would be expended.

LOOKING BACK, LOOKING FORWARD: EXAMPLES OF HOLOCENE CONDITIONS

Judging the present by looking back to early Holocene conditions, before the Anthropocene's maximum thermal limits supporting species life had been reached, there must have been sufficient water and moisture to support plants, animals and microorganisms.[5]

By contrast today as Earth's temperatures scale upwards, the lack of groundwater well may become fatal to many land species reaching their maximum thermal limits. Local episodes might then lead to ecocides of overlapping varieties, also bringing a final end to the Holocene sustaining human society.

Global warming produces either too little water or too much—as Hurricane Katrina proved, slamming into America's Gulf Coast on August 29, 2005. It was a category-5 mega-storm, the strongest there is, that had slowed to Category-3 levels by the time it made landfall. But its initial category-5 water-surge, raging in the Gulf of Mexico, dragged the waters of Lake Cochitrain ashore along an 170-mile front into New Orleans, which was already significantly below sea level. It broke the old, poorly-constructed levees, unleashed 10 to 20 feet of water, and destroyed all in its path. New Orleans poorer neighborhoods in the lower ninth ward became a rushing river of destruction, drowning unknown hundreds, displacing many thousands of others.

The city had already sunk about three feet during the 20th century, largely because levees built by the Army Corps of Engineers, to keep the Mississippi River from depositing silt that would have replenished and raised the underlying land mass and sea-absorbing barrier reefs. And as the storm brought the lake and sea over the levee-breach into the city, it put 80 percent of it under water, becoming a harbinger of what global warming would do to other coastal cities in the future.[6]

There is no doubt that global warming rose in the 20th century and so far has intensified in the 21st. The U.N.-sponsored Intergovernmental Panel on Climate Change (I.P.C.C.)—a peer-reviewed, international collaboration among thousands of scientists—has determined that hot Atlantic and Gulf vapor, merging with west-thrusting, searing skies, will likely launch hurricanes more frequently, more powerfully, that will be more devastating to all forms of life. There may be no way to stop the destruction short of worldwide action directed against the purveyors of fossil fuels.

BIOGEOGRAPHY: FATE OF SPECIES LIFE

Biogeography sends a deeper message, too, about how the Anthropocentric Age is altering natural cycles, though the fulminations of Earth are often unpredictable. Human misfeasance also merges with continental drift, ocean basins disturbances, and atmospheric gyres, together determining the evolution, development, devolution and fate of species life.

FROM beneath our footsteps, the planet is made up of several distinct layers. Its core is an amalgam of iron and nickel; its mantle of stone; its outer crust composed of varying materials lying at different depths.

The outer layer, called the lithosphere, is composed of plates that slide over the Earth's surface, riding on waves of semi-molten rock and metals, named the asthenosphere.

On these plates, Earth's outer crust or shell is formed, consisting of relatively light materials, separated from the dense underlying mantle (known as the Mohorovicic discontinuity). At the base of oceans, the crust is relatively constant in thickness, unlike considerable variation of continental crust, with its greatest depth beneath the weight of mountain ranges. Upon this crust of land and subterrain and sea-bed exist the conditions essential for species evolution, development and diversity—constituting Earth's biogeography.

The core of Earth affects both the composition and geologic profile of sea-bottom and *tierra firma*. Geologists have long known that Earth's core lies far beneath land cover, made up of a dense, chemically-charged ball of alloyed iron and nickel, roughly the size of Mars and every bit as alien from the rest of Earth. It is under the pressures of 3.5 million atmospheres, where temperatures reach 10,000 degrees Fahrenheit—hot as the surface of the Sun. Here iron circulates as both a solid and a fluid that twists and spirals like liquid toffee.

Earth's core itself is divided into an inner and outer layer at a depth from the planet's surface of between 3090 and 3170 miles (4980 and 5120 kilometers). The inner core is solid with a radius of about 800 miles (1300 kilometers); but the outer core is liquid. Above its throbbing heat is a separation shield (called the Gutenberg discontinuity) partitioning the dense core from lower mantle's partially molten rock; and above it an upper section of ferro-magnesian silicate materials.[7]

The torrid heat of Earth's inner core shapes and regulates its outer flows and geologic structures, casting tectonic plates floating far above it, building up mountains and gouging out sea-beds. The fulminations of the core's iron also generate Earth' magnetic field, blocking dangerous cosmic radiation, brightening northern skies with flashing aurora lights, and even guiding terrestrial travelers.[8]

So far as scientists now know, the core's thermal heat is overwhelmingly primordial, left over from the planet's gravitational formation and mostly trapped inside by the mantle's rocky structure acting as a muffler. As Earth meanwhile orbits relentlessly through frigid space, the core also obeys the second law of thermodynamics, gradually shedding some of its stored heat.

The speed of transferred heat depends on which one or both of two basic pathways: conducted straight outward, the way heat travels along a frying pan, or confected out in plumes, the way hot air rises in the atmosphere or boils and churns through Earth's layers.

Conduction wastes energy during transfer—heat moves under the second law, warming Earth from within. Convection, by contrast, is potentially positive, as its currents ripple through the mantle and shuffle the positioning of tectonic plates. And convection also fuels the geodynamic gyres that yield Earth's switching field of energy.

Perhaps the central core mechanism can be discerned from studying seismic waves generated by earthquakes. As John Vidale of the University of Washington explains, most earthquakes *originate* in the upper 30 miles of the globe (as do many volcanoes), and no seismic *source* has been detected below 500 miles. But the quakes' energy waves nonetheless radiate across the planet, detectably passing through the core.

The core does leave faint but readable marks on the surface, by way of the magnetic field that loops out from the vast chthonic geo-dynamo of swirling iron, permeating the planet and reaching thousands of miles into space. Magnetic particles trapped in neat alignment in rocks reveal that the field; and presumably the core structures that generate them, have been around for well over 3 billion of Earth's 4.8 billion years of existence. For reasons that remain mysterious, though, the field alternates. Every 100,000 to a million years or more, the north-south orientation of the magnetosphere reverses, an event often preceded by an overall weakening of the field. As it turns out, the strength of our current north-pointing field, which has been in place for nearly 800,000 years, has dropped by about 10 percent in the past century, suggesting we may be headed toward a polarity switch.[9] So far the effects on species remains unknown.

ATOMIC THEORIES

This is no surprise if one accepts early atomic fancy and later theory. In the first century B.C., the Roman poet Titus Lucretius Carus' 7,400 lines of verse, *On the Nature of Things*, rhythmically covers the thought of the Greek philosopher Epicurus, who taught his Athenian students three centuries earlier that life's purpose turned on beauty and pleasure that were the very "seeds of things"—eternal particles of matter, a modern metaphor, for all that exists is made of "atoms and void and nothing else," put in motion—constantly colliding and combining with one another in an inexhaustible variety of forms and substance.[10]

ATOMS and voids lead us down the path of the minutest forms of life existence and their evolution, devolution and extinction. In all species, cells may divide, DNA may replicate, growth then occurs, leading to cellular exhaustion, decay, disease and cancers of various degree, spelling finality for individual life.

Fitness to survive and reproduce meanwhile turn on species proprioception (self-awareness), behavior, and habits of self-preservation. "Foraging, stalking prey, hoarding food, searching for and finding a desirable mate, and nest building are all examples of activities that greatly enhance an animal's chances of survival and reproduction, or what biologists call fitness," writes U.C.LA. cardiology professor Barbara Natterson-Horowitz. "Animals are rewarded with pleasurable, positive sensations for these important life-sustaining undertakings."[11]

The vast turmoil of creation and destruction; the thermonuclear power comparable to the Sun at Earth's core; the explosive energy and its release of torrid heat traveling from the planet's outer core through the boiling rocks of the middle planet, fulminating plate movements on which continents float—these offer a portrait of where, how, and to what purpose species emerge, evolve, reproduce and morph their structures, organic shapes, metabolisms and devolving forms.

EARTH'S HEAT, SPECIES EVOLUTION

The broad conclusion is that thermodynamics posits not simply energy transfers from Earth's core to its crust, but also as heat ripples upward from the core, it raises both mantle and atmospheric temperatures, affect the environments for species evolution, development and extinction.

Earth's core heat also melts rock and circulates lava, shifting tectonic plates, altering the float and condition of continents essential for species life.

This heat sustains billion of microscopic species living 500 feet or more below Earth's surface, sustaining surface species in ways we barely understand.

Living matter literally is being squeezed and pressured between changing geological land and sea conditions on one side, and on the other an increasingly torrid atmosphere largely shaped by humankind.

BIOGEOGRAPHY AND INHERITED GENETIC VARIABILITY

Species evolution and fitness to reproduce are being threatened by a changed and changing ecosphere, possibly leading towards widespread extinction.

Searing Planet Earth; Species' Niche Environments; Imperfections of Natural Selection

How precisely will different species, including humankind, survive the oncoming sear of planet Earth?

Theories and myths about the planet's core and its affects on the surface are many, often canceling out one another. Our concern is how they affect species evolution; so it matters which conditions are present and changing that affect life, including human life.

Over the passage of eons of time, the traditional or home niche of aquatic, land and subterranean species sustained their continued existence, development, and evolutionary reproduction. Each species' hereditary, genetic code framed, and still ftrames, the limits of its physical, structural and organic variability to transform and morph its form, though sister species facing different environments often evolved into new varieties.

Together two factors, operate in tandem and set limits on an individual's ability to adapt to environmental change, especially rapid upheavals.

An *individual* within a species *might* adapt to a changing environment by transforming its feeding patterns or habits during the period of its *own lifetime*. But once an individual or a species comes up against its maximum thermal limits, it likely will expire before producing progeny.

If an individual alone or with a sexual counterpart reproduces, it may further the path of development and evolution. So *over many generations* a species might cumulatively preserve and develop favorable variations, or otherwise discard, change or destroy unfavorable or useless ones. Its structures or organs might adjust to *slow changes* in geography or environment and outlive, reproduce, and outlive those that do not structurally or otherwise develop.[12] Darwin said that.

HUMAN ENCODED GENETIC LIMITS

Humankind also has genetically encoded maximum thermal limits in adapting to environmental change.

The question is: Has evolution of our species kept up with the challenges presented by Nature and the needs of other species?

Selective evidence of the Anthropocentric thermometer may seem camp; but the Union of Concerned Scientists' analysis of six decades of summer weather data for five major Midwest urban areas (Chicago, Cincinnati, Detroit, Minneapolis and St. Louis) reveals that Midwesterners are living with more hot and humid summer days and nights, more heat waves, and fewer cool days that bring relief from heat. Particularly vulnerable to health risks associated with dangerously hot weather are urban residents, older people, children, and those with cardiac, pulmonary and kidney diseases. [13]

The dangers are only too evident, as earlier Union of Concerned Scientists' reports emphasize. Unchecked global warming could greatly increase the incidence of heat-related illness and death. Indeed, some 22 people died from heat-related illnesses during the June 2012 heat wave in the Northeast; and by July 9th, heat claimed 82 lives across the United States.

These may seem like small numbers. But at stake here are human evolutionary vulnerabilities, not only to their Maximum Thermal Limits, but also their organic and physical structures able to cope with the *rapid* onset of heat, lower oxygen levels, particle-infused air, and congested living conditions. [14]

EVOLUTIONARY BIOLOGICAL VARIABILITY

"Nothing at first can appear more difficult to believe than that the more complex organs and instincts [of species] have been perfected . . . by the accumulation of innumerable slight variation, each good for the individual posescessor," Darwin wrote in *Origin of Species*. For "all parts of the organization and instincts offer, at least, individual differences—that there is a struggle for existence, leading to the preservation of profitable deviations of structure or instinct —and, lastly, that gradations in the state of perfection of each organ may have existed, each good of its kind."

Dr. Randolph Nesse says Darwin was incorrect in interpreting species' past evolutionary variability in adapting to changing environments; and that that natural selection is not necessarily advantageous to a species' body structure. Imperfection of the genome, looking at the collection of genes, is not like a perfect, advantageous blueprint. And once set in place, natural selection cannot go backward, only forward.

Looking at a quadruped donkey, for example, all organs hang from the backbone. But if you compare the separate evolutionary origins, going from a quadrupled donkey to a human biped, one sees mixed-up placement of anatomical human organs handing downward, complicating their independent functions, tangling up intestines, and, bending forward, causing lower back-pain.

Other designs of the mechanisms of the human body are also flawed by natural selection. Natural selection may shape the body for maximum reproduction, often in a changing environment. But why did natural selection fail to build the body better?

As well, human evolution structurally allowed the half-circular rotation of legs and arms; but these advantages proved ill-suited for the human-design of tools and physical structures, leaving a legacy of investment in ill-shaped machines like computer keyboards that can injure users with maladies like carpal tunnel syndrome.

Another example reveals that the human eye jiggles constantly to avoid the hole in its back, creating a blindspot acquired by natural selection. By contrast, an octopus has a perfect eye for vision, without being forced to move its eye. An eagle eye, however, with a half-mile vision is not color-coded for discerning exactly what it sees. The point is this: natural selection by evolutionary transitions creates imperfections and can't go backward and remake the body design of any creature.

And evolution the human eye is modified, so single cells fire action potential in the different regions of the visual system responding to elementary stimuli, transmitting information by various relays extending from the retinal ganglion cells to higher-order visual areas of the brain.[15] Yet, other species such as big cats like leopards have inherited visual systems superior to ours, able to more rapidly to attack, retreat or freeze in confrontations.

NATURAL SELECTION AND ITS LIMITS

Human natural selection also has evolved body defenses against sicknesses, that may prove a saving grace for survival. Though bacteria and viruses evolve far more quickly, human responses and defenses to them have evolved over a much longer time frame. Natural selection awarded humankind antibodies as a sort of memory, say for measles, so if a population lives through the first exposure, it can't be infected again. As well, natural selection has given humans the capacity for pain, as by fever regulating the body temperature to save the vestige during sickness—that, contrary to modern medical practice, should not be offset by drug intervention to drive down internal heat.

There are definite limitations to natural selection, though: The Black Death, other plagues in Asia and the Americas, and purposefully disseminated diseases destroyed millions. These became stepping stones in both European migrations and blinded conquests of others.

There is also no way for the body alone to *rebuild* the heart or the brain; and without medical intervention, individuals likely might expire.

As well, DNA chromosome pieces get clipped at their ends as cells divide, so *telomeres* that would allow longer life and maximum reproduction fitness are cut short with humans aging.

Natural selection also builds up the potential for human suffering: Anxiety of brain activity bent on survival, or being successful among peers, can aid survival through alternative competitive strategies. But anxiety also brings on disease, leading to early death in taking actions to alternatively fight, take flight or freeze. Male competitive anxiety leads to earlier death than among females, who functionally reproduce or forego reproduction to collectively care for children and grandchildren, living longer for maximum group reproductive survival. And fortunately medical administration of medicines, designed to block negative anxieties, have evolved to allow individuals to reach their maximum capacities for competitive success.[16]

So there we have it: Natural selection produces bodies that have physical limitations and brains unable to release untapped potential, especially so as climate and social conditions change. And there is no going backward on the evolutionary tree of life to make corrections that would have better prepared humans or other species to cope with rapidly changing environmental conditions or climate, especially those bringing on extreme heat or cold exceeding the maximum thermal variation of species.

RAPID CHANGES IN BIOGEOGRAPHICAL CONDITIONS

Despite the genetic variability of an individual, a species, or its phylum, together they collectively may be annihilated by *rapid changes* in biogeographical conditions—a state that during the Anthropocene has posed increased dangers to species life.

Innumerable adverse or favorable conditions also may arise.

For a single species living through different geographic climate changes, developmental variations often occur, forever differentiating the structures and reproductive organs among its *own descendants*. Thereafter these descending varieties can no longer directly mate with their extant forebears to produce progeny, nor usually through their forebear's *other* descending genetic lines—thereby increasing species diversity.

Favorable conditions have emerged in our strange age of genetic bioengineering, as scientists seek to overcome barriers, often merging the genes of the same or different species' descending lines. Empowering the body to regenerate tissues by self-healing treatments of various proteins, fibers or cells, and cloning extra copies of semi-specialized stem cells of adult patients, the future bodes well for revolutionizing the treatment of damaged livers and neurodegenerative disorders like Alzheimer's disease, as well as supplying or restoring other organs.[17]

These measures certainly do not ensure human or other species against ecocidal conditions or deathly cascades of interdependent life forms, though these dangers might no longer be localized, leaping over the barriers of continents, islands, city-states or principalities.

The stability of regional or provincial conditions essential for species diversity, as well as entire ecosystems, today are being wrenched and decomposed by global warming, jeopardizing feeding chains and webs of interdependent existence. Worldwide, too, the sinews of Nature are being uprooted and destroyed on a scale endangering, depriving and disconnecting species from the sustenance and reproductive networks in which they have endured, evolved and developed for millions of years.[18]

It is no longer a simple matter of the time sequence during which each species, their genus, family, and order will expire; but, with the last of them gone in a particular phylum, there may be no precise moment when each will be extinguished in a seeming endless chain leading to an all-encompassing disclimax—consuming the inter-connection between the vast majority of species.

MOMENTARY STATES OF EXTINCTIONS

So consider the following 52,205 species in their momentary states of existence (based on the evaluation of the International Union for Conservation, excluding 9,709 species evaluated but found to be too poorly known to assign a category of "threatened"):

- Of 99% of known bird species. 8601 are not threatened, but 1253 are, making up 13%
- Of 85% of know mammal species, 3448 are not threatened, but 1138 are, making up 25%
- Of 70 % of known amphibian species, 2767 are not threatened, but 1917 are, making up 41%
- Of 29% of the known number of reptile species, 1961 are not threatened, but 772 are
- Of 23% of the known number of fish species, 5346 are not threatened, but 2028 are
- Of 5% of the known number of mollusk species, 1946 are not threatened, but 1673 are

- Of 0.02% of the known number of Arachnid species, 5 are not threatened, but 19 are
- Of 0.3% of the known insects, 2122 are not threatened, but 741 are
- Of 3% of known crustaceans, 924 are not threatened, but 596 are
- Of 3% of known non-flowering plants, 629 are not threatened, but 716 are
- Of 5% of known flowering plants, 3445 are not threatened, but 8525 are

Hope still breeds strong that so long as one species out of a multitude remains alive, their phylum will survive. Still no one knows just how long that one will continue to develop in the wake of environmental changes threatening extinction, initiating further cascades of species life.

AT least four conditions remain essential to secure species diversity: interdependence of species in ecosystems where one exchanges with and provides others food, nutrients, minerals, and raw materials; symbiotic regulation of climate and air, soil and water quality; collective buffers against natural hazards; and maintenance of soil fertility and pollination cycles, and, on land and in water, interlocked feeding and elimination chains.[19]

EVOLUTION AND DEVELOPMENT: SUDDEN ENVIRONMENTAL CHANGE AND THE SCIENCE OF EVO-DEVO

Many species are able to survive, evolve, reproduce, and develop resistance to *sudden changes* in temperatures and atmosphere, terrain and subterrain, seas and sub-sea beds. This is due to the viability and flexibility of species' natural systems that are built into their evolutionary tract of genetic complexity and variability.

Evidence of such variability has been handed down from 19th century scientists.

IN 1831 and 1846, veteran geologist M. J. d'Omalius d'Halloy wrote that it was more probable that new species had been produced by *evolutionary descent with modification* than they had been if separately created (by a so-called first creative cause).[20]

Herbert Spencer also wrote of species developed as organic beings, modified by changed circumstances, as well as gradual acquisition and modification of each of their mental powers and capacities.[21]

Expanding on these thoughts, Darwin penned:

> The process of modification must be slow, and will generally affect only a few species at the same time; for the variability of each species is independent of that of all others. Whether such variations or individual differences as may arise will be accumulated through natural selection in a greater or lesser degree, thus causing a greater or lesser amount of permanent modification, will dependent on many complex contingencies—on the variations being of a bene-

> ficial nature, on the freedom of intercrossing, on the slowly changing conditions of the country, on the immigration of a few colonists, and on the nature of the other inhabitants with which the varying species come into competition. Hence it is by no means surprising that one species should retain the same identical form much longer than others; or, if changing, should change in a less degree.[22]

His essential meaning, says psychologist Howard E. Gruber, is that *within the life of an individual*, death limits a species' modification and maintains its integrity. But if an *individual organism* indefinitely accumulates a habit, so the fortuitous modification takes a durable form in the individual's *own lifetime*, it transforms the habit into an instinct. And the individual thereby may or will lose its identity as a member of the original species; so if the changes are heritable, so too will its progeny lose the old species identity.[23]

A SINGLE individual among a surviving species may not have a time-sufficient lifecycle to make significant changes in either its structure or organs. But if this individual survives long enough to reproduce, its progeny might continue to make changes in its form to adapt to altered geologic conditions and geographic environments.

If we imagine it might takes 100 generations or even millions of years to advance the species' form to catch up with Nature's transformations of geoclimate, then it is possible to view fitness to survive and reproduce in the perspective of the *global time required for evolution and development*—evo-devo—of species forms.

In terms of survival, then, a species and the phylum to which it belongs would develop superior survival capabilities over those that in the same passage of time did not evolve its own fitness to survive and reproduce.[24]

BIOLOGICAL DISRUPTIONS AND EVOLUTIONARY MODIFICATION

Without micro-fauna, mega-fauna could not exist. Without soil, its interlocking micro-organisms, mushroom diversities enriching loams, worm varieties processing biomass and their symbiosis, all land life would be impossible and natural evolution and the development of its web of life would cease.

"Nature does nothing uselessly," Aristotle opened the door to our understanding the complexity of the web of species life.[25] Slow evolutionary development tightened the symbiosis, as Carl von Linné (Linnaeus) observed; for indeed, he said: "Nature does not proceed by leaps."[26] Darwin later added that Linnaeus' famous statement that the *genus* gives the characters of a *species*—and not that the physical species characters make the genus—implies "that some deeper bond is included in our classifications than mere resemblance," adding: "I believe that this is the case, and that community of descent—the one known cause of close similarity in organic beings—is the bond, which though observed by various degrees of modification, is partially revealed to us by our classification."[27]

Species evolutionary modification takes place in step with adaptation to their environmental conditions. Such development involves a forward-looking track-work that can be observed in various creatures.

Darwin cited the example of the tadpole of the common Salamander or Water-newt carrying traces of its ancestral modifications, that your author knows well on his rain-walks in Tilden Park in Berkeley, California.

As G. H. Lewes remarked in 1859, this creature possesses attributes having gills, passing its existence in water. It contrasts with the *Salamandra atra*, which lives high up in the mountains, never lives in water, brings forth well-formed young in a perfect state, though with rudimentary and quite useless parts. If one opens a gravid female, revealed are tadpoles with exquisitely feathered gills, enabling them to swim like tadpoles of the water-newt. Yet this physical aquatic attribute has no reference to its future life, nor has it any useful adaptation beyond its embryonic condition; but rather is solely a reference back to its "ancestral adaptations," expressing "a phase in the development of its progenitors."[28]

It remains a newt by any other name, revealing its evolutionary development beyond the watery pond, now living under different environmental conditions in the mountain reach. Evolution, development and modification are not new.

SOIL, PLANTS, AND ANIMALS MAKING UP THE WEB OF LIFE

As climate shifted out of an Ice Age and into today's interglacial period over the past twenty thousand years of the western-phase of the Holocene, the scattered fragments of Amazonian forest refugia expanded and coalesced; oak and larch began their northerly American continental march, following in the root steps of conifers—illustrating, says Richard Leakey, the need of species to move when climate changes, if they can.[29]

Soil sustains plants; plants sustain animals; among animals, our species could not survive without soil and both plants and animals.

"Essentially, all land life depends upon the soil," wrote my famous Berkeley neighbor Charles E. Kelogg. "There can be no life without soil and no soil without life; they have evolved together."[30]

Lynn Margulis discovered that soil bacterial, fungi and mysenials are base of the life of plants and animals, making up more than 90 percent of Earth's living matter, able to combine with other genetic material, draw on the DNA of other species, and breakdown soils from sedimentary rocks washed into the sea, producing breathable oxygen.[31]

Mycologist Paul Stamets, has also noted, for example, the important role mushrooms play in the survival and health of the Earth and human species. "The task we face today is to understand the language of nature. Nature is a force of good. And good is not only a concept, it is a spirit." says Stamets. Fungi generate soil that gives life. They form a powerful never-ending network underground, detoxifying and contributing nutrients as they move through the soil.[32]

Soil also requires political protection, though.

U.S. President Theordore Roosevelt, hoofing the mountain reach with John Muir, sought to politically preserve soil as the foundation of the natural domain; and President Franklin Delano Roosevelt, tried to protect the western range, both Roosevelts voicing similar concerns, FDR warning: "The nation that destroys its soil, destroys itself."

It is a broad edict. And yet, undercutting the soil and the functions of natural systems, have followed the domestication of land and bioengineering of species—revealing how human destruction of micro-species unravels the web of life.

CASCADE OF SPECIES

Other species might not survive, either. Transformation of geoenvironmental conditions have deprived certain species of the essentials for immediate existence, setting off a downward cascade of species within their webs.

Some among a chain of life-varieties have been obliterated; though others, genetically-evolved, may develop variability of structure and organs to survive.[33]

Species too, may be forced to vary their offensive and defensive structures in order to competitively sustain themselves, as Darwin emphasized. And you can view the species modification process with examples of dinosaurs and other creatures on the excellent second floor exhibits at Australia's New South Wales Historical Museum.

Darwin wrote:

> When many of the inhabitants of any area have become modified and improved …, on the principle of competition, and from the all-relations of organism to organism in the struggle for life, that any form that did not become in some degree modified and improved, would be liable to extermination. Hence we see why all the species of the same region do at last, if we look to long enough intervals of time, [they] become modified, or otherwise they would become extinct.[34]

Then, as a species disappeared from the face of Earth, Darwin added, there was no reason to believe that the *same identical form* would ever reappear.[35]

Some who doubted this found themselves not only wrong, but also caught in the dilemma of trying to go backward in the evolutionary process, rather than follow species development that over time marches forward.

As an example, by 1627, the last of aurochsen bulls that Julius Caesar and Black Forest hunters sought, were annihilated as a species phylum. Three centuries later, seeking pure German origins, under Lutz Heck, who took over the Berlin Zoo from his father in 1931, the Nazis sought to reverse the process using eugenic methods—concentrating on the resurrection of three pure-blooded, extinct species, supposedly untainted by racial mixing. By looting east European zoos and the wilds for the best specimens, Heck hoped to concentrate the genes of the aurochsen (the wild cow progenitor of all European cattle breeds), as well as Neolithic forest horses known as tarpans; as well as the European or "forest" bison.

Reasoning that an animal inherits 50 percent of its genes from each parent—and that even an extinct animal's genes remain in the living gene pool—Heck hoped to concentrate the genes of a last living ancient species by breeding together animals that most resembled the extinct one, so in the time of six or seven generations they would arrive at their purebred ancestor.

Heck scouted the Polish farms for mares, for example, preserving most tarpan traits, to mate them with several wild strains (including Shetland, Arabians, and Przywalskies) to breed them back to the fierce, ideal creature, nearly un-rideable, painted in ochre on Cromagnon caves. Despite the fact that tarpans still roamed Polish northeastern forests as recently as the 1700s, the last wild fertile mare died in 1880, and the last captive tarpan expired in the Munich Zoo in 1887, the species becoming extinct.[36]

SPECIES MODIFICATION IS SLOW

Again, in recounting the geological succession of organic beings, Darwin wrote that:

> The process of modification must be slow, and will generally affect only a few species at the same time; for the variability of each species is independent of that of all others. Whether such variations or individual differences may arise will be accumulated through natural selection in a greater or lesser degree, thus causing a greater or lesser amount of permanent modification, will depend on many complex contingencies—on the variations being of a beneficial nature, on the freedom of intercrossing, on the slowly changing physical condition of the country, on the immigration of new colonists, and on the nature of other inhabitants with which the varying species come into competition. Hence it is

by no means surprising that one species should retain the same identical form much longer than others; or, if changing, should change to a less degree.

HUMAN EVOLUTION ENDANGERS ITSELF

Evolution also altered human structures and organs, often out of step with changes in other species. Like other species, though, evolution inbred human instincts and the drive to survive and reproduce. It advanced humankind's intellectual and spiritual powers of wonder at the creative forces embodied in matter and organisms.

Humanity did not stop at wonder, however: It crafted tools and social mechanisms to comprehend, cope with, and take charge of the natural domain, and ultimately to geographically expand.

Increasingly elaborate tools and technologies ensured human reproduction and dominance over Nature and its species, revealing over time the naked face of its collective reality and inevitable consequences.

The dangers are apparent today. Thermal power is pushing Planet Earth towards entropy. However justified as expressions of "creativity," or solutions to the very problems technology and social disparity have created, we remain determined to secure our own species—often through pillage and at the expense others.

NOTES

1. John Donne, *Holy Sonnets*, no. 9, L. I The sonnets recall unknown forces and a grim future—not unlike those humanity faces today. Darwin, *Origin*, 365-66. See also *Guardian*, "Climate Change," faq, *EST*; December 10, 2010, 07.15.

2. Mark Hertsgaard, 'Climate Change: While Washington Slept," *Vanity Fair*, November 8, 2012.

3. Wynne Parry, "Climate Change Allows Once-Rare British Butterfly to Thrive," *The Christian Science Monitor*, May 25, 2012, http://ezorigin.csmonitor.com/tags/topic/United+Kingdom.

4. Hertsgaard, 'Climate Change," *Vanity Fair*, November 8, 2012.

5. Ecologist Dr. Ed Berg and colleagues examined Kenai, Alaska's Jigsaw Lake's sedimentary macrodeposits of wetland vegetation that accompanied an 10,000-11,000 years ago extreme decline in available water during the Holocene Thermal Maximum. Only thereafter between 9,000 to 8,400 years ago did the water level again rise. So the water draught had taken place well within Kenai's warm Holocene Thermal Maximum spanning 11.5 to 9 million calibrated years before the present. Either the vegetation that became peat bogs occurred *in situ*, meaning in place; or they were carried to today's lake bottom by water or shoreline floatation. But in any case they imply an extreme decline in the water once available, offering a clue to what might happen as the Holocene nears an end in the 21st century, at least in the Kenai. See Ed Berg, "Analysis of the Jigsaw Lake Sedimentary Peat Macrofossils, vers. 2.0," April 13, 2012.

Berg has worked on the cutting edge of investigations of such Holocene conditions sustaining life varieties in Alaska's Kenai, noting that bacteria living in peat likely are very different than terrestrial bacteria. The cause might be that peat macrofossils, living in a highly acidic environment, were not tuned to flesh consumption like their terrestrial counterparts. This might mean that peat bacteria would preserve, rather than consume, both vegetation and animal remains found in peat bogs, offering clues of how hot Holocene temperatures could become before peat was formed, dehydrating and eliminating species.

6. Mike Tidwell, *Bayou Farewell: The Rich Life and Tragic Death of Louisiana's Cajun Coast*, 2006. See also A Spike Lee Film, *When the Levees Broke: A Requiem in Four Acts*, an HBO Documentary Films Event, 2006. DVD Video. Aspect Ratio: 16.9/93973, Four Hours. Also Richard Krooth, "Interview with Photographer Lana Husser: Filming the Devastation," February 2013. And Richard Krooth and Hiroshi Fukurai, *Little Left: New Orleans in Katrina's Aftermath*, manuscript 2006.

7. John Pernetta, *Guide to Oceans* (Buffalo, New York: Firefly Books, 2004), 8-9.

8. Krooth, *Gaia*, "Introduction."

9. Natalie Angier, "The enigma 1,800 Miles Below Us," *The New York Times*," May 28, 2012. Citing David J. Stevenson planetary scientist (Caltech), *Nature*, 2003; Dario Alfè (University College London), et al., *Nature*, 2012; John Vidale, (University Washington) 2012.

10. See Stephen Greenblatt, *The Swerve*, 2011, elaborated by Lewis H. Lapham, "Ignorance of Things Past," *Harper's Magazine*, May 2012, 33.

11. Barbara Natterson-Horowitz. and journalist Kathryn Bowers, "Our Animal Natures," *The New York Times*, Sunday Review, June 10, 2012.

12. Charles Darwin, *Journal*, 1836 and *Origin*, 145-177.

13. See Union of Concerned Scientists, *Heat in the Heartland* (UCS: Climate Change and Your Health, 2012).

14. *The Air We Breathe, passim.*

15. Eric R. Kandel, *The Age of Insight: The Quest to Understand the Unconscious in Art, Mind, and Brain* (New York: Random House, 2012), 250.

16. See Dr. Randoff Nesse, *Evolutionary Medicine*; Evolutionary Medicine.com. See also Marilyn French, *Beyond Power: On Women, Men, and Morals* (New York; Summit Books,1985), 65, 128-29, *et passim*.

17. See The Editors, "Special Report: The Future of Medicine, Boosting the Body's Healing Powers," *Scientific American.com*, April 2003: 48-67.

18. See O. E. Wilson, *The Diversity of Life*, 191-194

19. See Wilson, *The Diversity of Life*, 191-194. See also Frijof. Capra, *The Web of Life*. (New York: Anchor Books/Doubleday, 1996), Part IV, "The Nature of Life."

20. M. J. d'Omalius d'Halloy, *Bulletins de l'Acad*, (Bruxelles,' 1831 and 1846), tom. xiii, p. 581.

21. Herbert Spencer, *Leader*, March 1852 (reprint 'Essays' 1858).

22. Darwin, *Origin*, 365-366.

23. See Howard E. Gruber, *Darwin on Man: A Psychological Study of Scientific Creativity*, Second Ed. (Chicago: University of Chicago Press, 1974, 1981), 185-186. Darwin thought that acquired, hereditary characteristics also were part of the process of the acclimatization of plants, saying for example that the evolutionary development of "Habit is hereditary with plants, as in the period of flowering, in the amount of rain requisite for seeds to germinate, in the time of sleep, etc. . . " Darwin, *Origin*, 139, *et passim*.

24. See Nova, "Australia: The First Four Billion Years," PBS.

25. Aristotle, *Politics*, Bk. I, chpt. 2 (British Library copy).

26. Lineaus. *Philosphia Botanica*, Sec. 77 (British Library copy, 1750).

27. Darwin, *Origin*, 452.

28. Darwin, *Origin*, 491.

29. Richard Leakey and Roger Lewin, *The Sixth Extinction: Patterns of Life and the Future of Humankind* (New York: Anchor Books/Doubleday 1995), 50-51.

30. See Charles E. Kelogg, *U.S. Department of Agriculture Yearbook*, 1939.

31. Charles Darwin, *The Expressions of Emotions in Plants and Animals*. (London: British Library manuscript copy, n.d.). See also, *Variation of Animals and Plants Under Domestication*. (London: British Library manuscript copy, 1868). And see http://www.nextworldtv.com/videos/inspiration-1/ fantastic-fungi.html.

32. See Kenneth Miller, "Mushroom Manifesto: A crusading mycologist says fungi are key to human and environmental health and can clean up everything from oil spills to nuclear

meltdowns," *Discover,* 07/08; 38-47. See also Richard Leakey and Roger and Lewin, *Sixth Extinction*, 49-51. Worms too consume biomass creating top soil in which new plant life evolves. For Darwin, see Krooth, *Gaia, passim.* Yet, on the forest edge, worms may consume biomass to the moment when the roots of old trees are deprived of the nutrients needed to reproduce.

33. See Carroll, *Endless Forms, passim. See also* Krooth. *Gaia*, 1-2, 25-27, 473, 583, 631-632. And Leakey and Lewin, *Sixth Extinction*, 49-51.

34. Darwin, *Origin*, 365-366; Richard Leaky corroborated this with anthropological proofs and script. See Leakey and Lewin, *The Sixth Extinction, passim*.

35. Darwin, *Origin*, 365. Richard Leaky corroborated this with anthropological proofs and script. See Leakey and Lewin, *The Sixth Extinction, passim*.

36. Diane Ackerman, *The Zookeeper's Wife* (New York, London: W.W. Norton & Company, 2007), 78-81, 317. See also Darwin, *Origin*, 365-66. See Krooth, *Gaia*, "Introduction."

II

Thermal Power & Entropy

> Russian scientists have discovered hundreds of plumes of methane gas, some 1,000 meters in diameter, bubbling to the surface of the Arctic Ocean. Scientists are concerned that as the Arctic Shelf recedes, the unprecedented levels of gas released could greatly accelerate global climate change. Igor Semiletov of the Russian Academy of Sciences tells the UK's *Independent* that the plumes of methane, a gas 20 times as harmful as carbon dioxide, have shocked scientists who have been studying the region for decades. "Earlier we found torch-like structures like this but they were only tens of meters in diameter," he said. "This is the first time that we've found continuous, powerful and impressive seeping structures, more than 1,000 meters in diameter. It's amazing."—Eric Pfeiffer. "Giant plumes of methane bubbling to surface of Arctic Ocean," December 14, 2011 [1]

Could a thermal disaster that boiled both the oceans deep and the surface of the planet 55 million years ago again possibly recur during the present Anthropocentric Era? Evidence from the Palaeocene-Eocene geologic boundary is still seen in the Norwegian Sea, revealing that a vast natural, gas-driven accumulation of fossil fuel—hydrocarbons consisting of methane gas—was buried in deep sea sediments. Along with icy methane concentrations (called clathrates) they likely were ignited in a massive submarine explosion by super-heated volcanic rock magna that squeezed through the Earth's crust.

The remaining methane then combined with and eliminated seawater oxygen, forming carbon dioxide that broke the ocean surface and aerated, with any remaining CO_2 turning sea depths so acidic that most sea creatures went extinct. (Possibly only new blooms of surface plankton re-absorbed the excess carbon over the following 20,000 years or so.)

Some 1,500 to 3,000 gigatons of carbon were injected into the atmosphere, pushing up carbon concentrations from around 500 to 2,000 parts per million of air molecules over a geologic eye-lash of decades, possibly lasting centuries. Oceans turned acidic, threatening tiny marine foraminifera at the bottom of the food chain, fueled a cascade of species leading to massive extinctions.[2] Thereafter, it took millions of years for the evolution of *new forms* of foraminifera along coasts and in open seas.

WORLD'S FIRST CONTINENT: AUSTRALIAN LAND AND EVOLUTION

On land, torrid heat dried soils, generating an evaporation and plant transpiration crisis that led to less rainfall, abrupt warming, and periodic migrations of Asian fauna over rough land bridges into Europe and North America. And what took place in Australia some 30-to-40 million years ago offers a future look at what the modern world might expect.

Australia had broken off from Gondwana, a giant world super-continent, taking fresh water with it, making it the first genuine continent.

Set adrift on Australia's floating continent, some 26 million years ago its edges moved towards the torrid Equator, still later drifting North, transformed into an Ice Age climate. During both climate swings, the continent simply dried out.

At first Central Australia was transformed from a massive inland sea covering two-thirds of the continent. Then under a reign of searing heat it became a spreading arid desert of stubble grasses, as the jungles of North Queensland dried.

Over millions of years, grass-grazing animals then fed over a vast area. Ancient possums had evolved into giant kangaroos with solid ankles bones that no longer rotated or twisted, enabling them to bound great distances across grasslands to feed, females reproducing in great number, carrying their young in a pouch. The architecture of evolution would only gradually change.[3]

Cassowaries developed from the rails of Gondwana's other fleeting continental segments, their alimental discharges now keeping 37 varieties of trees alive and well. Overall, though, the continent continued to dry out, its forests shrinking back, leaving the eucalyptus the dominant variety.

DEEP FREEZE

Then some 55,000 years ago, during an oncoming Ice Age, a deep freeze absorbed the seas. The remaining forests further shrank. Giant mammals called Rhino Bilateria and Equatine Alligators hid in caves to survive.

Some flora escaped arid conditions, too, birds carrying pollens and defecating seeds in their gut. Air currents helped distribute their essences. And furry animals transported their burs—throwing newly-arrived seedlings into competition to secure soils and resources, often driving remaining ancient plant forms to extinction, along with the creatures that thrived on them.

FROM ICE TO HEAT

In the long cycle of 5,000 to 55,000 years, the Ice Age ended and Hot House Earth began, following the path of the now-proven Milankovich cycle. In modern Australia, as most elsewhere on Planet Earth, thermal heat replaced most ice, Earth's other major alternative cycle.[4]

"For all the major greenhouse gases the concentrations are reaching once again record levels," the World Meteorological Organization (WMO) Secretary-General Michel Jarraud told a November 6, 2013 news conference in Geneva, presenting the U.N. climate agency's annual *Greenhouse Gas Bulletin*. The dangers, Jarraud said, were being ignored by the public and nations worldwide, though the accelerating trend was driving climate change, making it harder to keep global warming to within 2 degrees Celsius, a target agreed at the Copenhagen summit in 2009.

Summit agreements and annual bulletins had become quite meaningless without enforcement tools, of which there were few, though delegates from over 190 nations met in Warsaw in October 2013 for a U.N. conference to work on emission cuts under a new climate pact later signed in 2015, to come into full force only in 2020. So, Jarraud could only again report: "This year is worse than last year, 2011. 2011 was worse than 2010. Every passing year makes the situation somewhat more difficult to handle, it makes it more challenging to stay under this symbolic 2 degree global average." For the future, greenhouse gas emissions are set to be 8-12 billion tons higher in 2020 than the level needed to keep global warming below 2 degrees, the U.N. Environment Program revealed on November 5th, 2013.[5]

Carbon dioxide is very stable and is likely to remain in the atmosphere for a long time, Jarraud said, with the highest concentrations lasting for more than 800,000 years. So logically, if the world pursues its "business as usual" trajectory, carbon dioxide will probably hit the 2 degree mark in the middle of the century, Jarraud foretold, noting that this would also affect the all-critical water cycle, sea levels and extreme weather events.

He emphasized that the climate system was dominated by the ocean rather than the atmosphere, and the time needed to warm the seas meant the full impact of current carbon emissions would only be felt later. Futures sea levels would overwhelm: "Even if we were able to stop today—we know it's not possible—the ocean would continue to warm and to expand and the sea level would continue to rise for hundreds of years." So Earth's future awaits a death cycle: a human-created thermal furnace; the onset of inescapable increasing atmospheric heat; rising and heating oceans; and, we will see, certain extinction for many forms of species life.

The following two chapters review the contemporary transitions.

NOTES

1. Eric Pfeiffer. "Giant plumes of methane bubbling to surface of Arctic Ocean," *The Sideshow, Yahoo News*, December 14, 2011.
2. See Flannery, *Weather Makers*, 50-53.
3. Stephen Jay Gould, *The Structure of Evolutionary Theory* (Cambridge, Mass./London, England: The Belknap Press of Harvard University Press, 2002), 2 ff.
4. In some places in the desert have been and hopefully will remain oasis, fed by subterrain rivers formed when rare rains filter through sand (which does not absorb water) and cumulate in giant ponds, some over 2,000 feet below the surface. In Morocco, your author has visited oases fed by such river/ponds. Today Permaculture expert Geoff Lawton has take us on a virtual tour of a 2,000 year old food forest in Morocco. This is the real definition of an "oasis": intense lush green growth in the middle of the desert! Imagine helping yourself to date palms, olives, figs, pomegranates, guavas, citrus, mulberry, tamarind, carob, bananas, quince and more—a food forest, still stable after 2,000 years.
5. "Greenhouse Gas Emissions Reached New Highs Last Year. Again." *HuffPost Green*, November 6, 2013, Reporting by Tom Miles; editing by Stephanie Nebehay and Alistair Lyon. "Irrefutable Facts About Carbon Dioxide, Methance, and Nitrous Oxide: Nitrous Oxide." The WMO bulletin said the volume of carbon dioxide, or CO_2, the primary greenhouse gas emitted by human activities, grew faster in 2012 than in the previous decade, reaching 393.1 parts per million (ppm) of air molecules, 41 percent above the pre-industrial level. The amount of the carbon gas in the atmosphere also grew by 2.2 parts per million, higher the average of 2.02 ppm over the past 10 years. And carbon dioxide was not the only threat, as the second most important greenhouse gas, methane, continued to grow at a similar rate, reaching a global average of 1819 parts per billion (ppb) in 2012; while the other main contributor, nitrous oxide, climbed to 325.1 ppb.

Chapter Four

Entropy

Machines & Inanimate Power—
Destroying the Environment

Modern society has constantly shifted the base of technical equipment used in production, running parallel with the advance of inventions, tools and energy use.

The accelerated development of technologies has been led by engines powered by fossil fuels, filling the skies with gases and debris, workers producing commodities seemingly without end. But then the end has finally arrived, draining Earth's resources, aerating greenhouse gasses and effluents, and bringing on suffocating heat beyond the tolerance of many species, including our own, reaching the moment of environmental entropy.[1]

This chapter reveals the past sequence leading to this end.

INDUSTRIAL TIME-LINES

The struggle between Britain and France laid the early framework for industrialization: building the technologies and mobilizing fossil fuels to sustain the forces for production. Releasing carbon gases and debris to the atmosphere, ground cover and seas followed—foisted on European and colonial environments and populations.

Industrialization has had a brief, but powerful impact on Earth. From the 1750-75 quarter century start of the English Industrial Revolution, to the year 2014, atmospheric carbon dioxide rose from 270 to over 400 parts per million of air molecules. By the late 20th century, the annual burning of fossil fuels had already set free about 10 billion tons of carbon dioxide. Some 5 billion tons were recycled in soils, vegetation and ocean sinks—with the other 5 billion aerated, already having cumulatively heated the planet for over 270 years.[2]

SANDWICHED between the layer surrounding Earth's surface and the deep oceans below, live sea life and land plants and animals dependent on the cycling of moisture and oxygen, nitrogen, carbon dioxide, nitrous oxide, phosphorous and other gases. As we will see, these gases and species collided with each other, and came into the crossfire of significant technical, environmental and social changes.

RESTRUCTURING SOCIAL AND PRODUCTION RELATIONS

During the 18th and 19th centuries, England led, and France followed, the demise of feudalism and the sweep of rural populations into great and lesser urban centers—this demographic shift restructuring the groundwork for workers' development of new tools of labor, government-promoted inventions, and production output on a scale never before known.

Britain advanced the technological revolution by burning fossil fuels to generate steam power for production in practically every then-known type of engine.

Her industrial successes turned on these machines using these fuels, as well as oils to lessen the friction of gnashing factory machines, producing intense heat. Machines also generated massive quantities of carbon dioxide, methane and other gases and debris; and, to clean both tool shops and a grimy workforce required soaps made from animal and other fats drawn from farm production. As Britain industrially expanded she burned more coal for power, drew raw materials to process, and imported foods to feed her workforce and population from an expanding colonial empire. France would adapt her methods. Here resides the genesis of our anthropogenically torrid zone of planetary heat.

BRITISH SURGE: OFF THE LAND, INTO THE FACTORY

British expansion from an insular economy to a global empire was without parallel in history. The British surge followed the accumulation by merchants of wealth drawn from foreign societies and by those seeking profits from the labor of English peasants and workers. To build an empire, these aspiring capitalists first separated the English population from the land, turned them into an industrial workforce, and eventually deprived them of ownership of the tools and instruments of labor.

Expanding industries meanwhile attracted these workers. A series of discoveries in the 18th and early 19th centuries placed in English owners' hands a means to manufacture wool and (later) cotton. Until the 17th century English sheep-wool had been hand spun, the product then sent to Flanders for weaving. But in that century the spinning wheel was invented, run first by one hand, afterwards by the foot of the spinner. Although only one thread at a time time could still be spun, James Hargreaves soon produced a machine (1764) in which one wheel turned eight spindles, spinning eight threads instead of one. Seven years later, Richard Arkwright appropriated other inventions to draw out the spun threads by means of water-power. In 1779 Samuel Crompton combined the two previous inventions so a spinner could spin two hundred threads at a time. With so much thread, weavers now needed Edmund Cartwright's invention of of the power loom (1784) to throw the shuttle back and forth automatically, allowing by 1800 one man to weave more cloth than two hundred could do in 1700. And when Eli Whitney invented the American cotton gin, England also began to weave cotton fibers from United Sates plantations on a grand scale; for one slave could clean with the gin as much cotton of its seed as three hundred had been able to clean by hand. With James Watt's steam engine (1785), the way was cleared for mass production in owners' factories.

Soon new industrial centers like Liverpool, Manchester, Leeds and Birmingham drew spinners and weavers to their side. Over-grown industrial villages sprang up in surrounding country areas. Supplying wool fibers to domestic workers, merchants employed the rural populace, eventually drawing them out of the fields and placing them in spinning and weaving factories. And as more and more people were hurled into industrial villages and towns, the nation's population swelled two-and-a-half times in the central years of Britain's industrial advance from 1750 to 1840.

To protect Britain's superiority over foreign rivals in the cloth and other industries, waterpower was replaced by factories burning coal, evaporating funnels of carbon dioxide and particle matter into the atmosphere. Owners also used their government influence to impose severe penalties on the export of machinery and the emigration of skilled operatives. Alongside the old mill hands, more English working people were drawn to the cities, turned into a mass of poverty-stricken wretches, driven by factory owners and their mechanized production lines to unceasing toil, and unable to purchase or consume what they produced. Owners' exports of manufactures, imports of luxuries, flourished, calling forth improvements in the mining of fossil fuels to burn in industry, on rail locomotives and ocean transport, calling up London's banks for financing both domestic and overseas operations. England was the workshop—and the sweatshop—of the world.

FRANCE: REVOLUTIONS, ARMS, DELAYED INDUSTRIALIZATION

French self-sufficent feudalism, rural household production for cross-road fairs, and small-scale urban production under mastercraftsmen directing endentures workers were the dominant modes of production—delaying a foundation for industrialization and building an urban wage-work force.

Paris itself had small workshops, when it was consumed by revolutionary fire in 1789. The monarchy was bankrupted by social inefficiency in production, rural crop failures and court extravagances; then by populist demands, named the *Rights of Man*, more appropriately the rights of all.

Harvests had periodically failed, leading to the underproduction of grain and other produce to feed the population, also positioning the landlord-owners of estates to economically squeeze the serfs to secure their own living standards, as well as to pass the burdens to populations in villages and towns by raising the price set on the limited food supply. Food riots soon became commonplace throughout the countryside, also witnessing revolutionary urban movements to support the impoverished and workshop *sans-cullotes*, not only in 1789, but also in1830 and 1848.[3]

French revolutionaries and anarchists sought to wipe out all aristocratic privileges—calling for *liberté*—freedom for all to the fullest individual development; *égalité*—equal rights before the law; and *fraternité*, expressing the belief that each and every person with all others held a secure place within the commonality of cooperative, available civic privileges—stressing each man and woman's innate goodness, as Jean-Jacques Rousseau professed.[4]

But their collective defeat lay on the horizon.

There would be no unity of forces once political factions emerged. Consummating and consuming the potential revolution, at least one faction—opposing the spread of popular drives to democratically shape the steps towards social transformation—prevented the population from grasping power.

During 1793, the brilliant, strident lawyer, Maximilien Robespierre, with other leaders of the Revolution, had taken charge of the State Committee of Public Safety, amidst the vacuum of power left by the collapsing Monarchy. As the most powerful, if momentary, leader in the nation, Robespierre soon fell out with George Danton and other Jacobin leaders over the measures to be taken for ultimate success of their cause.

The very concepts of equity and justice soon were forgotten, as Robespierre set loose a Reign of Terror in which Danton and many thousand others were indiscriminately sent to the guillotine. As heads rolled, Robespierre's was next; and the bloody *régime* soon extinguished itself, replaced by a reactionary Directorate, then undermined by Napoleon Bonaparte's *coup d'etate.*

Napoleon then pursued a dual, if duplicitous, policy. At one pole, he weakly supported many of the rights articulated by the Revolution, with his *Code Napoleon* lying at the heart of culture and science. At the other, he used the revolution's momentary attack on feudalism as the foundation for consolidating the state and the pursuit of European continental markets under a government scheme: to safeguard owners' previous rights to hold and rent landed estates; to protect their rights to industry; to build roads and bridges to markets; to police the workforce; to control the relations of employers and workers regulating wage contracts, hours, and working conditions; to introduce a State-designed system of uniform weights and measures, as well as to educate the workforce in mathematics and science—effectively creating the conditions essential for the success of the nation's workshop owners and protection of their means of production.[5]

BRITAIN OUTFLANKS FRANCE

By 1800, British merchants had achieved a practical monopoly of the carrying trade between Europe, America and the Orient.

To displace continental markets held by Britain, in the late 19th century, Napoleon attempted to turn the Continent into an economic union under French control to the exclusion of British influence and commerce. At the height of his power in 1806, he declared a Continental Blockade, forbidding the subjects of France or any other power under French aegis to trade with England or with English colonies. The act threatened English manufacturers and merchants with ruin; for they relied upon European markets to absorb their surplus products, since England's impoverished workers could never buy with their meager pay what their own hands made. In retaliation to the French Blockade, British owners forbade all trade with France and her subject allies. forcing people on the Continent to pay exceedingly high prices for the limited supply of English contraband goods that slipped through Napoleon's iron curtain. So great was dissatisfaction accompanying the Anglo Blockade, so short was the supply which produced the enormous prices charged Continental people, that Napoleon's overthrow began—long before Waterloo.

Yet France devised an innovation which would be repeated by later military powers, to the cost of workers. To ensure quick victory, Napoleon had changed the human cannon fodder of war from mercenary to citizen armies.

For this Napoleonic France then called forth its minions of citizen armies, to die if need be, in the name of conquest. The Revolutionary Government had decreed, August 23, 1793, a *levée en masse* conscripting all citizens into a national army and mobilizing the entire nation's men and materiel. Within a year of the decree of '93, France boasted half a million men under arms in a still-growing popular army. Rather than deploy the costly mercenary armies of the past, the monetary outlay for citizen armies fell—at an appalling and escalating expense, as Napoleon called forth millions to die in the name of a French-led conquest, betraying what he promised—a European revolution.

But there came a limit to Napoleon's reliance on the biggest battalions. For British technology with her engines powered by burning fossil fuel producing steam, the working classes were able to build armaments and cannon, bringing supplies to the frontlines of Waterloo. England's fire-power of her infantry stopped Napoleon's human wave, winning a decisive victory which led to nearly 100 years of English continental, if not global, dominance, peace in Europe and colonial expansion.[6]

FRANCE ATTEMPTS TO COPY BRITAIN

In the wake of the British victory over France, State power was reorganized. New property relations were established. Machines and the quest to uncover Nature's secrets and resources uplifted the spirit and material foundation of ensuing technological and social ties in France and on the European continent.[7]

These forces captured the imagination of French soldier/scientist Nicolas Leonard Sadi Carnot. Seeking to emulate Britain's rapid industrial advance, Carnot built a model of a cycle and reversible engine to quicken France's production—inadvertently discovering the first principle of thermodynamics, explained in his *Reflexíons sur la puissance motrice du feu.*

HIS invention was a simple reciprocal engine. (I ask the reader to understand or skip over the following three paragraphs.)

Burning coal for heat compressed steam that entered cylinders by the action of piston rods. Acting alternatively on either side of a piston, steam forced it back-and-forth within a hollow cylinder. The piston was attached by a piston-rod to a cross-head, that in turn was linked by a connecting-rod to a flywheel that revolved, regulating the gyres of the engine's revolutions.

Through the action of valves, the part of the cylinder compressing steam in front was connected with an exhaust that released the pressure. The piston then moved towards the opposite end; and at the precise moment the steam was cut off, the piston finished its stroke, and the valves automatically changed direction, so the action was reversed.

The whole reverse-cycle engine was limited by metal grinding on metal, making it vulnerable to overheating. It would shut down when the fuel-energy and friction produced so much heat that the production of steam dissipated, causing the machine to slow-down, seize-up and lose power, so it no longer fed stream into the cylinders with adequate force.

Even in the best engines of Carnot's day only about 12.5% of the input of energy as fuel was delivered as mechanical energy; and 87.5% was wasted as inefficiency, changing its potential from work-power into heat and waste. This Carnot did not fully comprehend. Rather, he stated that all forms of motive-power are invariably inherent in burning Nature's fuels; and that such motive-power once produced is not destroyed, but only changed in form. He did not understand the second law of thermodynamics, whereby energy input dissipates in part as friction, heat and waste, replacing work-output and becoming the foundation of disorder—entropy. That revelation was left to others.

MIND OVER MATTER; MACHINES DRIVING ENTROPY

Could British and French factory owners burning coal and other fossil fuels not known that loading the atmosphere with smoke, carbons and debris not risk a future heat-crisis on Earth?

British biologist Sir Charles Lyell (1797-1875) had theorized about ever-increasing domination of mind over matter in the step of evolution during successive geological periods of life—from sensation and instinct to intelligence of higher mammals, improving the reasoning power of humans. Humankind, he said, was poised "to trace the events of indefinite ages before the creation of our own race, and is not even withheld from penetrating into the dark secrets of the oceans, or the interior of the solid globe; free, like the spirit which the poet described as animating the universe."[8]

Animating the universe under human direction would take a giant leap from the late 18th century, leading humankind to alter the biosphere with machines fueled by burning wood, peat, and coal, oil and gas, and much later igniting fissionable nuclear isotopes that they could not completely control.

In the course of such animation, machines became the leading tools of production; and machines put in motion by both animate and inanimate power made thermodynamics the soul of the scientific relations between heat, waste and mechanical, electrical and other expansive or contractive forms of energy.

From its invention, the odyssey of the steam engine became an early 19th century foundation for intensive study. A machine heating water raised its thermal units to the moment when the expansive power of steam could be transferred into mechanical energy as useful work and heat.

It was an old technology known before the Middle Ages, used during the early 18th century, when steam appliances were used for pumping water out of coal and other mines. In 1705 Newcomen built the first practical engine in which pressure was applied to a piston to pump water. In 1765, James Watt then invented the separate condenser of steam for driving machines, later making other mechanical improvements.[9]

At root, though, steam engines required workers to accurately construct and efficiently operate them. And through their operation, the workers graduated from being mere laborers to machine-tenders.

Before the days of the steam engine, a worker's energy had been assisted by animals, wind and water, supplying the power to raise and move objects, for operating farms and workshops. The steam engine relieved the human laborer of using muscle and mechanical power, supplying it a hundred-and a thousand-fold, making the worker the operator, director and tender of the machine.

As higher intelligence and knowledge often was required to direct the mechanical work of the early machine than was demanded in self-directed labor-power, training and some education became a necessary requirement for the workers' industrial efficiency in production.[10] And to build the machine for efficient use of fuel to produce power required still more knowledge.

FIRST AND SECOND LAWS OF THERMODYNAMICS AND ENTROPY

Long before the discovery of the notion of steam-power leading to work efficiencies and entropy, it was known that when two bodies of different temperatures were connected with each other by purely thermal links—by evaporation, convection, conduction, or radiation—heat always flows from the hotter body to the colder one.

In 1850 Rudolf Julius Emanuel Clausius, one of the most eminent of mathematical physicists of the 19th century, and professor of physics at Zurich, Wurzburg and Bonn, explained the second law of thermodynamics.

A predecessor of the famous Max Planck in Berlin, Clausius said the first law of thermodynamics offers only a partial picture, as it posits that energy can be neither created nor destroyed, only transferred and changed in form.

Looking more carefully, Clausius posited that in any irreversible process a small amount of heat energy is incrementally dissipated across and within a system's boundary. By his restatement of Sadi Carnot's principle, known as the Carnot cycle, he put the theory of heat on a truer and sounder basis. His most important paper, "On the mechanical theory of heat," published in 1850, first stated the basic ideas of the second law of thermodynamics: "Heat can not of itself be passed from a colder to a hotter body." And in 1865 he introduced the concept of entropy.

Clausius's second law of thermodynamics offered an irrefutable theory: a mathematical expression of time-based dissipation of the store of energy that can never be fully recovered in its original form. The world as well as the universe would always tend towards such increasing dissipation of energy as disorder, adding layers upon layers of entropy as a one-way time-clock running down.[11]

His experimentally reproducible distinction between heat dissipation and animated work creating output or product lies at the heart of the second law. Simply, as the second law of thermodynamics portends, the forces producing work-output are generated along with heat. Over time, heat will enlarge relative to both output and the value of "work". And that heat will move from the hottest points to the coldest points of a containment vessel, be it a machine, the planet, the universe or a living thing—eventually making heat relatively uniform, creating conditions for energy dissipation and inefficiency, and eventually shutting down the operating mechanism, machine, or the conditions for supporting life.

As the law posits that heat can spontaneously be conducted or radiated only from a higher-temperature region to a lower-temperature region, and not the other way around, the prime examples of irreversibility are in the transfer of heat by conduction, convection, or radiation.[12]

The concept of disorder later was developed in response to the observation that a certain amount of functional energy released from combustion-reactions is always lost to dissipation or friction and is thus not transformed into useful work.

Early heat-powered engines such as Thomas Savery's (1698), the Newcomen engine (1712), and the Cugnot steam tricycle (1769) had already proved their inefficiency, converting less than two percent of the input of energy into useful work output; and a great deal of useful energy also was dissipated or lost into what seemed like a state of immeasurable randomness.

THERMAL POWER LEADS TOWARDS ENTROPY

Thermal power as a driving force actually changes external environmental conditions, and increases the output or value of workers' industrial efficiency—yet loses part of its work power to friction, heat and waste.

Over two centuries, physicists investigating this puzzle of lost energy elaborated the concept as entropy. As Max Planck himself was taught as a student, this can best be explained by the human hand grasping, moving or making an object that, in theory, changes its place or physical form but does not destroy it. A laborer lifts a heavy stone to the roof of a house under construction. Storing his labor-energy, the stone remains on the roof until one day it falls back to Earth, releasing the energy the worker put into the original lifting, but not changing its solid mass to, say, a gas or a liquid.[13]

This is a partial picture, though, a half-way point between the first and second law of thermodynamics. In this simple instance, the energy expended by the worker to lift the stone to the roof is actually dissipated in part as heat, as the worker's muscles burn up oxygen and protein, building up lactic acid in his muscle-mass and, as he perspires, dehydrating his body. What remains of work-energy expanded in originally lifting the stone, once it falls to the ground, illustrates a step towards the second law, though the stone itself has not changed form.[14]

THE SECOND LAW; CONDITIONS CREATING AND ABSORBING CARBON DIOXIDE

During operations of the Second Law of Thermodynamics, energy sources produce ever-less work output and more and more heat, gases and waste, together escaping to cooler sectors of the planet.

Some of the waste is stored, temporarily minimizing planetary heat. Blue sky, green Earth and deep sea are the three principal storage sinks for carbon dioxide, methane and other greenhouse gases. As each storage space has been loaded to near capacity, however, the internal pressure and balance of the conditions of each has been altered by natural implosions and exogenous pressures.

Together these have led to overlapping entropies, their gradations helping to explain the moments of their origins and expiation, especially in the step of technological advances.

As we will see, timing is everything in the long reach of using energy sources producing work or a work-product in relation to the build-up of labor-exhaust and machine-friction, and the escape of heat, carbons and waste. Such sequences often strikingly appear during changes in the technical means of production and the types of burned fuels; as well as in the advances in means of transportation, marketing, and our species self-righteous fight for control over parcels of global geography.

So long as too little heat and waste can permanently be stored on the planet, or can escape into space, entropy on Earth is avoidable.

CLOSED SYSTEM; PLANETARY DISORDER

The Second Law of Thermodynamics posits that within a closed physical system, heat and waste eventually will escape from the highest thermal points to lower and the lowest ones. So, for example, the buildup of atmospheric heat, due to the burning of fossil fuels, converges with solar radiation striking Earth's surface re-radiated skyward, eventually distributing the heat to cooler regions, creating Hothouse Earth, that dissipates relatively little heat into space.

Thus is the Sun's radiant energy reaching Earth at ground zero from outside the planet's physical system reradiated to and absorbed by the atmospheric makeup of carbon dioxide, other greenhouse gases, particle matter and moisture.

Over decades, the continual buildup of such atmospheric heat then leads to planetary disorder, as natural systems spread torrid temperatures to cooler atmospheric regions, to green ground cover and species, and into the colder sea.

The new earthly conditions and species then reorganize themselves in unpredictable ways—as entropies confronting and upending Earth's previous natural systems' self-regulating and self-organizing flows of energy between the Sun and Moon and Earth's elemental surface, subterrains, oceans and atmosphere, that together once supported and redirected evolution of the planet's surviving species.[15]

HEAT IMPACTING SPECIES LIFE

Earth's geography and temperatures play a major role impacting species.

In the northern hemisphere, the gyre of Earth's rotation around its axis in relation to the Sun frames seasonal plant photosynthesis absorbing carbon dioxide and releasing oxygen.

Today this seasonal swing has merged with the anthropocentric release of greenhouse gases and particle matter from the burning of fossil fuels, together creating conditions fostering Hothouse Earth and exposing each species to its maximum radiation endurance.

In themselves, the natural cycles of the planet account for at least half of the release of atmospheric carbons and thus potential heat. This is because the northern hemisphere landmass is geographically larger than that of the southern hemisphere, with northern temperature swings conditioning evolution of predominantly deciduous plants.

In Fall and Winter, as the axis of the planet creates an acute angle between the Sun and Earth in the northern hemisphere, plant life loses it leaves or slows its growth—its uptake of carbon dioxide falling, excess carbons and methane concentrating in the atmosphere. In Spring and Summer as the planet circles its star, with Earth's axis at an ever-wider angle *vis-à-vis* the Sun, it favors plant rejuvenation, photosynthesis absorbing carbon dioxide from the atmosphere and releasing oxygen.

ONSET OF HUMAN-CAUSED ENTROPY

The onset of human-caused entropy changes this rhythmic expectation.

Into the natural cycling of the northern hemisphere's Fall and Winter sky, human-generated greenhouse gases know no season and continue to evaporate, despite reduced Fall-Winter natural plant absorption, enlarging Earth's overall aerated carbon dioxide blanket.

During Spring and Summer, plant life should normally absorb and minimize the animal release of carbon dioxide as cloud cover. But that is not what has happened in the past few decades, when the overall concentration of released gases has shifted the year-round seasonal cycle of carbons steadily streaming skyward, its continuance threatening a planetary calamity.[16]

Indeed, today the warning signs of entropy are overwhelming, though rough calculations about the release and storage of CO_2 and other greenhouse gases are about the best one can do.

What is certain, though, is that during the industrial age release of carbon dioxide absorbing the Sun's radiation, the glaciers are fast melting; the seas are rising far faster than any time in the past 2,800 years; and continued release of greenhouse gas pollution is leading to about 22-to-52 inches of sea rise in the 21st century, swamping islands and coastal territories.

Far beyond humankind's awareness or motives for their actions, their deeds and projects have set and reset this carbon balance, poisoning the only home we will ever know—this and following chapters painting a testimonial picture of the process.[17]

NOTES

1. Krooth, *Arms & Empire, passim*; Marvin Harris, *America Now: The Anthropology of a Changing Culture* (New York: Simon and Schuster, 1981), Chapter 1. Today the division of labor functions continues among production-line-workers—as advertisers produce "fetishes" that all labor-toil is equal—though workers often can not identify their own labor in the commodities they collectively produce and market.

2. See Michio Kaku, *Physics of the Future: How Science Will Shape Human Destiny and Our Daily Lives by the Year 2100* (New York: Doubleday, 2011), 227.

3. Krooth, *The Great Social Struggle*, Vol. III, 7.

4. The rise of the French State and Anarchism stood at opposite poles of the social and political spectrum. Eighteenth and Nineteenth Century French society had been divided both spatially and by class. Geographic partition of city from countryside produced a rising urban bourgeoisie apposed to the social/political interests of rural, small-scale farming. Populations of the rural domain viewed city classes as an alliance designed to reduce their standard of existence. One affected the other in many ways, as a rise in the costs of farming soil brought rising prices to both urban workers and the emergent middling classes, while the reduction in craft income cut both the demand for and prices of farm produce.

The aristocracy bred to sumptuous court life around weakening kings meanwhile effectively discipated revenues or taxes from their own large rural estates, that also was contrary to their interests in raising funds for their own security through a growing military force—periodically used to suppress and control discontented rural populations and petty farming communities.

The military protecting the aristocracy and emergent bourgeoisie also took on a life of its own by suppressing alike owners of small urban workshops, master craftsmen and their apprentices, who were threatening revolutionary change that might upset all other social classes.

It was a complicated social patchwork that could not last the test of strife.

Seeming *déclassé* populations of artists, writers, musicians, and members of intellectual circles periodically shook the urban stage, took to the cultural barricades, supporting the injured, deprived, and exploited. They would support the Paris Commune and protests that held sway for limited moments. Utopian ruralists, emergent communists, and anarchists alike would organize, carrying the flag of equality, fraternity and justice in their hearts. Among them, Russian Prince Peter Kropotkin believed in work equity and the bread of life for all. His intellectual lead brought an army of great thinkers to the fore.

At base, though, their Utopian principles turned on the mastery of coal and petroleum energy to map and sustain the future of French society.

Thus did the 19th century mobilization of fossil fuels for production employing an army of French workers jell with an imperial empire in Asia and Africa, securing the remaining French aristocracy and emergent bourgeoisie, that sustained state authority they held, maintaining their affluence and cultural decadence.

5. Richard Krooth, *The Great Social Struggle*, Vol. III (Santa Barbara: Harvest Publishers, 1980), 11-12.

6. Krooth, *Arms & Empire*, 7-8, *et passim*.

7. Emanuel Chill, trans., *Power, Property and History: Joseph Barnave's Introduction to the French Revolution and other Writings* (New York: Harper & Row, 1971).

8. Sir Charles Lyell, *Principles of Geology*, Vol. 1 (London: British Library, 1830), Chapter 13.

9. Steinmetz, *America and the New Epoch, passim*; S. Lilley, Men, *Machines and History: The Story of Tools and Machines in Relation to Social Progress*, Rev. enlarged ed. (N ew York: International Publishers, 1965), Part One, *et passim*. See also Krooth, *Gaia*, Part VII.

10. Steinmetz, *America and the New Epoch*, 45-46.

11. http://en.wikipdia.or/wiki/ Nocolas_L%33%A9onard_Sadi_Carnot. See also "http://en.wikipedia.org/wiki/Carnot_cycle; http://en.wikipedia.org/wiki/Theory_of_heat; http://en.wikipedia.org/wiki/Second_law_of thermodynamics; and http://en.wikipedia.org/wiki/Thermodynamics_entropy. See also William R Johnston, The Austrian Mind: An Intellectual and Social History, 1848-1938 (Berkeley, Ca.: University of California Press, 1972), 144, et passim.

12. Allan Janik and Stephen Toulmin, *Wittgenstein's Vienna* (London: Weidenfeld and Nicolson, 1973), 94, *et passim*.

13. See http://en.wikipedia.org/wiki/Entropy. The same principle is illustrated in thermochemistry dealing with the absorption and release of heat in chemical compositions. An equal amount of heat is released by the formation of a given weight of a particular substance or compound under common standard conditions. The decomposition of any compound then requires the expenditure in the form of heat of exactly the same amount of energy as was liberated by its formation. I.e., for compounds formed by the absorption of heat, the exact same amount of heat is liberated when the compound decomposes. Richard Krooth, Training in the drafting offices, City of Chicago, Division of Bridges and Viaducts, 1954-1958.

14. Or, for example, you are at the Arctic Circle and use your labor-energy to hang clothes you wash in the morning on an outdoor line. The wind freezes your shirt to a solid, then aerates the solid into a gas, drying your shirt in time for dinner. Your labor-energy is stored, then released by environmental conditions transforming its state. But your energy is not lost, only changed in form from a wet shirt into a solid mass, and then into an aerated gas, leaving the detritus of an altered state—a clean shirt—behind.

15. Compare Krooth, *Gaia*, and Al Gore, *The Future: Six Drivers of Global Change* (New York: Random House, 2013), xvi-xvii. Gore's disbelief in entropy stems from his interpretation of LIya's Prigogine's view of Earth as an open system, importing flows of energy from outside the planet's system then into it, though it and out again—the continual flow of outside energy into the system forcing the open system to reorganize itself at a higher level of complexity, as a self-organizing process, changing nature. It escapes Gore that this is the essence of the Second

Law of Thermodynamics: one moment of entropy followed by the loop of natural, or Gaian, self-reorganization of conditions for evolutionary reproduction of surviving species life. This is also the essential logic of the exhibit "The Future is Wild," Sydney, Australia, Sydney Aquarium, 2012-13.

16. Gore, *The Future*, 312-314. To better understand the First and Second Laws of Thermodynamics, consider chemical combinations producing or absorbing heat:

- The quantity of heat developed/emitted by the formation of a given weight of a particular substance under standard, uniform conditions is always the same.
- The decomposition of any compound requires the expenditure, in the form of heat, waste or otherwise, of exactly the same amount of energy as was absorbed by its formation.

For some compounds created by the initial absorption of heat, exactly the same amount of heat is liberated when the compound decomposes.

17. See Seth Borenstein, "Seas are Rising Way Faster than Any Time in the Past 2,800 Years," *AP*, Feb. 22, 2016. Studies by Bob Kopp (Rutgers Univ. Prof. Earth and Planetary Studies) and Ben Horton (Rutgers Prof. Marine Sciences), *Study One*; and Anders Leverman (Researcher Potsdam Institute), *Study Two*, Proceedings of the National Academy of Sciences, Feb. 22, 2016.

III

Isolines and Isotherms

During the remains of the Anthropocene, the destiny of species on Earth will turn on isolines and isotherms. The following chapters 5 and 6 explain why.

Chapter Five

Cosmos: A Holistic Concept of Earth

> Descriptive botany, no longer confined to the narrow circle of the determination of genera and species, leads the observer who transverses distant lands and lofty mountains to the study of the geographical distribution of plants over the earth's surface, according to distances from the equator and vertical elevation above the sea—Alexander von Humboldt

In 1845, Friedrich Heinrich Alexander Baron von Humboldt, a world traveler and naturalist, published the first volume of his famous *Cosmos*, a vast and comprehensive survey of natural phenomena. His scientific adventures began with the dawn of the 19th century, exploring the Orinoco River and the upper portion of Rio Negro embracing the great tract between Quito and Lima, Peru, then probing the region between the City of Mexico and the Gulf of Mexico, as well as the island of Cuba.

Venturing high in the Andes, in 1799 Humboldt led an expedition, starting from Caracas, Venezuela, reaching the then largely unknown climate of plant diversity in the *Páramos*—a fabled *alto plano* colony where Fibonacci's mathematical equations of fast evolution (see Chapter 14) are illustrated by grasslands high above the equatorial zone. Here he found a vast kaleidoscope of unique vegetation growing at altitudes between 9,200 and 14,800 feet (2800 and 4700 meters) above sea level, covering some 13,500 square miles among the Andes, crisscrossing the skies of Venezuela, Ecuador and Colombia—most of its 3,431 species of vascular plants still found nowhere else on Earth.

Von Humboldt discovered that this neo-tropical, high elevation ecosystem was largely grasslands lying far above the forest tree-line, no animals to be found—no bats, bees, butterflies or hummingbirds as a force of pollination driving plant reproduction—yet it was favorable to rapid species evolution and eventual demise.

It was a climate unlike any known—sandwiched between subterranean forces far below and the aerial ocean surrounding the globe. Thin air pressure and powerful winds, rains and snow at such ethereal mountain heights favored self-pollinating plants, minimizing the stress placed on the plants' vascular tissues which conducted water and nutrients among flowering flora, ferns and their fern-allied phylum.

Humboldt thus viewed what he named the great Colossus of the Andes, the *Cordilleras*, as an inexhaustible source of fertility and plant variety, its climate zones distributing plants according to their niche environmental needs—each responding to its place beneath the heavens and volcanic reaches, as well as to other climates pervading the planet's lower levels and interiors. He wrote:

> There [in the *Cordilleras*] the depths of the earth and the vaults of heaven display all the richness of their forms and the variety of their phenomena. There the different climates are ranged the one above the other, stage by stage, like the vegetable zones, whose succession they limit; and there the observer may readily trace the laws that regulate the diminution of heat, as they stand indelibly inscribed on the rocky walls and abrupt declivities of the *Cordilleras*.[1]

Here dense fogs brought sufficient moisture. Constant winds carrying pollens, seeds and insect eggs also provided available nutrients and means for germination; and daily impulses of sunlight sped and intensified photosynthesis.

From decaying biomass, the strands of vascular tissue on plant stems and leaves (phloem on the outside, xylem on the inside) quickened the intake of methane and carbon dioxide, as the garden of foliage then exchanged oxygen sent into the atmosphere to marry methane, accelerating the collective evolutionary drama.

SYNTHESIZING THE UNITY OF NATURE'S FORCES

Based on his mass of fresh knowledge of geography, climatology, geology, botany, zoology, meteorology and almost every other branch of natural science, Humboldt synthesized the unity of forces controlling the various manifestations of Nature. He argued that it was far superior to perceive connections between them than to study isolated environmental facts, thereby providing a holistic concept of the universe, as if it were a "painting of Nature" (*Natufgemälde*).

With his knowledge broadly focused on the cosmos—its volcanoes, stratigraphy and physical geography—and thus its topography, atmospheric pressure and chemical composition; its humidity, as well as internal heat; and its overriding climate and weather—he added to his *Essai sur la géographie des plantes* (1807-08) with a "Tableau physique des Andes et des pays voisins"—an iconic cross-sectional profile of South America from the Pacific to the Atlantic at the altitude of Mount Chimborazo (20,498 feet) in Ecuador, detailing the window of appearances of different plants at different altitudes.

LAW OF ALTITUDE AND LATITUDE

Later von Humboldt joined to his *Nova genera et species plantarum* (Vol. 1815) a table entitled "Geographiae plantarum lineamenta," that depicted the vertical, zoned distribution of plants on three mountains, one in the tropics, another in the temperate region, and a third in the polar region—demonstrating Humboldt's famous "law" that the changes in plant distribution by altitude matched the ones by latitude.

Concentrating on South American topography, Humboldt revealed plants thriving in eleven zones by altitude, changing temperatures, and soil composition. He compared those of Chimborazo to other mountain ranges and their plant distribution worldwide, from the Poles to the French Pyrenees, Swiss Alps and Tenerife (located on one of Spain's seven Canary Islands). Such plant diversities also revealed something else.

CONTRASTING EXCHANGE OF GASES AT SEA LEVEL AND AT ETHEREAL REACHES

Placed historically in the northern hemisphere at sea level during the Holocene Era, Earth's optimal climate for life had largely depended on biological feedback controls that set the atmospheric pressure of oxygen and nitrogen, as well as concentrations of gases, like water vapor, and carbon dioxide that absorbed the Sun's infrared energy.

Plants exchange approximately 21 percent oxygen for concentrated carbon dioxide. Carbon dioxide is exhaled by animals and released by decaying vegetation. This process has evolved as a system of natural biological balance and regulation, removing excess carbon and oxygen from the atmosphere. Carbon dioxide absorbs the Sun's slowly elevating heat level that otherwise would have dehydrated plant and animal life. And plants release just enough oxygen, preventing spontaneous forest and grass fires that otherwise would destroy the food supply for birds and other animals.

STORAGE VAULTS FOR CARBONS

Meanwhile, our species began cutting forests and domesticating land to occupy, plant and husband, which meant they no longer could act as a major sink for carbons. This left the oceans and atmosphere as the major storage vaults for carbon dioxide on the planet—momentarily relieving heat levels as the Sun's radiant heat reaching Earth's surface was re-radiated to the moisture-laden cloud cover. At ethereal reaches, by contrast, environments and species were vastly different. Here lay the essence of Humboldt's storied account.

COSMOS

In *Cosmos*, Humboldt's aesthetic-holistic epistemology is superior to the meaning of "cosmos" in Homeric times when it meant "ornament" and "elegance." For Humboldt described "cosmos" as the balanced order or harmonious arrangement of the world. And after 1834, Humboldt spoke of the universal laws of nature: "to comprehend the phenomena of physical objects in their general connection, and to represent nature as one great whole, moved and animated by internal forces."[2]

In his own summary of *Cosmos*, speaking in the third person of his adventures and research, Humboldt challenged others to replicate his work:

> Descriptive botany, no longer confined to the narrow circle of the determination of genera and species, leads the observer who transverses distant lands and lofty mountains to the study of the geographical distribution of plants over the earth's surface, according to distances from the equator and vertical elevation above the sea. It is further necessary to investigate the laws which regulate the differences in temperature and climate, and the meteorological processes of the atmosphere, before we can hope to explain the involved causes of vegetable distribution; and it is thus that the observer who earnestly pursues the path of knowledge is led from one class of phenomena to another, by means of the mutual dependence and connection existing between them.[3]

By designating isolines (geologic topography) interlocked with the distribution of heat extremes (isotherms), Humboldt discovered the causes of the distribution of vegetation across the planet. This means that distance from the equator and vertical elevation both determine temperatures; and together they define the geography of vegetable distribution and sustainability. Thus delicate species like skunk cabbage and mushrooms could grow in wet muddy terrains; mosses and lichens would appear below the snowline; at the tropical base of mount Chimbozo and other mountain ranges, palm tree varieties would grow tall; in temperate zones hundreds of varieties of ferns and even sturdy oaks could grace the soils; above the tree line, grasses and uncountable *Páramos* varieties of flowering and other low-lying surface plants would pour forth in glory.

Viewing the cosmos as a layered, interlocked system, Humboldt also discovered that Earth's volatile interiors framed changes in its geological surface formations, together changing life appearances and structures:

> The boundaries of sea and land, of fluids and solids, are thus variously and frequently changed. Plains have undergone oscillatory movements, being alternately elevated and depressed. After the elevation of continents, mountain chains were raised upon long fissures, mostly parallel, and in that case, probably contemporaneous; and salt lakes and inland seas, long inhabited by the same creatures, were forcibly separated, the fossil remains of shells and zoophytes still giving evidence of their original connection. Thus, in the following phenomena in their mutual dependence, we are led from the consideration of the forces acting in the interior of the Earth to those which cause eruptions on its surface, and by the pressures of elastic vapors give rise to burning streams of lava that flow from open fissures.[4]

THE GEOGRAPHICAL DISTRIBUTION OF PLANTS

When he returned to Europe in 1804, Humboldt quickly put pen to paper. In his classic study, *The Geographical Distribution of Plants according to the Temperature and Altitude*, he wrote that plant organisms resulted from rapid adaptations to extreme environmental conditions that had evolved during three to five million years, when the Andes reached an altitude that was capable of sustaining unique vegetation.[5]

We now know the Andes started forming tens of millions of years ago, but it wasn't until the last 2.5 to 5 million years that the northern Andes rose above the elevation where trees can survive. And this treeless space led to the rapid evolution of diversity of other plant species bathed in the radiant energy of year-round, intermittent daytime sunshine. This rapid evolution was either despite or because of harsh ultraviolet rays, icy temperatures, and sweeping weather changes, alternating between precipitation and rivers, dense mist and torrid sunshine.

As a source of raw materials and a storage sink for waste disposal, *Páramos* plants were reliant on the atmosphere. Under extremely low air pressures, plant-evolved stomata pores that exchanged their production of oxygen for carbon dioxide. All excess carbon dioxide was absorbed by weathered sedimentary, igneous and metamorphic rocks, a combination of wind, ice, and precipitation leaching their particles down mountain sides and waterfalls to be deposited as sediments in rivers, lakes, valleys, deltas, continental shelves and ocean floors.

Continual compression of eroded particles then formed sedimentary rock waiting to be uplifted to the mountain fastness and once again eroded, turned into the mountains descending sediments; metamorphosed or re-melted in Earth's engines of fiery, crushing and folding volcanic magna. Facing such pressures and conditions, *Páramos* plants began to evolve into newly formed species, shielded by occupying specialized niches, natural rock cover, and dry hillsides among the grasslands, also sustaining themselves in damp bogs and among protected stands of modest shrubs. All this we know from Alexander von Humboldt's holistic accounts.

DNA, EVOLUTIONARY BIOLOGY, AND FIBONACCI'S VASCULAR BEAT

Does evolutionary biology function differently in the Andes than at Earth's surface terrain? On the planet's surface face, plants have evolved as a major driving force of animal life, especially as flowering plants developed and their nectar was gladly taken by bats and bees, butterflies, hummingbirds and a *mélange* of other birds on the wings of life, as in Hawaii and hundreds of islands, and on continents.

Millions of microscopic organisms live just below the surface, producing nutrients, releasing gases, purifying water, building web-like networks, supporting the feeding and elimination chains of one another and supplying surface organisms with moisture, trace minerals, and oxygen-laden loams. Unnumbered ground animals also consume and process plants, releasing their own gases, water and wastes, defecating seeds, giving burs and seedlings a ride bound up in their coats and furs.

At sea level, plant roots systems rely on a complex system of bacteria, fungi and minerals in the sands and soils. Older roots anchor a plant or tree to the ground and enable it to take up water and nutrients from the groundcover. The surface area of the root system is increased by millions of microscopic projections of root hairs growing just behind the tip of each root, the tip functioning like a plough for the delicate younger roots to take up moisture and minerals giving the organism energy.

EARTH'S EQUILIBRIUM

Clearly organisms affect soils, water, gases and atmospheric composition. But could organisms making and exuding gases affect global climate by creating dynamic balances between methane oxidized by sunlight and oxygen produced by plant life?

Ice core analysis reveals both have been fairly constant in the atmosphere over the past million years or so. As these gases and nitrogen, carbon dioxide and nitrous oxide are made by organisms, used by them, processed by them and eliminated by them as waste, is it possible that they keep Earth's equilibrium favorable for life itself?

One remains in a comparable environmental milieu while climbing hills and elevations up to the tree line, the complexity of vascular plants advancing along Fibonacci's fast-evolution time lines. Beyond that point, moving into the *Cordilleras*' corridors of plant evolution, a completely different environment prevails. On reaching the ethereal heights of the *Páramos* Andes, lacking animal life, DNA alterations among sister plant species reveal each coping with different resource and waste elimination chains, as well as niche stresses, explaining the lightning speed of their evolutionary modification.

In the late 19th century, my Peruvian grandfather lived in the Andes within fifteen degrees south of the Equator, heights reaching 12,000 to 20,000 feet, spelling alternating levels of heat, moisture, and unique "jungle" plants. In this environment more than five tobacco species still thrive, their decay of leaves with other tissues of vegetation upwardly spiraling methane and other gases dissipating slowly, keeping oxygen levels well below 17-to-21 percent of the atmospheric ceiling required for respiration of most earthly mammals.

Yet Andean plant food production remains plentiful in this atmosphere of less than 21 percent oxygen, hard-working, mountain-defying farmers supporting a relatively modest population—living at 10,000 to 14,000 feet above sea level. They still live and labor on slopes of more than 65 degrees from the horizontal, with the distance from the Pacific shore in most places a scant 75 miles, many less than fifty. The wide range of topographies and altitudes within a short distance still leads to supporting not only relatively dense human communities, but also to micro-climates in the short transverse from oceans to mountains, passing through 20 of the world's 34 principal types of environments.[6] And like the *Páramos*, the Peruvian Andes still retain a spectacular diversity of wild and domesticated plant species that have rapidly evolved, gracing hundreds of niche environments.

RAPID EVOLUTION OF SISTER SPECIES

Rapid evolution is the unmitigated byword in the vast reach among the Andes, the modified DNA of species living in each niche environment measured by a few generations. Nature here is weathered, too. The longer the time frame that two sister species diverge from a common ancestor, the more moments each sister's lineage has to accumulate vascular mutations in slightly different climates. Young, new species can be compared from Fibonacci's starting line (see Chapter 14), for they have accumulated relatively few mutations.

GOING GLOBAL

Winds of stratospheric strength also have carried pollens and seedlings worldwide, as to-and-from the North American deserts and the *Páramos*—evolving into such strange variations as the daisy *Espeletia ibei*, that grows as tall as trees. This daisy develops defenses against the elements by growing white hairs on its flowers to protect them from damaging ultraviolet rays, also covering its stem with a thick layer of dead leaves to keep it warm, just as do palm trees at ground and sea level. In fact, the wind-swept, organic powers of Nature have carried *Páramos* plant pollens and seeds to similar niche environments and climate zones across oceans to continents and island with like elevations and temperatures—leading to their further evolution.

Understanding the processes that cause such speciation is a key aim of evolutionary biology. Lineages or biomes that exhibit recent and rapid diversification are ideal model systems for determining these processes. These biomes reported to be of relatively recent origin, i.e., since the beginning of the Miocene era, include Mediterranean ecosystems, such as the California Floristic Province, oceanic islands such as the Hawaiian archipelago, and, today, the scientifically-reconsidered Humboldt neo-tropical high elevation ecosystem of the *Páramo*s.

NATURE'S WHOLE

Humboldt's interest in Nature consisted in the "gradual development of the reciprocal action of natural forces constituting one natural whole"—partly revealed in the geographic, topographical, and temperature-related arrangement and cultivation of exotic vegetable forms, their animated delineations, and even representations in von Humboldt's *Naturgemäide* landscape painting.[7]

According to this thinking, everything in Nature is seen as interconnected, as a living whole. It's like looking at Séraphine Louis, dit De Senlis's, oil painting of spectacularly decorated, interlocked flowers symbolizing an entirely new conception of a singular bouquet of life. As a visionary, she created rivers and trees as a symbolic, universal natural ordering of all species as a composite tree of life.[8]

Or consider that von Humboldt himself was upstaged by the still mysterious holistic representation of cosmos inscribed on the Bronze Age "Berlin Golden Hat," displayed at Vorderasiatishches Museum zu Berlin: Antkensammlung (Pergamonmuseum); Museum für Vor-und Frühgeschichte (Nues Museum), Berlin Bodestraße, 2000). In modern terms, von Humboldt discovered that the cycles of the universal, natural order turned on forces that operated together in uniform patterns, as if joined at the hip.[9] *Cosmos* is the finest study of Gaian cycles you may ever read.

PROPERTIES OF LIFE FORMS

In the body and margins of Darwin's original journals, one discovers his reformulations of von Humboldt's factual findings and theories. So too have hundreds of other writers since penned Darwin's thoughts into their work. Collectively they hope to uncover the strands clarifying the properties and forces shaping evolution.

Turning a page from species biological inheritance of organic forms—whether or not they are able to cope with changing environmental conditions—many properties of species life turn on physics as the architecture of molecules that, of their own nature, produce patterns of existence, quantifiable as numerical formulations.

This is to say, evolutionary instructions encoded in plant or animal genes and molecules both guide and constrain their physical structures and properties. They are able to be expressed by the physics of natural phenomena or by mathematical equations.

DNA molecules, for instance, form seemingly artistic, architectural knots and coils. Inorganic matter and organic spheres also are linked together by patterned viruses, like the tobacco mosaic virus; so that the momentary artificial separation of these viral components in a test tube, once freed of restraint, spontaneously reassemble into a mosaic whole that replicates their original life form. These properties of matter also border Lynn Margulis's concept of evolutionary symbio-genesis—merging two or more separate organisms to form an interdependent, holistic, single new organism that cannot function without its components.

Is this not akin to plants in the *Páramos*, or like the biological process that forms carnivorous plants like the Sundew that grows in Northern California's bogs and wet meadows, their leaves bristling with stalk glands dotted with droplets of a "glue-like substance"? As small insects touch the glands, they become ensnared, the stalks bending down over them, to be digested by enzymes.[10] Yet to be explained is exactly how (*à la* Lynn Margulis), and at which (Fibonacci's) sequence of complexity these stalk glands developed and fused with the glands producing enzymes during the evolution of the singular Sundew organism.

The above are examples of bionome evolution. By contrast, primitive forms of slime mold—the soil amoeba *Dictyostelium discoideum*, as well as the division of embryonic cells—follow Ian Stewart's discovery of definite development patterns that can be approximated by mathematical equations.[11]

Perhaps one day we will discover that Ian Stewart and Lynn Margulis were following Fibonacci's scientific yellow brick path, tracing the mathematical complexity of new life.

Unanswered questions remain. Is there an atomic subtext or subatomic structure of these developmental patterns? What are the deeper forces that bring them about and lead towards entropy and the decline of species—still only vaguely understood by modern day scientists, especially biologists, physicists and mathematicians?

Do these questions hinge on the very beginnings of life on Earth? And what ultimate role have humans so far played in bringing down this living house of global cards? We have yet to discover the answers.

NOTES

1. Friedrich Heinrich Alexander, Baron von Humboldt, *Cosmos*: *A Sketch of the Physical Description of the Universe*, trans., E.C. Otté, (Baltimore and London: The John Hopkins University Press, 1997), Vol. I, 83.
2. Humboldt's "Author's Preface," *Cosmos*, Vol. I, 7.
3. Humboldt's "Author's Preface," *Cosmos*, Vol. I, 7, 8.
4. Humboldt, *Cosmos* Vol I., 161.
5. Friedrich Heinrich Alexander, Baron von Humboldt, *Journey to the Equinoctial Region of the New Continent; On the Geographical Distribution of Plants according to the Temperatures and Altitudes*, 30 Vols., Paris/Berlin, 1804-1824. For a detailed account of Humboldt's explorations and studies, see Andrea Wulf, *The Invention of Nature*, (New York; Knopf/ Random House, 2015). In fact, the development of a holistic concept of Nature had its origins among Bronze Age thinkers. See "Berlin Golden Hat," Vorderasiatishches Museum zu Berlin: Antkensammlung (Pergamonmuseum); Museum für Vor-und Frühgeschichte (Nues Museum), Berlin Bodestraße, 2000). Grecian philosophers, led by Aristotle, also viewed the cosmos as interlocking terrain, atmosphere and sea.
6. Krooth, *Gaia*, 229.
7. Humboldt. "Author's Preface," *Cosmos*, Vol. I, 10.
8. Hans Körner/Manja Wilkens, *Séraphine Louis 1864-1942,* Übersetzt von/Traduit par Annette Gautherie-Kampka (Berlin; Reimer/Benteli, 2009). See also Patrick Guillot's pictorial review, Séraphine Louis, dit De Senlis, on the web; and Wikipedia, "*Séraphine Louis.*"
9. Alexander von Humboldt, *Cosmos,* trans. O. C. Otté, intro. Nicolaas A. Rupke, Vol. I; trans. O.C. Otté, intro. Michhael Dettelbach, Vol. II; Friedrich Heinrich Alexander, Baron von Humboldt, *Journey to the Equinoctial Region of the New Continent; On the Geographical Distribution of Plants according to the Temperatures and Altitude*, 30 Vols., Paris/Berlin, 1804-1824. To depict the *Páramos*, scientists comparing net diversification rates of lineages in fast-evolving biomes, have used 73 dated molecular phylogenies. Based on their sample, they demonstrate that average net diversification rates of Páramo plant lineages were faster than those of other reportedly fast-evolving hotspots; and that the faster-evolving lineages are more likely to be found in Páramos than the other hotspots. Páramos today represents the ideal model system for studying diversification processes. Most of the speciation events that they observed in the Páramos (144 out of 177) occurred during the Pleistocene era, possibly due to the effects of species range contraction and expansion that may have resulted from the well-documented climatic changes during that period. Understanding these effects will assist with efforts to determine how future climatic changes will impact plant populations.
10. Ruth Kirk, Editor, *The Enduring Forests: Northern California, Oregon, Washington, British Columbia and Southeast Alaska* (The Mountaineers & Mountaineers Foundation, n.d.), 35.

11. See Ian Stewart, *Life's Other Secret* (New York: John Wiley, 1998; Penguin paperback, 1999), 22, 66, 89, *et passim*. See also Public Exhibits at the Wisconsin Institutes for Discovery, The University of Wisconsin, Madison, 2011; and the San Francisco Botanic Garden, Ancient Plants, Golden Gate Park, 2015.

Chapter Six

Loss of Control Over Global Heat

> Climate change is a reality. What we are responsible for and the consequences we'll have to face up to is that because our emissions of CO_2 are accelerating the process, things are happening much faster. One of the biggest impacts is the reduced protection of our coastlines due to corals dying and sea level rising. [Anticipate] increased storms and flooding; and millions being displaced around the world.—Jean-Michel Cousteau, April 3, 2012

> Whatever we do as geo-engineers is unlikely to stop dangerous climate change or prevent death on a scale that makes all previous wars, famines, and disasters small; but to continue "business as usual" [burning fossil fuels and decimating forests] . . . could be worse and would probably kill most of us during the century. . . . We can adapt to climate change, and this will allow us to make the best use of the refuge areas of the world that escape the worst heat and drought. . . . Parts of the world, such as oceanic islands, the Arctic basin, and oases on the continents will still be habitable in a hot world. We need to inhabit them and see that they have sufficient sources of food and energy to sustain us as a species.—James Lovelock, *The Vanishing Face of Gaia*, 2009

As Earth's orbit around the Sun alternates from circular to elliptical and back again over timelines of 5,000-to-55,000 years or so, it changes the planet's cycles of heat intensity (isotherms) impacting life-sustaining zones to and from the Equator; a well as topographical elevations (isolines).

In our moment of the Anthropocene, Earth's orbit remains circular.

Yet, human intrusion into practically every activity on Earth has produced heat that is being transferred hither and yon as radiation. As well, the Sun's power striking the globe is re-radiated by thermal convection to the atmosphere. And in turn atmospheric heat is transferred to the cooler parts of Earth.

This should have offered our species a historic signal: loss of control over global heat; its amplification by the destruction of forests naturally absorbing methane and carbon dioxide; and the reduced availability of oxygen for millions of breathing animal species, also testing their maximum thermal limits.

By themselves, these conditions have forced species to search for alternative escape routes towards cooler climates, moving along existing isolines to avoid dehydration, organic stress, starvation, and rarity—either by moving to higher elevations or towards existing polar-regions and the seas.

From regions of torrid isotherms varied species already are moving to higher elevations and into mountain ranges, along isolines away from the Equator, as well as plunging into wetland fringes, and relocating along or near inlets of the seas.

1. CONTRASTS BETWEEN PAST AND PRESENT

The contrasts between the past Holocene and the present Anthropocene are stark. The Holocene Era, covering about 200,000 years, had centered on species moving along isolines for sustenance, resources, and the disposal of wastes without poisoning themselves; enabling them to crawl, slither, walk, fly, or emerge from the seas—enduring in equable surroundings and moderate isothermal conditions, able to evolve their fittest to reproduce.

With the onset of the human-driven Anthropocene, species movements centered on avoiding rising and spreading climatic heat and the increasingly torrid equatorial regions. Heat was advancing across Earth's surface, into the atmosphere, and seas; as well as from hot to cooler regions, as we have seen the Second Law of Thermodynamics predicts. And to avoid reaching their own maximum thermal limits, if they could each species also began moving towards these low-temperature areas surrounding the remaining polar regions, and towards higher elevations; as well escaping to areas near or even in the sea.

Life on the Planet Increasingly Was Being Placed in Jeopardy

The obvious ways to stabilize existence on a torrid Earth should have been to stop burning fossil fuels evaporating carbon dioxide, other greenhouse gases and particle matter into the atmosphere; as well as to replant the globe with forests absorbing carbons and transpiring oxygen and moisture; and to regulate and divert human population movements away from already-overheated urban centers, where machines aerating greenhouse gases degraded living conditions.

But these measures were not and are still not being taken. And looking back in time, perhaps we can uncover the sources of the present human blind spot.

How the Past Offers Clues to Future Species Destruction

As the technological sphere had advanced over nearly three-and-a-half centuries between 1600 and the war-threatening year 1940, it increasingly had impinged on and overcome the biological domain. There were many known sources of the gradual degradation of the ecosphere.

From the early 16th through 18th centuries, the mercantile powers had copied each other's methods. Their empires struggled for a favorable balance of trade and the accumulation of gold and silver as outlays for the implements of war and the pay of soldiers to guard colonial resources from other Powers. They almost completely ignored the biological domain under foot.[1]

Colonial treasure from exchange or looting financed seemingly eternal wars. Against Spain, Holland and France, Britain fought a total of 219 years from 1558 to 1815. For over 103 years she battled Spain, a former ally alienated by English piracy and slave-trading, the execution of Queen Mary of Scots, and the English help to Protestants on the Continent, finally bringing the invading Spanish Armada to heel in 1588.

For 33 years she fought Holland, a country she had helped against Spain, taking over Dutch colonies and trade after clashing on land and sea. And after the Hundred Years' War, she again fought France for 83 years, eliminating her last great Continental rival after Waterloo.[2]

The casting of and firing of cannon, the smoke and debris circling sailing ships in battle, produced relatively little heat compared with effluents yet to come from the "cannon kings" in Europe and gun-ships built by the Carnegies in America.

Still the great Colonial Powers continued to battle with one another for colonial territories under the convenient, if fictitious, doctrine that non-European affairs were outside the sphere of diplomacy, enabling the merchant companies themselves to threaten and fight for empire by military means, while their home governments remained at peace. Portugal maintained a standing army at a heavy cost. Spain invaded the Americas and Caribbean islands. The Dutch took plantations by disease, gun and sword. All the Great Powers established Plantations or trading *depôt*s throughout Asia. And later, England and France fought militarily for dominance in India and North America.[3]

Between them, the dominant power-balance changed hands, shifting the center of European economic, political and military control. Venice held center stage in the 15th century, Spain in the 16th, Holland in the 17th, France in the 18th, and Britain in the 19th. *Pax Britannica* then ruled the waves for a century.[4]

Species lying in the path of advancing colonial empires either fled or were simply destroyed. One estimate postulates that least 100,000 plant species and 20,000 animal species were completely destroyed during these five centuries: some 100-to-200 lost *vis-à-vis* every 1,000 remaining today; compared with one-in-a-million lost before humankind appeared on Earth! Species that survived were often reduced in number, weakened, became rarities, or were completely destroyed, as Darwin observed—leaving the straggling few endangered or threatened by human encroachments on plant and animal habitat.[5]

The Great Powers remained in charge, colonizing great swaths of the globe, captivating its peoples, resources and species, mindlessly unraveling Nature's symbiotic web.

Human Exodus, Machine Society, Trophic Cascades

Recent history of human population movements along isolines preceded intensely rising thermal temperatures associated with the onset of industrial society about 1750 and its most expansive period of colonialism after 1882.

There were few warning of future heat intensity and effusions of waste, elevating atmospheric temperatures—later causing glacial melt, rising sea levels, inundation of islands and continental coastlines—and exceeding the maximum thermal limits of millions of species.

Even considering the second law of thermodynamics, with the burning of fossil fuels, most 19th and 20th century concerns centered on the inefficiencies of machines producing too little work-output, with most energy-inputs being lost to heat and waste.

Nonetheless, during the height of capitalist machine-based society and colonialism through 1939, plantation production and moving agricultural products, mining metals and raw materials for western production, witnessed the global thermometer actively rising.

There was little doubt that heat followed groves of trees removed for agriculture; cutting terrain for mine operations; laying out transportation routes and railway beds; building cities and port facilities—together opening corridors for intense sunlight reaching and dying Earth's terrain. These measures sharply disrupted earlier biogeographical conditions, leading to destruction of species' previous feeding niches, webs and waste-disposal chains.

The catalogue of destruction of Nature and its species speaks volumes. Tearing down forests for farms planted year-in, year-out, in single crops reduced natural plant transpiration as well as species diversity. And machine production and burning fossil fuels sustaining metropolitan and colonial centers emitted greenhouse gases, heat and waste, filling the airwaves, lessening the proportion of oxygen for breathing life. Steam engines and coal-burning ships evaporated ever-more carbons and minute particles of debris into the moisture-laden atmosphere—still increasing in our moment.

Heavy technologies directed by cost-conscious, often brutal, administrators connected the infrastructures for colonial production sent to western markets—laying gage railroad ties; importing powerful locomotives, cranes and other portal lifting engines; docking ships at port facilities; and building warehouses to sleep and feed gangs of workers.

These machines, production and transport complexes alike displaced woodlands, savannah and species, disrupting and chemically changing the circulation of gases and wastes they emitted—clouds of carbons and debris blocking the Sun also impacting germination and transpiration of plants, and the respiration of animals generating heat and releasing wastes.

Many thousands of species and varieties succumbed, unable to acquire the necessary sustenance, breathable oxygen and other means of life. And even if opportunities to rebuild species symbiosis had emerged, the restructuring of their networks would have consumed many hundreds of years. Bioengineering could not and did not restructure the biosphere.

Ignoring the Biosphere

As the technical means of production, trade and consumption advanced, the biosphere was either ignored or plundered. Focus was placed on the expanding map of empire, accumulating wealth and power.

British trade had trebled from the beginning of the 17th century to the start of 18th century. After Napoleon's defeat, the volume of British production demanded wider markets still, for output soon outdistanced the limited markets that its old mercantile trading policies had established. The free traders then successfully used parliament to authorize British commerce with other nations, overturned the limiting Navigation Acts, funded the Royal Navy, and facilitated the import of cheap foreign grains and other foods for consumption by wageworkers.

The lifeblood of England had become overseas commerce, her food largely imported in British ships—though her island fastness was able to survive starvation but three months in case of a naval blockade. Yet, not blockade, but the break point of the Irish potato famine brought want and starvation—leading not only to the repeal of the Corn Laws protecting domestic farmers, but also heralding unlimited free trade, further leading to Britain's vast export of capital and the exclusive administrative rule over colonies to the exclusion of other Powers.

The other Powers followed suit, ignoring the environmental domain unless it spelled profitable opportunities. Global territories were soon divided into colonies and spheres of influence. The continents were entirely parceled out among the Great Powers. The desire for land to occupy, a main motive of nineteenth century colonial annexations, left scarcely any vacant areas—no noman's land—which would-be emigrants could conquer. For every Power was engaged in a mad scramble to build up a colonial empire for its own plantations, as well as for logging and mines by carving out new domains in the Americas, Africa, Asia and the Pacific islands.

Lost in the scramble were John Locke's original concept that human beings possessed certain fundamental, inalienable rights, among them life, property ownership and political equality. The very idea of property ownership on a grand scale was extended in the form of the Great Powers taking colonies and trampling on the rights of other peoples, also displacing evolving species.

Nature's Wonders Ignobly Forgotten

This proved no climate for worrying about poetic admirations of Nature's wonders, nor much else than spreading colonies to sustain mother countries with food and raw materials, and to sell their metropolitan manufactures.

In 1899 Rudyard Kipling captured the public imagination, as well alike of English soldiers and administrators sent abroad to checkmate colonial aspirations for freedom and independence—as if their captives needed British oversight:

> Take up the White Man's burden,
> Send forth the best ye breed—
> Go, bind your sons to exile
> To serve your captives' need.

With colonies under administrative, armed guard, British wealth surged. The sum of trade and earnings abroad was reflected in the accelerated accumulation of capital: in the 63-year period 1812-75 British wealth increased £5,848 million, as compared to the total accumulation of £7,924 millions in the following 37-years (1876-1912). The most rapid accumulation of capital came from the large-scale expansion of colonies after 1875 and especially after 1882.

Obviously, the colonies, where British manufactured goods were exported and raw materials and foodstuffs were obtained, offered investment and loan rewards for British financiers, about one-half of Britain's increased wealth from 1875-1912 seeking reinvestment abroad! Without such capital exports, the colonial production of raw materials and foodstuffs for the British market would have fallen; and imperial earnings from overseas operations would have been miniscule.[6]

Such imperious activities did not comport with looking after environmental integrity supporting the Tree of Life.

Interlude in North America

By contrast in the United States, President Theodore Roosevelt's December 3, 1907 Message to Congress, sought to offset wanton plunder of Nature at home:

> To waste, to destroy, our natural resources, to skin and exhaust the land instead of using it so as to increase its usefulness, will result in undermining in the days of our children the very prosperity which we ought by right to hand down to them amplified and developed.

Still unregulated deeds already had overcome such hopes. Frontiersmen backed by Federal militia had subdued Indian tribes, cleared trees for farming, and altered two-thirds of the entire continental United States of its natural waterways, forest cover, and indigenous species.

From the Alleghenies and Monongahela River to the Colorado River system, the geography surrounding the Mississippi River had witnessed the largest washout of deep-soil micro-organisms and larger species in recorded American history.

Dense forests from Minnesota, Wisconsin and other states along the twisting waterways feeding the Mississippi already had been cut for lumbering, to supply America and the world timbers for beams and studs, building elegant furniture and carved banisters, window-casings and pre-hung doors.

Now without root systems to hold terrain in place, precipitation during Winter, Spring and Fall liquefied top-soils and muds, flowing into rivulets connected to lakes and rivers, pouring into the Mississippi, heading for the Gulf of Mexico.

Where once were trees, expanding farmland periodically dried; the Sun striking the ground was re-radiated to the cloud cover creating domes of heat—temperature rising during once-variable Spring, Summer and Fall. Deepening Mississippi waters then spread into a widening alluvial plain, inundating thousands of farms, towns and cities in a 2,000-mile long cascade to the Gulf.

And as the water periodically subsided and the plain dried, the mud was dredged and ploughed for Southern cotton and other crops that already had ruined the natural fertility of the soil for lack of crop rotation and animal fertilization.

There was no going back to save trees, soils, micro-organisms, or large bear, deer, beaver, silver fox or other forest and river animals. As now blindsided, dams were built to keep out periodic flood waters; and the Federally-financed Tennessee Valley Authority built giant dam-sluices to generate electricity for surrounding states, factories and communities.

Yes, Pare Lorentz carefully tracked this path of "The River" in his 1937 classic documentary film, with Virgil Thompson's original scores, among the most famous ever composed for a film. And yet few seemed to be listening, comprehending the destruction of Nature's web.

Full-Flower Empire: Pacification of Nature

European empires meanwhile viewed Nature's pacification through a different prism. By the late 19th century, the colonial system already was in full flower, having been extended by mobilizing finance, heavy industrial equipment, overseas administrators, and occupying armies.

Within the home territories of old world empires, giant industries came together. Some brought out competitors or combined in the same or similar fields, controlling production so it would not exceed market demand. This was done by competitors dividing production levels and market territories, thereby eliminating competition, and exchanging information. Others large industries secretly agreed on identical market prices set by and for all producers. Still others interchanged technical information on commercial and financial matters; exchanged patent rights; mutually-held each others' stock; interlocked their directorates; and allotted among themselves investments made in machine production.[7]

Other competitors created not only national, but also international cartels. There were some 100 global cartels in 1910, but 321 by 1931, mostly in three industries accounting for two-thirds of all global cartel arrangements. Fifty cartels were in the ferrous metal industry (16 percent of all cartels), 51 in chemicals (another 16 percent of the total), and 80 in transport and communication industries (24 percent of the total).[8]

These and other cartels together controlled a substantial portion of world production, as Table 6.1 shows.[9]

And as cartels multiplied, they concentrated on and allotted between themselves exclusive colonial territories, that brought raw materials and food supplies to England, France, Belgium and the other industrial powers.

An expanded, more rigid colonial system and a vast effusion of imperial capital now supplemented the free trade of earlier years, heedless of the effects on species life.

Indeed, capital exports were the dividing line between mercantilism and imperialism. For large-scale industrial techniques, the integration of finance with industry, the monopolization of industries, cartels and supporting State policies pushed forward capital exports and the export of capital goods for production abroad.

Britain, Germany and the United States together sent abroad as capital goods 26 percent of all exports in 1800, but 39 percent in 1900 and 46 percent in 1913—imperial earnings, representing an excess of revenues over capital exports. Especially in the form of means of colonial production, these became the birthmark on the face of the new imperialism, which replaced the

Table 6.1. Cartels

	Year	Per Cent of World Output		Year	Per Cent of World Output
Rubber	1940	97	Sugar	1937	85
Phosphates	1937	92	Tin	1939	83
Diamonds	1939	+90	Artificial Silk	1929	70
Copper	1939	+90	Synthetic Nitrogen	1932	67
Cement	1937	92	European Lumber	1926	75
Potash	1939	91	European Steel	1929	32
Bottle Glass	1932	91	European Steel	1936	45
Electric Lamps	1939	90			

[Source: Richard Krooth, *Arms and Empire: Imperial Patterns Before World War II* (Harvest Publishers, 1980), page 85.]

old colonial plantation system and harnessed the labor of the peoples of the world[10] —in process destroying species in their path.

Empires Battle; Little Place for Species of the Wilds

So complete was colonial expansion for resources, species of the wild had little place to extend their geographic range and track.

Together the cartelization of industries, export of finance capital, and colonial domination took over 109,744,000 square kilometers of the globe by 1939, subjugated 1,779,342 people in 200 colonies, and left Britain's the dominant power controlling 39 percent of all colonies and 35 percent of all colonized people.[11]

Between 1913 and 1935, however, the extraction of resources and produce varied as the Great Powers created exclusive trade areas with their own colonies, keeping out competitors. As Table 6.2 shows from the Caribbean, South America, Africa, Asia and Oceania came all rubber, about 73 percent of all colonial produce, some 56-60 percent of all seed oil, almost 50 percent of all textiles, some 34-35 percent of all cereals and other foodstuffs, 24-28 percent of all fertilizers and chemicals, and 17 percent of all cereals.

If similar resources from Eastern and Central Europe were added, as well as those hidden in the North American figure as derived from Canada, Alaska and Greenland, a growing portion of these raw materials and foodstuffs obviously came from outside Western and Maritime Europe and the United States. Only in fuels and wood-pulp did the imperial nations produce more than half world production, meaning even at home their pollution of the

Table 6.2. World Production of Raw Materials, Foodstuffs and Rudimentary Manufactures (In percentage of Aggregate Production)

Product Group	Year	Eastern & Central[a]	Rest of Europe[b]	North America	Total Colonies[c]	Caribbean	South America	Africa	Asia	Oceania
All Raw Materials, Foods, etc.[d]	1913	17.7	16.5	26.8	29.8	1.5	3.9	2.1	20.7	1.6
	1925	15.2	15.0	28.7	32.6	2.2	4.4	2.5	21.9	1.6
Raw Materials excl. Foods[e]	1913	12.8	21.3	35.9	24.7	1.0	5.2	2.4	13.9	2.2
	1925	10.0	17.8	28.7	30.0	1.5	5.1	2.9	18.3	2.2
Cereals:	1913	18.7	14.0	33.1	16.9	1.1	4.3	2.3	8.1	1.1
	1925	16.7	14.3	37.0	16.7	0.7	5.6	2.5	6.7	1.2
Cereals and Foodstuffs:	1913	21.5	12.7	19.7	33.8	1.9	2.9	2.0	26.0	1.0
	1925	19.6	12.6	20.1	34.8	2.8	3.8	2.2	24.8	1.2
Colonial Produce[f]	1913	5.4	4.1	15.0	72.6	8.4	29.6	3.4	31.0	0.2
	1925	3.7	4.8	15.7	73.7	6.9	28.7	9.2	28.7	0.2
Oil seeds, etc:	1913	2.4	16.8	20.1	54.4	0.3	5.4	9.2	38.4	1.1
	1925	1.8	13.9	17.8	60.3	0.3	9.9	10.8	37.7	1.6
Textiles:	1913	3.1	7.3	30.2	49.3	0.5	6.0	5.1	30.1	7.6
	1925	4.4	10.1	31.4	47.4	0.4	4.7	5.1	30.4	6.8
Wood pulp:	1913	23.2	29.0	47.0	0.8	0.0	0.0	0.0	0.8	0.0
	1925	15.4	27.9	52.9	3.8	0.0	0.0	0.0	2.8	0.0
Rubber:	1913	0.0	0.0	0.0	100.0	0.0	39.5	15.8	44.7	0.0
	1925	0.0	0.0	0.0	100.0	0.0	7.1	1.4	91.5	0.0
Fuels:	1913	16.7	27.4	45.0	7.1	0.8	0.1	0.6	4.6	1.0
	1925	13.6	21.4	50.8	12.0	2.8	0.9	0.8	6.6	0.9
Metals:	1913	18.3	29.3	42.5	5.2	0.5	0.4	0.1	3.4	0.8
	1925	14.4	24.6	50.8	8.4	0.6	1.6	0.7	4.5	1.0
Fertilizers & Chemicals:	1913	25.9	30.1	15.9	27.7	0.0	23.7	1.6	2.0	0.4
	1925	24.7	31.1	19.6	24.4	0.0	17.1	1.7	3.5	2.1

Source: Richard Krooth, *Arms and Empire: Imperial Patterns Before World War II* (Harvest Publishers, 1980), page 67.

a. Excludes USSR.

b. Includes Western and Maritime Europe.

c. Includes Caribbean, South America, Asia, Oceania. Excludes colonial and subject areas in North American group ie., Canada, Alaska, etc.

d. Includes tobacco and hops, and all other items listed.

e. Excludes cereals and other foodstuffs.

f. Includes tobacco and hops

atmosphere with greenhouse gases continued, along with cutting back forests that could no longer absorb and store carbon dioxide.

As colonialism extended its domain, it mined and developed almost every useful fossil fuel and mineral for industry. It demanded crop uniformity reducing species diversity—destroying perhaps hundreds of thousands of species, concentrating domesticated species in confined areas, and turning people, animals and plants into a zoological infirmary of life, destined to remain underdeveloped to this day.

Future Division of Global Landmass; Species Transfers and Evolution; Diversities and Conflict

On the cutting edge of species diversity, British biologist Charles Elton wrote in *The Ecology of Invasions by Animals and Plants* (1958) that:

> A disadvantage in describing [species] invasion only by examples, however famous, is that this does not quite convey the tumult and pressure of species that have been and are escaping from confinement of their ancestral continents to range the world . . . . Whenever we know the history [of a species] it starts with a very small nucleus of population, growing into an "Autumn Rivulet" and then not infrequently to a flood. And when the population has got that far, its movement is seldom absolutely checked except by natural limits of the environment.

Elton illustrated the historical movement of crop pest invasions by six maps compiled by the Commonwealth Institute of Entomology, showing several kinds of stages in the fungal spread of diseases to crops and trees. As we will see, this was particularly significant, as colonialism and post-world-war neocolonialism focused on agricultural crops to feed and supply domestic populations and a hungry world.

Many crops had been confined to their original domain on continents and islands. But others had dispersed with colonial crops, along with their crop-pests, together moving abroad.

- Thus the original Palaearctic, East-Asian Japanese-beetle, that attacks garden and shade trees, rather than native forests, reached North America by ship.
- The European spruce sawfly, *Gilpinia hercyniae*, also a Palaearctic species, spread to the eastern part of North America.
- The Colorado potato beetle reversed the track-work, traveling by ship to Europe with the crop.
- The licerne flea, *Sminthruus viridis*, not a flea but a pringtail, reached the Antipodes, but not North America.
- The small cabbage white butterfly, *Pieris rapae*, also a Palaearctic species, spread in the middle 19th century across North America, solidly establishing itself in Bermuda, Australia, New Zealand and Hawaii. The Australian fluted scale insect seems to have traveled to almost every country that it could occupy.

- The classic case of the invasion of the cotton bollworm was bound up in history, its transfer from the American South's slave-driven cotton plantation culture reaching Caribbean island planters, and thereafter Egyptian brokers at Alexandria and other ports of call.

A small brown moth whose larvae are colored pink, this cotton-pest, probably originated in India, or perhaps Southeast Asia, living on wild and cultivated cotton. It may even have been in Central Africa before its world-wide spread began.

But it is certain the U.S. pink cotton bollworm, *Pectinophora gossypiella,* invaded Egypt during the American Civil War, when the Northern states militarily interrupted southern cotton shipping from New Orleans. English cotton mills turned to Egypt for raw cotton supplies, using American cotton seeds, accompanied by the boll weevil, together carried in British ships to Egyptian suppliers to plant, deploying *corvée* labor.

Later it may also have been brought to Egypt on imported cotton or cotton-seed from India in 1906, noticed there in 1911, then spread to East and West Africa; also spread to China, the Philippines, and Hawaii; from there to the West Indies, and to Mexico in 1911; and on Mexican cotton-seed to Texas by 1917; then much later carried to Brazil and Australia.

Species invasions, Elton concluded, most often reached cultivated soil, or to land much modified by human practices. A minority of invasive species penetrated further into natural waters and woodlands, and into communities rich and varied, even if they also suffered the results of human occupation through fire, forest succession after lumbering, water control, or channel drainage.

Seeds, species transfers, and crop-infestation arrived through commerce and transport systems, too; under government designs; by inland waterways and seas; in ballast; by animals and plants crossing borders; as well as birds on the wing; and were carried on animal burs and by human travelers with equipment.

Charles Elton's major concern was bringing long-isolated species into contact in distant or local environments, impacting all species.

"If we look far enough ahead," he projected, "the eventual state of the biological world will become *not* more complex [in terms of the biological number of different species in the world], but simpler—and poorer."

Similar species might spread world-wide, though on a regional or provisional geographic scale, only *temporarily* flourish.[12]

On a local scale, the species battle for dominance would intensify everywhere once-isolated species spread. Darwin's concept of species battling one another for survival and reproductive success would intensify, leading to some species becoming rare and expiring.

Long-Term Decline in Global Species Diversity

This past offers clues to how a future consolidating or dividing global landmass might shape species diversities—creating a second, virtual Pangaea through natural processes or human transport.

Colonialism initially brought massive invasions of wild and domesticated species into too little geographic space, especially along coastal fringes—eventually leading to too few natural resources to sustain all; too little nutrients or room for their regional foraging; even leading to a surfeit of their elimination of wastes, poisoning air, water and terrain; and too often breeding so many, resulting in provincial conflicts among them for survival and reproductive fitness.

Source of Decline of Global Species Diversity

Species decline had already happened in the wake of sea transport by caravels, then sailing ships and great coal-burning steam vessels carrying microorganisms in ballast; or as seeds, or plants and animals, brought together as an agglomeration of species.

Ocean transport and territorial conquest and consolidation effectively renewed Pangaea-conditions, as if a single continent. In this sense, a virtual, continental world threw long-isolated species into combat for resources and territories, leading to renewed struggles to survive, eliminating many—the future leaving perhaps one-third as many mammal species as once existed and half as many bird species.

Your author's study of *Gaia and the Fate of Midas: Wrenching Planet Earth* documents how this transpired over time in the expanding territories of Greece, Rome, and Islam; Galilee, the Fertile Crescent and the East; Europe, the Americas, Britain, and Ireland; Australia, New Zealand, and smaller Pacific Islands; as well as China and India, climbing up through the Himalayas and down to the Bay of Bengal—foretelling on-going extinctions, entropies and regional collapse.[13]

2. A VIRTUAL PANGAEA: COMPARISONS OF SPECIES DECLINE

A virtual Pangaea had thus emerged, throwing species dispersed to almost every continent and island into competition for survival, reducing their diversity—providing a grim account of long-term species decline over several eras.

During the more equable Holocene Era, first appeared Iberian sea trade in horticultural species, then Holland's Hortus gardens, its ships on the bounty importing and spreading plants worldwide through the Dutch East India Company. Hortus was followed by the British government's funding the

Royal Navy and merchant marine, as well as ex-officio Kew Gardens, its enlarging plant center becoming the international hub for herbarium transfers.

Britain conveyed plants wholesale around the world, expanding exponentially in the mid-nineteenth century as its grasp extended to Australia and Canada. Spread across much of Africa and Asia, Kew botanists sorted out the world's flora. By 1865, Kew was at the global crossroads, deploying its herbarium to create new agricultural crops in overseas territories, setting up botanic gardens in Calcutta, Peradeniya in Sri Lanka (Ceylon), Jamaica, Singapore, Sydney and Mauritius—with the thought that improved species could be acclimatized and tested as new potential crops.[14]

In Hawaii, a major example where the British set its boot, its takeover of the land, destruction of forests, and creation of vast plantations with domestic and foreign seed served English commercial and financial interests. It came, though, at the expense of island flora, fauna and human populations.

Britain's herbarium lasted in Hawaii until 1893, when U.S. army incursions deposed Queen Liliuokalani. Occupation led to many indigenous animal and bird species being eliminated, one after another, especially avian honeycreepers, whose lifelines as pollinators and nectar-gatherers had evolved over millenniums.

Rising Island Isotherms Drive Population Isolines

During these and other invasions, species diversity plunged on a global scale. Temperature changes also hinged on the Sun's increasing intensity; agricultural and wetland outpour of methane; and the onset of Europe's industrial surge, at first burning wood and peat fuels, then black coal firing mine pumps, factory and marine boilers and engines—evaporating carbon dioxide and particle waste into the air.

Now the balance of the atmospheric was no longer *regulated* or balanced by species symbiotic production of gases and wastes—one plant species, for example, fertilized on the rumen of a bovine animal; the cow or ox in turn releasing essential methane and carbons produced by methanogens digesting cellulose in its gut; the plant absorbing carbon dioxide, photosynthesizing oxygen released to the atmosphere for breathing species, that in turn gave off their own carbon dioxide, that the plants also absorbed.

The future then lay in the hands of our species bent on destroying forests for farming; building urban infrastructure; extracting and burning masses of fossil fuels aerating carbons and particle matter; burning tires, garbage, nuclear fuels, and evaporating other chemicals and increasing volumes of greenhouse gases into the atmosphere.

We may never know the precise moments when rapidly rising global temperatures sped the relocation of assorted species populations—when torrid isotherms became the predominant force pushing species along new isolines to higher elevations and towards remaining polar ice caps and cooler seas.

But aerated carbon dioxide, methane, and particle waste stored in the moist heated atmosphere sharply rose during the Anthropocene, especially after two world wars exploding and burning fossil fuels, along with two dozen or more lethal regional conflicts, powerfully heated the planet, continuing to our moment.

These increases in heat accompanied a rapid rise in global populations extracting and burning or using fossilized oil, gas, coal and coal tars—filling fuel tanks of millions of vehicles and engines.

3. ATMOSPHERIC WARMING: DANGERS TO SPECIES LIFE

Dangers to species life from atmospheric warming belittle my efforts using solar collectors and ice-defrosters on every house I have ever lived in.

Proof positive is Keeling's Curve of carbon dioxide scaling into the atmosphere in both the past and now:

- During the pre-industrial age, the average concentration of carbon dioxide amounted to 275 to 280 parts per million of air molecules
- By 1959 the average concentration had risen to 316 parts per million
- In 1970, it rose again to 325 parts per million
- In 1990 it was 354 parts per million
- In 2000 350 parts per million
- In 2010 395 parts per million
- In 2014 400 parts per million
- And by 2050 an estimated giant leap to 708 parts per million

In a brief 55 years, the discharge of microscopic carbon waste and carbon dioxide into the atmosphere rose from 316 parts per million in 1969 to over 400 parts per million by mid-2014. And as we have seen, carbon dioxide is very stable and is likely to remain in the atmosphere holding the Sun's heat re-radiated from Earth for as long as a century or more, with 2014 concentrations of carbon dioxide the highest for more than the last 800,000 years.

What level of heat can we expect in the future? Looking ahead one hundred years, as if the biological world lived on a fictitious Pangaea reassembled as a single continent, we might see isotherms up 8° Fahrenheit or more, glaciers completely melted, ocean levels 90-to-180 feet higher, and coastlines and islands completely inundated.

Floating on tectonic plates, a great continent might once more coalesce; but then again might separate as isolated smaller continental fragments and islands. New species also might evolve from those surviving and radiating from those dispersed—having once walked, been blown, swum, or flown around the globe.[15]

Once more, speculation remains rife, especially after discovering an eleven-story dinosaur in South America; Australian anthropologists foreseeing the future appearance of avaricious trees as big as elephants that move through the forests for sustenance; and fish surfing the sea by evolving sails.[16]

Human Populations Impacting the Conditions Essential for Species Fitness to Reproduce and Radiate

The past nonetheless offers hope for a few species to survive, reproduce and radiate during the waning days of present Anthropocene. Recent anthropological evidence links the decline of species diversity to habitat loss and human population movements; disappearing landscapes; the human reallocation of remaining territories and resources; internecine wars; communicable diseases —and rapid global warming.

Population size and movement are generally known, though unpredictable, dependent on food supplies and distribution; wars and plagues; tectonic plate movements and volcanic explosions; human species dispersions and the ways they survive, as well as the gases and wastes they emit.

Broad estimate of human population size from 7-to-15 billions, with upper and lower limits, and the rise of carbon dioxide in the atmosphere underline the imbalance of sources of sustenance; the operation of market forces requiring means of exchange to buy necessities; political and economic crises and wars; poverty, malnutrition, sickness, disease, pandemics and mortality—still affecting 20-40 percent of the world's population today.[17]

Wars without End

Major wars alone have swept away millions of people, billions of insects, birds and mammals. World War I battle casualties took nearly 20 million, with their loss of future potential labor-power worth $33.5 to $37.5 billion; and civilians killed at $33.5 billion in future labor-worth. Civilians' property losses on land equaled $30 billion. Production losses at $4 billion; shipping and cargo losses $6.8 billion; losses to neutral populations $1.75 billons; and war relief $1 billion.

World War II led to the death and injury of almost twice as many. From the start of the war in 1939, some 45.5 million soldiers were ultimately available for field service in the British Empire, France, Germany, Italy, Japan and Russia, recruitment having hardly begun in the U.S.[18] Civilians became targets for invasions, pillage, destruction of infrastructure and life.

The restructuring of economies for warfare mobilized labor and resources, almost completely disregarding the conditions of species lying in the path of armed forces. Speed of mobilization was essential to support armies at war, as national industries produced several times their normal capacities.

With large sections of former working populations in the military, with transport disorganized and the supply of raw materials for civilians diminished, rigid controls were placed on all national resources. Waste and luxuries were eliminated for most; the gold supply to purchase necessities was conserved; production was planned by the state to ensure an ample supply of vital commodities; and workers' efforts were quickened, populations being relocated and redistributed to achieve maximum labor-efficiency.[19]

There was little protection for species in the crossfire of the battle lines or along the home front, though there are no accurate estimates of species destruction. But in 1939, the home and colonial territories of the major Powers that were disrupted or torn by war covered 109,744,000 square kilometers.[20] Some 17,000 Frenchmen died in the first day of the Battle of the Somme. And if each 1,000 square kilometer lost but two species in the course of the war from 1939 to 1945, some 219,488 species would have disappeared from the face of Earth—comparable to the kill of a minor extraterrestrial comet invasion.

The land was scorched. Trees, forests, farmland, whole cities were reduced to rubble. Birds by the millions were killed; insects multiplied if only for the moment until their food supply ran out; and wild animals were parched, flared, massacred, or eaten by starving populations.

"Never counted in the 'costs' of war," rails author Arundhati Roy, "are the dead birds, the charred animals, the murdered fish, incinerated insects, poisoned water resources, destroyed vegetation. Rarely mentioned is the arrogance of the human race towards other living things with which it shares the planet. All these are forgotten to fight for markets and ideologies. This arrogance will probably be the ultimate undoing of the human species."

In our moment, these facts, figures and prospects largely are forgotten. Yet state focus is still placed on ceaseless wars and eliminating populations—as well as not allowing them to conceive and reproduce.

Studies that call for human "reproductive choices" claim that the carbon-legacy of an average U.S. female for each newborn child adds some 9,441 tons of carbon dioxide, equivalent to 5.7 times her lifetime emissions.

The Optimum Population Trust in Britain insists that family planning programs in poorer countries should be treated as non-coercive, "legitimate candidates for climate change funding."[21] But paying women to use medical or other means not to have children as a population control policy—designed to reduce potential carbon emissions by each and every newborn—carry the prospect of state enforcement.

Rather than attempting to reduce the extraction and burning of fossil fuels, such programs stress absolute population reduction, especially in non-industrialized and impoverished nations—implicit with ethnic and racialist overtones.

Planetary Life: Fragile and in Jeopardy

Clearly the human presence and rising number on Earth have altered the equation for all species' survival in a vulnerable global ecosystem.

In place of Earth's orbital periodicity, alternating frigid and torrid environments, and vulnerability to dangerous beasts of the wild, humans and other species now face an even greater peril: discharges of lethal gases and effluents from burning fossil fuels—so much so that the very environmental conditions, that species in the past created and altered as foundations sustaining life, are being undermined, placed in jeopardy, and face an uncertain future. The air is so rich in carbon dioxide and other greenhouse gasses and debris that it is disrupting the biosphere and choking life's force.

This is largely because the technological sphere has displaced, degraded and destroyed once-existing environments—outpacing the renewal of the biosphere, creating a deficit in new viable soils, as well as blocking plant and animal evolutionary exchange of minerals, nutrients, pollens and gasses—that otherwise might have sustained the future longevity of the web of life.

By itself planetary life remains fragile; so please allow a scientific moment to explain. Earth's atmospheric ozone layer blocks most of the potentially species-harmful, ultraviolet radiation—seen with special instruments or films. (Radiation is revealed in the part of the electromagnetic spectrum, where wave-lengths are just shorter than those of ordinary, visible violet light, but longer than those of x-rays.) If the ozone layer is repeatedly punctured or disappears due to human creation of certain gases elaborated below and elsewhere, most surface species will be severely damaged, suffer cancers, or prematurely be destroyed.

All gases on Earth, except rare, chemically unreactive argon, helium and xenon, are either made by organisms or processed by them. Over a million years, this is how Earth's atmosphere has maintained a fairly constant presence of oxygen and methane—respectively, oxygen at today's 400-plus parts per million of air molecules, comprising 21 percent by volume; and methane at 1.5 parts per million making up about .08 percent by volume—the disequilibrium between them maintained as radiant sunlight oxidizes methane by 67 percent over approximately each decade.[22] Normally that would bode well for oxygen-breathing life.

But the flow of oxygen is being blocked in part. Its supply is jeopardized as the atmosphere is increasingly being loaded with methane that combines with oxygen to form carbon dioxide—now expected to rise to over 708 parts per million of air molecules by 2050.

As carbon dioxide accumulates in the moisture-laden atmosphere, the Sun's heat re-radiated from Earth is raising temperatures exceeding the maximum thermal limits of many species, undermining the previous symbiotic exchange of plants providing oxygen and absorbing animals' discharge of carbon dioxide as waste.

So Questions Remain

Can species sidestep and adapt to humans uprooting forests as a source of oxygen and as a sink for carbon dioxide? Can they adjust to humans tearing down mountains releasing toxins and effluents into the airwaves and polluting essential groundcover, aquifers, and other water resources sustaining them?

Humans are invading subterranean lodes of shale to extract gas, as well as drawing on aquifers left over from the Ice Age, poisoning water sources and causing earthquakes—also bringing toxic residues to the surface destroying nutrients, trace minerals and soil fertility—the very elements supporting the evolution of various species drawing sustenance and maintaining their fitness to reproduce.

Earth may not even be able to feed and sustain more than half its 2009 population of near 7 billion before the end of the 21st century; yet there is no stopping its growth to 11 billion except by massive contraception programs, says the Optimum Population Trust, failing to mention it would have to enforce uniformly throughout the word.[23]

The alternatives are to face a torrid climate, failing crops, wars for food and resources, and the possible death of billions of our own species from starvation, dehydration and plague.[24]

4. SOME SPECIES MAY ADAPT: EPIGENETIC SURVIVAL

The stepping stones to such an end nonetheless may see some species adapt to rapid environmental upheavals. With changes in terrain and land use, as wild species have moved into cities and towns, animals' habits and mental agilities have been changed. Natural selection and evolutionary success may require their enlarged mental capacities—the need for their brains to become significantly larger and more prescient, enabling different survival methods and the epigenetic development of new habits.

Humans, too, possess neural mechanisms linked to daily rituals and habits, requiring their struggle to go beyond inherited capacities to break old ways—essentially an epigenetic battle for survival as environments change, for example. "It is as though the habits are there, ready to be reeled off, if the neo-cortex determines that the circumstances are right," two brain science researchers say.[25]

As an example of the neo-cortext breaking out of habitual methods of foraging for food, in two known species—the white-footed mouse and the meadow vole—the brainpower of animals escaping to cities or suburbs has apparently evolved, becoming about 6 percent larger than the brains of animals living in farm or other rural areas, says Dr. Emilie C. Snell-Rood, a University of Minnesota biologist. And in rural parts of Minnesota, two species of shrews and two species of bats have developed an increase in brain size as well.

The brains of all six species have gotten bigger, Snell-Rood says, as humans have radically changed the environment in Minnesota. Where once there were pristine forests and prairies, now there are cities and farms. Animals colonizing cities and towns have had to learn how to find food in buildings and other places their ancestors hadn't encountered. In this disrupted environment, animals better at learning new things are more likely to survive and have offspring.[26]

The questions remains, though, whether or not the first generation of invading creatures immediately enlarged their cranial capacities; or if their epigenetic and evolutionary traits only emerged slowly after many generations have passed, as they successfully reproduce and radiate.

Humans Pull the Plug

So which wild species can survive during the Anthropocene? If humans pull the plug, plucking out one species in a feeding or waste elimination chain, the entire trophic structure might well collapse.

Consider salt marshes that shield coasts from flooding, pull pollutants from water, and are nurseries for many fish species. Fishing that removes the top predators like striped bass and blue crabs from salt marshes in New England's Narragansett Bay, Rhode Island, for example—as well as destroy animal prey species like marsh crabs feeding on plants—might wipe out marsh cordgrass with roots anchoring the soil, causing the entire marsh itself to erode and collapse.[27] And, says Wade Elmer, plant pathologist at the New Haven, Connecticut Agricultural Station, a new species of fungus that attacks and weakens cordgrass also may prevent them from creating chemical defenses to ward off the marsh crabs.

From these and other causes, a trophic cascade—destroying feeding-chain, gas-release and waste-disposal relationships between different types of organisms in an interlocked cycle—can cause wetland collapse and the loss of coastline defenses.

Restructuring the Web of Existence

As we have already seen, human shortsightedness can undercut biological inheritance and species epigenetic adaptations—creating barriers to and limiting reproductive success. As our species has restructured the web of existence, creating a torrid global environment, fungus, bacteria and viruses have become dominant strains of microorganisms.

Yet, over generations, humankind has developed genetic resistance to some of these viruses, altering their own DNA for survival fitness. Under a human, invasive-protective umbrella, lore has it that viral microorganisms of smallpox once were once spread by Spain, England and the U.S.; leprosy by Germany; dysentery by England and France; venereal disease by the American colonists; and tuberculosis by Japan.

And on the cusp of territorial expansion, cutting down jungles, deforestation, and eating so-called bush meat, our species has succumbed to threatening natural barriers, microbes, and mass extinctions, as documented by many experts, the best among them including Ivan Illich, *Medical Nemesis* (1976), and Dr. Rob Desalle in his concise study, *Epidemic: The World of Infectious Diseases* (1999).

Over time, some species, ours included, may develop epigenetic defenses and adapted cellular resistance to Ebola and 75 percent of infectious diseases that originate in other animals—stopped short by failure of human adaptation of defenses.[28]

The latest mosquito-carrying Zika virus may well be the cause of women in Brazil and elsewhere giving birth to fetus with abnormally small heads. But so far absolute proof by tropical disease experts is still in its formative stage, demanding extreme caution in the fight against enzootic diseases.

5. THE EMERGING PARADIGM

Like an unfolding Greek drama, human-designed technologies are destroying Nature's quintessential interdependence of species.

Our species meanwhile has retained its myopic vision, burning increasing quantities of fossil fuels and bioengineering species to the moment of Nature giving way to human will.

Geoscientist James Lovelock says that: "Global heat would not have happened but for the rapid expansion in numbers and wealth of humanity; that if we fail to curb global heating, the planet could massively and cruelly cull us, in the same merciless way that we have eliminated so many species by changing their environments into one where survival is difficult."[29]

Proof is not hard to find. Forests are torn down for farming, their timbers milled to build furniture, homes and office infrastructure. Prairies sustaining diverse wildlife are cleared, providing space for towns, cities and extended networks for transportation, mining, and cattle ranches.

Seas become dumping grounds for chemical fertilizers, leaking oil from tankers, used medical equipment, splinters of degraded plastics, together creating fetid oceanic islands of waste—further acidized with chemically-laced rain and moisture-laden anthropocentric greenhouse gases, jeopardizing the essential symbiosis of ocean plants and animals.

Bioengineered species and high-powered chemicals combat or leave still-born such naturally-evolved species as rootworm beetles or others creatures hostile to domesticated crops—human chemistries also leaving behind effluents and poisons seeping into and degrading underlying soils and contaminating foods that animals and humans consume.[30]

Toxic ash and gases from coal-power plants are released into the atmosphere. Government pollution-controls fail to regulate industrial evaporation of gasses, heat and particle matter to the airwaves, fallout causing congestive heart failure and a host of diseases to children, pregnant women, and oxygen-breathing animals.

Agricultural runoff of phosphorous, nitrogen, and other acidic effluents invade lakes, rivers, and wetlands, leading to lifelessness. And the U.S. Everglades, Earth's last remaining river of grass, is flooded with urban and farming pollutants, causing a massive trophic cascade, jeopardizing the life condition of hundreds of diverse species.[31]

As well, there is growing use and abuse of public common lands, drilling for oil and fracking for gas by cutting through shale and poisoning subterranean artesian wells and groundwater aquifers, along with rivers and wetlands that animals and humankind depends on for fresh water. Mountaintops are carved off for coal-, copper-, iron- and uranium-mining, leaving behind taconites and lethal wastes open to precipitation-runoff, jeopardizing soils, drinking water, and river networks once sustaining wildlife, farms and cities.

Into remaining wilderness and jungles invade human hunters for sport and bush-meat-adventurers for profit—setting fires as they go to clear land for planting of colonial crops like palm oil trees and bioengineered soybeans, cotton, corn, wheat and rice.

Corporate planners and poachers mobilize well-armed labor gangs to take over whole forests—to build logging-roads, cut thousand-year-old trees deploying chainsaws, massive tree-cutting-cables and logging trucks—completely upending the natural order of jungle life. And the destruction of forests is stripping the globe of the major source of oxygen and storage reservoirs for carbon dioxide.

Extinction Learning Curve

Once upon a time, when Alfred Russell Wallace wrote his magnificent 1869 Pacific garden island classic, *The Malay Archipelago*, it reflected a near-identical theory, elaborated by Charles Darwin in *Origin of Species*, from his county house half a world away. Both argued that in the course of evolution, favorable variations in plants and animals would *gradually* adapt to environmental change and be preserved for survival and fitness of their progeny; and unfavorable variations eventually would tend to erode and be destroyed.

It was a neat little circle that assumed that less-improved forms of species would succumb in the struggle for life; yet it largely excluded catastrophes from extraterrestrial invasions violently altering Earth's crust and conditions, determining future patterns of life survival and extinction.[32]

Scientific knowledge was running far slower than actual evolution, speciation, and the forces of extinctions under the long-term cycles of torrid and glacial environments—with interglacial, meta-stable conditions like the Eocene marking the dawn of large mammals. Already very hot by today's thermal standards, some 55 million years ago a geological accident (possibly caused by methane released from seabed entrapment in frozen clathrate crystals, or by molten lava exploding beneath Arctic Ocean petroleum deposits) released between one-and-two tetratons of carbon dioxide into the air—each teraton equal to one million, million tons. So much carbon compounds were pumped into the atmosphere that they raised Earth's temperatures by 15 degrees in both temperate and polar regions; and between nine and 15 degrees in the tropics.

Over the following 200,000 years or so, atmospheric conditions returned to their previous state; and the time-line of the ocean's deep water alkaline mixing in with surface water neutralized acidity, allowing the survival of many shell-forming organisms. But not quite so today, as humankind has injected a comparable quantity of carbon dioxide into the atmosphere in two and a half centuries since the start of the Industrial Revolution. Earth itself now lacks sufficient storage vaults for carbons in decimated forests, overloaded sea floors and the atmosphere—and may once again aerate another one-to-two tetratons of carbon dioxide. The rapid accumulation of carbon dioxide in surface ocean water also is making the seas too acidic for shell-forming organisms to survive.

The Sun's radiant heat and accumulated carbon dioxide in the atmosphere is now more severe than it was 55 million years ago, so there is little time for species to adjust to the advance from the moderate heat of the Ethocene to building torrid climates of today's Anthropocene.

Then the Sun was 0.5 percent cooler than today; there was no human-driven agriculture on Earth decimating forest cover; and natural vegetation regulated the climate. Then the atmosphere was better able to cool Earth, unlike human-made aerosol pollution today that limits cooling by three to five degrees.[33]

Often-repeated warnings now are issued from the United Nation's Intergovernmental Panel on Climate Change (IPCC). "Science has spoken," U.N. Secretary General Ban Ki-Moon said in Copenhagen at the November 2, 2014, launch of the IPCC fourth and final report. "There is no ambiguity in their message. Leaders must act [to stop greenhouse gases produced by burning fossil fuels]. Time is not on our side."

The IPCC assessment, which incorporated the findings of three earlier reports, reaffirms with 95% certainty that global climate change is both real and a mostly human-made problem. The entire project, which reviewed approximately 30,000 studies about climate, also suggests that if greenhouse gas emissions continue without intervention, there could be "irreversible" impacts, such as rising sea levels, more frequent heat waves and even a change in the human population's male-female ratio as gender lines change due to heat causing male heart problems shortening their longevity.

Meteorologists reported in October 2014, that the year could be the hottest year on record. With 2015 and estimates for 2016-17 increased heat, if climate change continues, billions of dollars worth of inundated seaside property could be destroyed, while some states could see crop yields drop by as much as 70%.

Is it simply conjecture that climate change could cause a rise in various infectious diseases, world hunger, respiratory problems, and heat-related illnesses, such as cardiac arrest and heat stroke? And among non-human animals and herbaceous plants and trees, will they be forced to adjust to new environments, face rarity and become extinct?

Rising isotherms clearly were now in charge on Planet Earth, driving species for survival and reproduction along isolines to cooler, if still unstable regions.

Which Emergent Species Will Recolonize Earth?

Which species were able to colonize Earth in eons past?

Since the first living cells appeared on Earth, they were accompanied by parasitic viruses. These cells and viruses together made up Earth's early mosses that had colonized the land and drawn carbon dioxide out of the air.

These viruses may even have predated their hosts, literally inventing cells as a renewable resource they could exploit to make new viral particles—a forced form of a biological feeding system. Since then, the viral universe, or virome, appears to have infiltrated and replicated in the cells of all living animals, plants, blue-green algae and molds, larger viruses and bacteria.

The Sun's radiation provided the energy for the reproduction of all these biogenic forms. But its emissions hitting Earth were re-radiated to an atmosphere holding far less heat than today—leading to falling planetary temperatures, periodically causing the extinction of emergent, warm-weather plants, animals and viruses.

Iceberg Earth slowed biogenic formations, too. The changing orbit of Earth's gyre every 5,000 to 55,000 years from circular to elliptical (egg-shaped), periodically leading from earthly lethal heat to frigidity and freezing seas, also froze organic life, leaving a trail of ancient tundra—*organic soil* frozen to the bottom of a still-surviving 2.7 million year old ice sheet. This is signified by millions of atoms of the isotope beryllium-10, revealing this organic soil had been preserved beneath two miles of ice in Greenland, the second largest remaining glacier on Earth.[34]

There was still another unknown freezing the planet: Perhaps planetary extinctions had followed the passage of our solar system through the arms of the Milky Way—a twenty-six million year old channel through the so-called Oort cloud, regularly or periodically producing comic showers that rained destruction on Earth. If the larger body of Earth distorted its surrounding gravitational space-time field, pulling smaller astronomical objects into the distortion, Einstein's General Theory of Relativity might set the stage for repeated extraterrestrial invasions, undercutting emergent planetary life.[35]

Other possibilities are still pending, linked to the idea that glaciation caused a mass extinction at the end of Ordovician Era.

In fact there have been five-and-a-half *known* mass extinctions:

- Eocene (About 55 million years ago, age of giant beasts)
- Cretaceous (About 65 million years ago)
- Triassic (About 208-210 million years ago)
- Permiam (About 245 million years ago)
- Denovian (About 360-365 million years ago) and
- Ordovician (About 438-440 million years ago)[36]

Of the first of the Big Five species annihilations on Earth, culminating at the *end* of the Ordovician, the oceans emptied out, eighty-five percent of marine species died off, and the animals and plants that survived became the foundation of the modern world of existence.

Only a few graptoltes species survived during this first great extinction 440 million years ago, today identified by their body images on successive layers of slate rock found at Dob's Linn, seen in the Southern Uplands of Scotland.

Able to be calibrated in the short period of a human lifetime, long before our species even existed, they represented an entropic change from what was a habitable seafloor to an uninhabitable one, impacting some 650 different graptoltes species—contoured by their dominating V-shaped images—into a handful of survivors with a streamline body plan, destined to repopulate the seas of the Silurian age of ancient fishes during the Paleozoic Era.[37]

Two Eras: Holocene and Anthropocene Constricting the Field for Nature's Success

Why do these evolutionary spirals and their demise matter?

They offer evidence of Earth's future for sustaining life. They help to judge the possibility of a sixth great species extinction, focusing on the Anthropocene and its immediate predecessor, the Holocene.

During the Holocene, species, without ethereal pre-design, had created, processed, used, and reimbursed Earth's gaseous balances, as well as periodically altered feeding chains. Lasting well over 200,000 or a million years the outflow of their own gases maintained an equable climate for life.

Species moved along isolines, not atmospheric heat intensity; so there was little if no inconsistency of these forces with the relatively small hominine or human populations burning peat, wood, coals and coke—together barely raising isotherms.

By contrast, during the Anthropocene humankind has transformed natural environments to the moment when technologies and genetic alterations have transformed life's condition and essence, as well as burned fossil fuels, rapidly advancing material production, generating heat and waste.

Current evidence of the consequent destruction is overwhelming. We have destroyed forests where wild species once flourished; bioengineered species' essences; and burned fossil fuels, forwarding production by aerating heat-absorbing greenhouse gases and particle debris.

At the same time we have eliminated the storage vaults for carbon dioxide in jungle, woods and seas.

We have moved species between continents and islands, rural and urban areas, lowland and uplands, undercutting the biogeographical sustainability of once interlocked species and their feeding and reproductive chains, that together stabilized global isotherms.

The resulting enlarged outflow of warmth and waste has spread globally to polar regions and higher altitudes, jeopardizing species' maximum heat tolerances and undermining their conditions for life.

Thus are modern methods clearing and burning forests, relocating species globally, and burning fossil fuels very different from those of the Holocene.

Today a rapidly growing population is igniting massive quantities of oil, gas, black coal, and nuclear isotopes, yet developed only small alternative power sources drawn from the sea, air, and geothermal subterrains.

The fundamental difference between the two eras, Holocene and Anthropocene, materializes under the Second Law of Thermodynamics. For during the Holocene, the outpour of heat and waste was less intense, warmth slowly spreading from the planet's tropics and temperate zones to the frigid polar regions of vessel Earth. By contrast, during the Anthropocene, today's burning great quantities of fossil fuels and nuclides loads the worldwide atmosphere with torrid heat and acids on an unprecedented scale—overwhelming species maximum thermal limits, inimical to their survival and reproductive fitness—effectively narrowing the field for diversity.

Simply put: The technological sphere evaporating lethal gases and particle debris is destroying the organic composition of matter required by most species for survival, as well controlling the ecosphere—diminishing the effectiveness of species' symbiotic release of their own gases and wastes, that once collectively maintained and reimbursed the global environment.

Who's to Blame?

This has happened because our species has elevated its power to create advanced tools, burn fossil and nuclear fuels, and cognitively reason above the ability of other species. We have taken effective control over animals, plants and their physical environment—changing the biosphere and species diversity.

Examples of industrial and mining polluters are well known.

In 2013, the Colorado-based Climate Accountability Institute found that 90 companies had been answerable for almost two-thirds of global greenhouse gas emissions since the start of the Industrial Revolution. The top 90 emitters included 50 investor-owned energy companies like BP, Exxon, Mobil and Shell, along with 31 state-owned companies and some proprietary nation-states themselves. Some 83 of the 90 were coal, oil and gas producers; and the remaining seven were cement manufacturers.

This group has been responsible for a release equivalent of 914 gigatons of carbon dioxide, which account for 63 percent of all industrial carbon dioxide and methane emissions between 1751 and 2010. And half of the greenhouse gas emissions were emitted in the past 25 years—enough to cover Earth's entire surface several times.[38]

As these institutions and corporations have advanced, their hubris, accumulating wealth regardless of social costs and species life, they have created still unimaginable future dystopias. Nature and species already have suffered . . . and likely will continue to do so until the technological sphere is totally reorganized.

No Exit

Now there seems to be no exit. The only hope to save species diversity is to completely stop cutting forests as a sink for carbon dioxide, and to stop the burning of fossil and nuclear fuel in order to prevent further heating of the atmosphere and acidifying the sea.

Our species is literally eating the proverbial seed corn that can sustain future life. We can invent new technologies, chemicals, and bioengineered seeds and animals to enlarge agricultural production to feed and supply increasing billions. But we can only do so by elevating heat in the atmosphere, melting the remains of the last Ice Age, raising sea levels, inundating islands and continental escarpments, and destroying coastal cities and both seaside and inland farm crops and domesticated animals.

We can send remaining species towards polar caps and up 20,000-foot mountainsides, where their biological preconditioning will be unable to cope with the new environments. We can witness the falling numbers of life forms upon Earth. But the technical sphere will continue to overwhelm the biosphere. And the tragedy of human arrogance likely will not be reversed in time to save ourselves or other species.

In our moment, every continent, territory, region and island fastness lies beneath a heavy atmospheric ceiling of debris and toxic gases, sharply differing from the past in volume, degree and danger to those living upon some branch, twig, leaf or vein of the tree of life.

All this has led to emerging global climate disorders, further blocking evolution and changing the inherent character of species life. As our species has pursued its own well-being at the expense of the natural domain and the life forms sustaining the environment, *the dominant pattern of species global evolution, speciation, and extinctions has been sharply altered.*

6. OLD PARADIGM; NEW VISAGE

Though a philosophical lens, the present evolutionary paradigm shift means old forms of species relationships are being gradually, if not suddenly, displaced by new ones.

Once the old dominant natural relations of plant and animal symbiosis have been interrupted, interceded and altered by human ones, both intended and unintended consequences follow.

The old paradigm turned on evolving strands connecting the seeming separate realms of environments and species; by their mutual exchange of gases, chemicals and nutrients; by creating conditions for natural, sexual, and biogenic selection. Through evolution, this turned on the biosphere they collectively created and inherited, their interwoven interplay dependent on feedback loops shaped by natural forces.

Today, the major pathways to understand the conditions essential for life and species reproductive fitness may well turn on a new visage: the periodic change in the orbit of Earth alternating torrid heat and extreme cold; possible astroid invasions destroying species life; the human technological sphere overwhelming the biological sphere, changing the pace of evolution, speciation and complexity—combining different species under the dominant nucleus of a single biological roof; and the epigenetic and super-structural alteration of the DNA double helix.

In my view, the new paradigm would calibrate species evolution on a global scale; so what seem to be independent life forms are viewed as entwined and interdependently locked in a web drawn ever tighter—as Gaian theory postulates.

The Present Evolutionary Paradigm

As the dominating force today, our species has diminished and annihilated others; mixed genomes of different species; intra-converted life forms; constricted the biogeographic web of existence; and twisted feeding chains among species. Some of these processes are well-known and documented.

As the genome of different species have been experimentally punctured, severed and recombined, they are structurally and organically worn threadbare—to the moment when evolution itself has been interrupted, altered, even quickened. Along Fibonacci's numerical track-work, they have neared a moment of biological overreach and forced devolution, where fruits die on the vine, animal limbs fail, and organs atrophy.

Parts and genetic pieces of the organic make-up of one species have been carved-up, ground to bits, and fed to other species in a whirlwind of replications. These in turn have stunned, jeopardized, and sometimes put to death those with weakened immune systems, unable to resist and overcome invading, swiftly reproducing, often intermingled, bacteria and viruses. These micro-organisms bring on intestinal and other organic conflict and disorder, affecting the host's embryonic progeny and the metabolic functioning of surviving species—a cyclonic leap towards a disclimax of biotic life itself.

Together these processes make up the essential ingredients of the present *Evolutionary Paradigm*. Functioning within the double helix of both animal and plants, we now anxiously await the *future* pace of Fibonacci's organic complexity, bringing on human discord with the natural realm and massive species destruction.

Yes, revolutionary destruction on the surface of Earth has happened before: glaciation as the endgame of the Ordovician; global warming changing ocean chemistry at the close of the Permian; asteroid showers in the final moments of the Cretaceous.

Coming within centuries, we now face the paradox of a near Sixth Extinction caused by our species overstepping and transforming Earth's ecological, biological and geochemical landscape. Collectively we already have forever closed the evolutionary pathways for thousands, maybe millions, of other species; and are now blocking the evolutionary gateways for ourselves. "Homosapiens might not only be the agent of the sixth extinction, but also at risk of being one of its victims," anthropologist Richard Leakey warns.

With remaining forms of evolutionary life, the tragedies that we have foisted on other species and ourselves have diminished the thrust of the unified forces of Gaia's cycles of energy—so much so that only polar and *alto mesa* mosses and sub-surface micro-organisms, along with chemosynthetic forms of deep sea life, ultimately may survive the present anthropocentric age of destruction.

Entropies Impending; Disclimax Possible; Speculation Abounds

By contrast with the millennial struggle of plants and animals to survive and evolve on a global scale with limited means to sustain them, the present Anthropocentric Era elaborates contradictory forces jeopardizing existence. On one tier is the production of the means of sustenance and commodities. On another, lethal gases, torrid heat, and perilous outpours of particle debris overwhelming Earth's biosphere, climate, and the ability of the vast majority of life forms to survive, epigenetically modify their organic DNA structures, and reproduce.

The Second Law of Thermodynamics sets up this inescapable disequilibrium: For the cycled spread of concentrated heat to the cooler portions of the planet's atmosphere, ground cover, sub-terrain and seas encloses all life forms in Earth's pressurized vessel—over time exceeding the maximum thermal limits of most species varieties, only a few able to slip through the poisoned veil at the highest elevations and in the deepest seas.

The contrast of Holocene and Anthropocene is thus sharply defined. During the Holocene, the forces of the distribution of vegetation and animals turned on *isoline* elevations and longitudinal distance from the Equator, interlocked with geographic distribution of *isotherms*. Distance from the Equator and vertical elevation calibrated low temperatures; and together they defined the geography of vegetable and animal distribution and their sustainability.

During the Anthropocene's overlapping moments of disequilibrium bordering on entropy, species move along isolines to escape the spreading blanket of heat, among them the largest expiring by the wayside for lack of adequate nourishment and oxygen, unable to reproduce.

The Rest is Speculation.

Might surviving, less-complex sea plants and animals move towards evolutionary sea-beds, blocked from moving further by the boiling vents of sulfur dioxide? Could heat-stressed species move towards ethereal elevations, the tree-line restricting animals conditioned to breathing levels of oxygen between 17-to-21 percent of the atmosphere—only a few able to survive and reproduce at 14 percent or less? Those species that could flexibly or epigenetically adapt to the new climates might then gradually exchange their own gases, chemistries and nutrients, generating a new equilibrium or disequilibrium cycle, potentially changing their mutual survivability in the new conditions.

These forces may well transform the atmospheric balance of species breathing oxygen to a mixture of their own effusions of carbon dioxide and methane. And the new conditions also might raise the maximum thermal and chemical tolerances for pelagic creatures.

Surviving life forms that generate their own surrounding conditions would then symbiotically consume what they produce and dispose of their own wastes by exchange with other species.

Back to Our Future

The 2014 United Nation's IPCC report predicted negative impacts of global warming. Crop yields will be outweighed by any potential rise in thermal heat. Glaciers will continue to shrink as the climate warms, which will have major impacts on rising sea levels and potable water supplies.

- Species on land and in the sea might shift their range in response to warming; and some might face an increased risk of starvation and extinction. The seas will continue to acidify, destroying the shells of billion of sea creatures constructing coral reefs.
- Following heat waves and floods in low-lying areas, hunger, starvation and plagues are possible. And violent conflicts between species, especially humans, will exacerbate the effects of global warming.
- All these are signs of massive disorders—overlapping entropies, setting battle lines for survival fueled by the onrushing technological sphere—altering, degrading and cutting away at the need of species for adequate food resources, nutrients, and habitat.

As this progression has advanced from one sphere to another, the technological infrastructures at first seemed to be all-powerful and dominant, drawing on the support of human populations. For worldwide, populations sought and still seek stability, resources and birth-to-grave security—often regardless of the damage using technology may cause to Nature and other species.

Those working the soil, eking out subsistence or engrossing profit—from peasants and working family-scale plots to employees of corporate agribusinesses—are pressured by exogenous, competitive market forces to enlarge output and cut costs, using pesticides and bioengineered seeds and animals, regardless of species' peril.

From government employees to the vast armies of street, hospital, shopping, service and other workers, there is little way to get to and from work or function on the job or in their personal domain without using micro-technologies, as well as motor vehicles, electric trains and buses, using countless inputs of gasoline, oil and carbon-based energy—evaporating into atmosphere, waterways and seas funnels of carbon dioxide, methane, nitrous oxides, and particle matter.

Manufacturing, from micro-devices to giant machines and aircraft, use the natural airways as their dumping ground for every conceivable effluent. And lobbying groups, representing each of these and other factions and fractions of the national, corporatized and religious social orders, manipulate political bodies to favor their own interests and goals at the expense of one another and the natural, evolutionary cycling of species, energy and matter.

Herein are social power relationships and control mechanisms deployed to realign national, regional and global policies—shaping Nature and circumscribing species survival.[39]

Human Awareness and Indisputable Climate Change

Human self-awareness still turns on Aristotle's concept of the natural desire to know through use and delight of our senses, most of all in the sense of sight.[40]

But surrounding us, in almost every aspect and dimension of our lives, we cannot fail to see the human-driven technological sphere invading and afflicting the biosphere, spreading disorders throughout the environment and weakening the condition of species life.

Now before our very eyes, drawing on great scholars and devoted researchers from around the world, we can read the Intergovernmental Panel on Climate Change's second of three reports on the causes, consequences of, and solutions to climate change.

The first report, released September 2013, in Stockholm, found humans were the "dominant cause" of climate change, and warned that much of the world's fossil fuel reserves would have to stay in the ground to avoid catastrophic climate change. The second report declares: "In recent decades, changes in climate have caused impacts on natural and human systems on all continents and across the oceans."

Some parts of the world could soon be at vulnerable points, facing environmental entropy and species collapse. For others, that tipping point has already arrived as—"Both warm water coral reef and Arctic ecosystems are already experiencing *irreversible* regime shifts."

This report for the first time also looks at the effects of climate change as a series of risks altering the conditions on Earth over the last 600,000 years – with those risks multiplying as temperatures continue to warm. The so-called 'death rate" of how many species face extinction each year out of a million species is largely due to the human population increase and human-altered conditions—causing loss of habitat. The pre-human rate of species extinction on Earth was about one to a million; but is now 100 to every 1,000—the fast death rate placing the world on the brink of a sixth great extinction, as Richard Leaky said over a quarter of a century ago.[41]

Futures Known

Considered together, as explained in this study, the following forces shaping and impacting Earth have been further identified and described by a handful of brilliant scientists:

- Astronomer Milutin Milankovich estimating the changing global cycle around the Sun from circular to elliptical every 5,000 to 55,000 years, during which Hot House Earth will end and a new Ice Age will begin, transforming planetary conditions for species life.
- Fibonacci's time-based mathematical progression, applying to species evolution, increasing their biological complexity and pace of successful reproduction.
- Linnaeus rules classifying species, the most basic biological category being a group of living things that look alike and normally mate only with each other.
- Alexander von Humboldt's holistic concept of the *Cosmo*s, synthesizing the unity of Nature's forces, equating isotherms with isolines in discovering the causes of the distribution of vegetation. This technique interlocks topograhical elevation and distance from the Equator (*isolines*) with the heat distribution of planetary *isotherms*. This proves that both distance from the Equator and vertical elevation shape the distribution of temperatures; and together they define the worldwide geography of vegetable distribution and sustainability.
- Rudolf Clausius' Second Law of Thermodynamics, effectively stating that "heat cannot of itself pass from a colder to a hotter body." In the context of vessel Earth, this means that heat will move from the hottest geographic regions to cooler ones—so Hot House Earth will eventually become uniformly torrid, testing various species' maximum thermal limits to survive.
- Charles Darwin's and Alfred Russell Wallace's concept of the struggle for existence among species—leading to their evolving favorable variations and structural and organic specializations that tend to be preserved, and unfavorable ones to be destroyed—leading to the formation of new species (speciation) with and through reproductive fitness.
- Stephen Jay Gould, Richard Leakey, Roger Lewin, Richard Dawkins, Edward O. Wilson, Carl Zimmer and Australia's Tim Flannery and Richard

Smith—elaborating how evolution of sister species, each turning on genetic, epigenetic or other adaptations to environmental changes, together lay the foundations for radiation of new species.
- Hugh Falconer's metaphoric prediction stating that future evolutionary speciation would and will be reconstructed with a different superstructural architecture.
- Lynn Margulis's research revealing that single species' endogenous interdependence on other species, creates a multi-structural web of mutual existence under the roof of a dominant, single, encapsulating species.
- Fritjoff Capra's detailing that exogenous species' interdependence, creating a web of life.
- Sean Carroll's investigations of species adaptations to sudden environmental change by precise, if still partly unknown, biological functions of the structure of DNA's nucleotide bases, turning on connecting the two backbones of the DNA's helical structure.
- The British Statigraphical Committee's findings concluding that the present Anthropogenic Age is signified by forest destruction and human worldwide transfer of species.
- And Gaian theory and James Lovelock postulating that species themselves have collectively created the conditions for life-sustenance and for storing their wastes by symbiotically emitting materials and gases into the ecosphere sustaining one another.

Today, technical impositions by our species, enlarging momentary controls—tearing apart, unraveling, disrupting and restructuring the filaments of life existence itself—are best described by David Quammen and Al Gore; and matchlessly viewed by Bernoit Mandelbrot through the prism of our own species' technical barriers imposing on Nature's evolving, reproductive patterns; and today pursued by thousands of biotechnical experimenters.

Will the Anthropocene bring greater changes still?

Read on.

NOTES

1. Krooth, *Gaia,* Chapters 29-30, *et passim*. See also Krooth, *Arms and Empire,* Chapters *III-IV, et passim*.
2. Krooth, *Gaia,* 268.
3. D. C. Somervell and Heather Harvey, *British Empire and Commonwealth* Rev. ed. (London: Christophers, 1959), 24-25, *et passim*.
4. Krooth, *Gaia*, 269.
5. Prof. Pimm and Clinton Jenkins, Institute of Ecological Research, Dry Totugas, Brazil, *Science*, May 29, 2014. See also Kathy Kirtland, *AP* via AEI-CASC, "Endangered Species Gallery," blog.sfgsate.com; Berkeley Resource Group, *Species Survival Rates*, Box 9515, Berkeley, CA., 94709. In 2014, some 300 species in California were on the endangered list; the 2010-2014 four-year drought enlarged the number.

6. Maurice Dobb, *Political Economy and Capitalism* Rev. 2nd ed. (London: Routledge & Kegan Paul Ltd., 1940), 229-230. See also Krooth, *Arms and Empire,* 20-21.

7. Charles P. Steinmetz, *America and the New Epoch* (New York and London: Harper & Brothers, 1916), 120-122, *et passim*. See also Morgan J. Rees, *Trusts in British Industry, 1914-1921* (London: P.S. King and Son Ltd, 1922), *passim*. And Krooth. *Arms and Empire, Chapter III. Also* James Harvey and Katherine Hood, *The British* State (London: Lawrence & Wishart, 1958), Chapters II, XIV. For further historical references, write Richard Krooth, regards *The Devolution of Capital; 1700-1970*: Unpublished Manuscript, Box 9515, Berkeley, CA. 94709.

8. See John A. Hobson, *Evolution of Modern Capitalism: A Study of Machine Production*, New and rev. ed. (New York: Charles Scribner's Sons, 1907), Chapter VIII. See also J. A. Hobson, *Imperialism: A Study* (New York: James Pott & Co., 1902), Chapters IV, V, VI, VII. See Wegenfür, "Statistisches Kartelle," *Allg. Statistisches Archiv*, 1932. BD. 22. H.II.S. 252. Percentages computed.

9. Kurt Wiedenfeld, *Cartels and Combines*, League of Nations: Economic and Financial Section. (1926.II.70), (Geneva, 1926), 25, 31. Compare with Earnest Mandel, *Marxist Economic Theory*, Vol. II, trans. Brian Pearce (New York and London: Monthly ReviewPress, 1968), 467-468. And Krooth. *Arms and Empire,* 67.

10. Maurice Dobb, *Political Economy and Capitalism*, 229-230. See also Krooth, *Arms and Empire,* 21.

11. League of Nations, *Statistical Yearbook*, 1941-42, 12-22, Table 12. Table constructed. See also Krooth, *Arms and Empire*, 121.

12. Charles S. Elton, *The Ecology of Invasions by Animals and Plants*. Reprint. (Chicago: University of Chicago Press, 2000), 50-51, 62-65. See also Kolbert, *The Sixth Extinction*, 210-213. And Krooth, *Gaia,* Chapter 22.

13. Elton, *The Ecology of Invasions*, 64-76. See also Krooth, *Gaia, Part VII, et passim*.

14. "Two-way traffic," *Kew*, Summer 2003, 15. See also Thor Heyerdahl, *Kon-Tiki: Across the Pacific By Raft*, trans. F.H. Lyon (Chicago/New York/San Francisco: Rand McNally & Co., 1950).

15. See Richard Dawkins, *The Ancestors' Tale: A Pilgrimage to the Dawn of* Evolution (Boston, New York: A Mariner Book, Houghton Mifflin Company, 2005*), passim*. See also Richard Leakey and Roger Lewin, *The Sixth Extinction: Patterns of Life and the Future of Mankind* (New York, London: Doubleday, 1995), *passim*. And see herein Krooth, *Darwin's Walk*, Chapter 9.

16. Sidney Aquatic Museum, *The Future is Wild, 2011-12*. See also Richard Smith, PBS Series. And herein Krooth, *Darwin's Walk,* Chapter 9.

17. See U.S. Census Bureau, "Historical Estimates of World Population," International Programs, http://www, census, gov,ipc//www/worldhis.html. See also *National Geographic*, "The New Age of Exploration: Pathfinders: The Arc of Exploration," 2014. And see the following: Keeling Curve at Al Gore, passim; Krooth, *Gaia, passim*; Krooth, *Arms & Empire, passim*.

18. Compton's Pictured Encyclopedia, *War Supplement: An Alphabetic Reference Book of the European War* (Chicago: F.E. Compton & Company, n.d.), 5-6, 19, 20, 32. See also A. Cloyd Gill, "Mr. Hoover's Debt Plan, *American Bankers Association Journal*, July 1931.

19. See also Compton's Pictured Encyclopedia, *War Supplement: An Alphabetic Reference Book of the European War* , 5-6, 19, 20, 32. See also, Krooth, *Arms and Empire, passim.*

20. League of Nations, *Statistical Yearbook*, 1941-42, 178-179, Table 92.

21. Optimum Population Trust, "Tackle Population Growth to Beat Climate Change," *News Release*, August 17, 2009.

22. James Lovelock, *The Vanishing Face of Gaia: A Final Warning* (New York: Basic Books, 2009*), 160, et passim*.

23. Optimum Population Trust, *News Release*, December 15, 2009.

24. Krooth, *Gaia*, Chapter 31, *et passim*.

25. See Ann M. Graybeil and Kyle S. Smith, "Good Habits, Bad Habits," *Scientific American* (June 2014) Vol. 370, No. 6: 38-43.

26. Carl Zimmer, "As Humans Change Landscape, Brains of Some Animals Change, Too," *The New York Times, Science*, August 27, 2013.

27. Mark D. Bertnesss, Caitlin P. Brisson. et al., "Herbivory Drives the Spread of Salt Marsh Die-Off," *PLOS ONE*, March 20, 2014. DOI: 10.1371/journal.pone.0092916.

28. Ivan Illich, *Medical Nemesis* (Toronto/New York/London: Bantam Books, 1976). See also Dr. Rob Desalle, *Epidemic: The World of Infectious Diseases, passim.* And *see* Lynne Peeples, "Why Scapegoating Bats Is a Big Mistake to Human Health," 10/31/14. http://www.huffigntonpost.com/green. Also Natalie Angier, "Ebola and the Vast Viral Universe," *The New York Times*, *Science*, October 27, 2014.

29. Lovelock, *The Vanishing Face*, 156-157.

30. See Robert B. Semple Jr., "Concealed Weapons Against the Environment," *The New York Times*, July 31, 2011, SR10. See also Krooth, *Gaia,* Chapters 29, 30, 31. And see Bill McKibben, *Eaarth, passim*.

31. Richard Krooth, "Travels in North America and the Everglades," *Notebooks*, 2010-13.

32. Kolbert, The Sixth Extinction, 94.

33. Lovelock, *The Vanishing Face*, 153-155.

34. Lovelock, *The Vanishing Face*, 153-155.

35. Richard Krooth, "Interview with Dr. Edward Berg," Tempe, Arizona, August 2014. See also Kolbert, *The Sixth*, 103. Astronomical bodies distort space-time, making it curve, creating gravitational fields. The larger the body and subsequent distortion, the greater is the gravitational field produced. As other astronomical bodies and light itself travel into this time-space distortion, their paths are also curved as the gravitation field literally pulls them in. See Gerald Tauber, ed., *Albert Einstein's Theory of General Relativity: 60 Years of its Influence on Man and the Universe* (New York: Crown Publishers, 1979). If we understand this theory, we can perceive Earth subject to violent cycles of extraterrestrial comets that surely will invade in the future and lead to the extinction of some, if not all, species.

36. Richard Leaky and Roger Lewin, *Origins: The Emergence and Evolution of Our Species and Its Possible Future* (New York: E.P. Dutton, 1977), *passim*. See also Wilson, *The Diversity of Life*, 29. See also Lovelock, *The Vanishing Face*, 153-154. And Richard Zeebe, et al., *Nature Geoscience*, April 2008, comparing the European Project for Ice Core in Antarctica data and carbon saturation in the deep sea for the last six glacial cycles regulating carbon dioxide.

37. Krooth, *Gaia*, 596-597. See also Kolbert, *The Sixth Extinction*, 96-97.

38. Lovelock, *The Vanishing Face of Gaia, passim. See* "Are 90 Companies Responsible," *HuffPost Green*, Nov. 22, 20, 2013.

39. Krooth, *Arms and Empire, passim*; Krooth, *Gaia, passim*. See also Marvin Harris, *America Now: The Anthropology of a Changing Culture, passim.*

40. Aristotle, *Metaphysics* 1.1, 980a20-25. British Library copy.

41. Leaky and Lewin, *Origins: The Emergence and Evolution of Our Species and Its Possible Future, passim. And see Elton, The Ecology of Invasions,* Chapters 2, 7. And Pimm and Jenkins, Institute of Ecological Research, "Dry Totugas, Brazil," *Science*, May 29, 2014.

IV

Natural Selection

That many and serious objections may be advanced against the theory of descent with modification, I do not deny . . . Nothing at first can appear more difficult to believe than that the more complex organs and instincts have been perfected . . . by the accumulation of innumerable slight variations, each good for the individual possessor. . . [For] all parts of the organization and instincts offer, at least, individual differences—that there is a struggle for existence leading to the preservation of profitable deviations of structure or instinct—and, lastly, that gradations in the state of perfection of each organ may have existed, each good of its kind.
—Charles Darwin. *Origin of Species*, 1859

Chapter Seven

Battle for Reproductive Success

> We can see why there should be so striking a parallelism in the distribution of organic beings throughout space, and in their geologic succession throughout time; for in both cases the beings have been connected by the bond of ordinary generation, and the means of modification have been the same. We see . . . that on the same continent, under the most diverse conditions, under heat and cold, on mountain and lowland, on deserts and marshes, most of the inhabitants within each great class are plainly related; for they are the descendants of the same progenitors and early colonists.—Charles Darwin, *Origin of Species*[1]

> Quite a large number of species are able to achieve a world-wide distribution as it is, either because the ecological barriers that hold in others are not barriers to them, or because, which is partly the same thing, they have exceptionally good powers of dispersal . . . But a great many other plants and animals never had the opportunity of ranging over the whole world. —Charles S. Elton, *The Ecology of Invasions of Animals and Plants*[2]

The search was on; and the eminent British ecologist Charles S. Elton was looking at extraterrestrial impacts on Earth and their geological disruptions of species, turning to the descriptions made in Darwin's time, not by Darwin, but covering Alfred Wallace's Realm.

Wallace had written that as species became cooped up, as it were, in various geographic regions for long enough, they had profoundly changed biologically and left a permanent evolutionary mark on the composition of flora and fauna.

To this were added, Elton said, two other processes that increase the complexity of species distribution patterns as viewed in modern times.

The first was that those species groups that were once very widespread had retreated to isolation, to be found only on a single continent or island, or even in a single lake.

The second, that after the long isolation in Tertiary times depicted as Wallace's Realm, there was some periodic remingling of faunas, before man came on the scene to carry the process abruptly further.[3]

THE UNFOLDING

Under the pressures of such geo-biological changes, the underlying question thus remained and continues—which species have already survived in isolated geographies and may continue to win the battle for reproductive success?

Here we review the three periods in the Wallace Realm, demarcating evolutionary patterns (biogenesis) and the consequences for species survival.

Where and at which biological moments are species to be found at different geographic regions on Planet Earth?

Over many millions of years, we know species have evolved and, by changing planetary conditions, appeared in many landscapes, mountain regions and seas.

Species survival and reproduction have remained conditional, too. For earthly species have diversified with the unfolding—

1. of their ancient distributions and concentrations throughout the entire world;
2. of their seclusion by the breaking up of the super-continents like Pangaea and Gondwana, creating virtual archipelagoes of world territories—isolating evolving regional species groups in confined geographic zones; and once more
3. of the random dispersal of these regional species groups over the entire world, which continues in our moment, especially by the human propensity for redistributing life on Earth.

The initial phases of biogenesis and planetary orbital gyrations were covered in earlier chapters. This chapter largely describes the second and third stages—species isolation and their elaboration and variation through evolution as a global configuration.

1. BIOLOGICAL MOMENTS

In his time, Charles Darwin emphasized global distribution of evolving species:

> On this same principal of former migration, combined in most cases with modification, we can understand, by the aid of the Glacial period, the identity of some few plants and southern temperate zones; and likewise the close alliance of many others, on the most distant mountains, and in the northern and

southern temperate zones; of the sea in the northern and southern temperate latitudes, though separated by the whole intertropical ocean.

He concluded that species of the same original genera found in widely separated geographies might vary:

> Although two countries may present physical conditions as closely similar as the same species may ever require, we need feel no surprise at their inhabitants being widely different, if they have been for a long period completely sundered from each other; for as the relation of organism to organism is the most important of all relations, and as the two countries will have received colonists at various periods and in different proportions, from some other country or from each other, the course of modification in the two areas will inevitably have been different.[4]

Today this is amply demonstrated by still living species, and applies to humankind as well.

We may never know exactly where went the 800,000-to-1-million old northern European human hunters and gatherers leaving footprints in estuary mud at Britain's east coast of Happisburgh—the most ancient evidence of humankind found outside of Africa—in a climate somewhat cooler than today, where they were beset with now long-extinct mammoth, ancient horse and giant vole.

Could they have crossed to the continent during the last glaciation, ice covering the English Channel, changing their physiology and mental attributes while living in a partitioned Europe as Neanderthals for 100,000-to-250,000 years; then journeying through frozen or tempered ice floes to the Americas and possibly Australia some 60,000-to-100,000 years ago?[5]

The ending of the last Ice Age, separating and isolating continents, confined species to regional conditions where they modified their physiologies and habits. For the developed Homo species, their turn from hunting to agriculture for survival supposedly dovetailed about 11,700 years ago. But many melting glaciers, still frozen parts of Greenland and Antarctica, and Arctic sea-ice and permafrost still linger today; and domesticated plants were being cultured and grafted by humankind at earlier dates in many parts of the world.

Patterns of Biogenesis: Worldwide Spead of Species

The Cenozoic Era beginning 50 million years ago and featuring the earliest great apes, continued into the start of the Holocene Era up through the present Anthropocene, that has fueled the ongoing spread of species worldwide.

Human progeny created the means to emigrate carrying plants and animals around Earth. The current stage of the global spread of species and their reassembly as new independent communities are passing under the veil of destruction of forests releasing enzootic life and causing planetary warming—brought on by humans burning fossil fuels, raising isotherms, bringing the Ice Age to a decisive close.

This transition has unsettled earlier geographic conditions and longitudinal scales of heat, with sharp contrasts between past and present.

Previously these zones were partitioned, distinguished, or overlapped into one another—as north and south polar regions; and as the torrid equatorial zones *vis-á-vis* the semitropical or moderate climatic regions. But today, these longitudinal lines of isotherms have been gradually erased by intense human-generated land, atmospheric and ocean heat, spreading from warmer to cooler regions as the second law of thermodynamics envisages—placing almost all species in the crossfire of either adapting to new environmental conditions or fleeing to maintain their fitness under threatened heat exceeding their maximum thermal resistance.

Endemic species with a given biological inheritance, or adaptability, or habits that were able to withstand extreme weather conditions on continents or islands—and where they once held an advantage over invasive species—are now challenged by rising temperature and new life forms better able to adapt and reassemble feeding and elimination chains.

These endemic species now face an uncertain future—either endure to the moment their maximal thermal limits are reached; or, if they can, flee to cooler upland geographies or to polar latitudes with more modest thermal readings.

Anthropocene upon Us: Altered Atmosphere, Land and Sea

Encapsulated under the drift of a premature torrid climate—that was not expected to arrive until 5-to-55 million years from the close of the last Ice Age—today's ongoing global invasions by plants and animals is destined to intensify their competition with endemic species on continents and islands.

The past five alternating heat and ice ages may prove prologue, too, indicating the consequences for different organic forms of both invasive and endemic species life in the present Anthropocentric Era—variously reaching their maximum thermal limits by the alterations of torrid heat in the atmosphere, on land and in the sea.

Earth already has suffered extraterrestrial bombardments, volcanic eruptions, frozen clathrate (methane) explosions, world wars, extensive deforestation, and the massive burnings of fossil fuel—that have led and are still leading to mass extinctions.

And we humans apparently have already selected the means of our own demise and extermination and that of most other species, though precise datelines of the Sixth Extinction remain uncertain.

But how exactly will any surviving species evolve, then lead to new biological forms, and assemble their communities?

That will depend on the resources available to them, as well as their geographic range, geological conditions, and regions restricting their expansion.

Paul Crutzen, the Dutch chemist who coined the word "Anthropocene" describing the recent epoch, wrote in journal *Nature*, that humans had altered the composition of the atmosphere through the combination of fossil fuel combustion and deforestation—raising the concentration of carbon dioxide in the air by forty percent over the last two centuries, and more than doubling the concentration of the more potent greenhouse gas methane—32-times as much of a heat-storage-blotter as carbon dioxide.

The stratigraphy committee of the Geological Society of London later agreed that the Anthropocene will be marked by a unique "biostratigraphical signal"—a product of the current mass extinction episode on one hand, and on the other the human propensity for redistributing life on Earth. This signal, their committee of experts summarized, will be permanently inscribed "as future evolution will take place from surviving (and frequently anthropologically relocated) stocks."

Richard Leakey said the same twenty-five years ago. And James Lovelock predicts the future face of Gaia will vanish as species that created Earth's atmosphere and equable temperatures are circumscribed or eliminated by our species burning fossil fuels, aerating greenhouse gases, cutting forests and acidifying the seas—destroying Earth's storage vaults for carbon dioxide and methane.

Tree of Life—Lacking Living Intermediaries

Casting backwards to uncover the root and branch of the tree of life, we may be able to better judge the future.

It may not be crystal clear, but Stephen Jay Gould explains that it turns on understanding evolution: "a process of branching, not of linear transformation, as the brilliant 'tree of life' metaphor. . . . Few, if any, living species are the unaltered forebears of another modern form; rather, any two sister species have branched and diverged from an ancient common ancestor. Therefore, we do not expect to encounter living intermediates between extant species for, although transitional forms surely existed, they died long ago and should only be found in the fossil record."[6]

Living intermediaries missing, the fossil record itself may not be as complete as some paleoanthropologists would like. And also missing are those fossils that did not survive the rags of time. They can be made up by known continental movements carrying still living sister species to divergent geographies, as well as by discovering past species forms through genetic analysis of extant animals and plants.

2. PANGAEA, GONDWANA & THE CYCLES OF SPECIES DISTRIBUTION AND EXTINCTION

The cycles of existence and extinction are anything but uniform and clear.

Darwin's belief in gradual evolution of species and their demise certainly were more usual than extraterrestrial bombardments that snuffed out billions of organisms in a great explosive flash.

Extinction might also mysteriously occur as one species radiated into another, so the earlier intermediate forms were either nowhere to be found, or only partial remains appeared in the fossil records that an intrepid paleoanthropologist or biologist discovered.

And extinction of microorganisms might never be suspected or even acknowledged in the archives of leading scientific laboratories and museums.

Then there were known alterations of the skin of Earth during the so-named Tertiary periods—mountain ranges and sea beds that should have left behind the remains of rocks, minerals, shells, and appendages and residues of living matter. These our species are still excavating to discover what we might learn to stop extinctions' repeat performances.

We now know there were at least three moments of species' distributions or restrictive boundaries across the face of Earth.

During the first period, about 400-to-350 million years ago or more, species distribution was signified by marine life pressing on land and fresh water. This occurred as early as Silurian and Denovian times of ancient fishes and the earliest land plants.

The Silurian and Denovian sandwiched the time-frame over the next 300-to-230 million years to the end of the Crestaceous Period.

And over the longer time-stretch of the Carboniferious Period between 400-to-300 million years ago, the earliest reptiles made their *début*.

Moving into the Jurassic Period from 200-to-100 million years ago, watch the earliest birds and flowering plants evolve with extreme modifications Though there were substantial regional differences between species in different climatic geographies in Northern and Southern Hemispheres, during climatic changes like the Permian Ice Ages, and dry and wet periods exposing greater or lesser land surface—as each of the major species group evolved, they eventually spread throughout the entire world, to all or nearby areas that are today present continents, as well as to some that are now islands like

Madagascar off Africa. Together these represent, ancient, "metropolitan" species distribution throughout the entire world.

During the second period that began during the Mesozoic Age, particularly in the middle and later parts making up the Jurassic and Cretacious Periods, a more uniform and warmer climate than today led to luxuriant forests in Greenland and Spitsbergen, and an Arctic Ocean filled with abundant ammonites and other marine forms.

As the sea flowed onto the land, it broke the land connections between many parts of the world, as by the Panama Isthmus, the Bering Strait, and connections to Australia, transforming the continents into an archipelago, isolating a great many groups of plants and animals that had already evolved towards their present state.

By the end of the Cretacious extinctions, primates appeared and many modern genera of plants like oak, popular, beech, sycamore, magnolia, laurel, pine, spruce and cedar flourished. Mammals, birds and fresh-water fish were poised for rapid evolution. And when the continents separated, each floating parcel held a quota of mammals, plants and birds, some carrying with them fresh water fish, some thereafter evolving as separate creatures.[7]

The separation of continents and islands thus led to the separate evolution of disparate regional species groups, each within their own respective domain.

During the third period, once again, the partial random dispersal of these regional species groups covered the entire world, though, as we will see in our moment, the fragmenting of continents and species remains incomplete.

Evolutionary Moments; Onto the Anthropocene

The geologic order of the so-called Tertiary period in the history of life on Earth are demarcated by the Eocene, Oligocene, Miocene and Pliocene—followed by the Ice Age and the post Ice Age florescence of life, the Holocene—the period of receding ice and its continual melt setting the framework for species reproductive success, rarity and extinction in our era—the Anthropocene.

After Gaia's single global landmass of Pangaea was surrounded by the single sea of Thalassa, its southern block, called Gondwana, edged over the South Pole causing extensive glaciation and biological devastation. The water froze absorbing the sea, glaciers reappeared, and tundra iced over.

Australia likely became the first *reappearing* continent, its drying reach leading to the evolution of a new tangle of species life. The living kangaroo, we know, survived on widely-separated grass ranges through evolution of a single, fused foot and ankle, powering its leaps across arid deserts to reach the great plains of grass, females carrying their offspring in a pouch ensuring reproductive success.[8]

Last Half Billion Years

The major periods covering evolution and extinctions on Earth during the last half-billion years or so include unnumbered instances of species radiation, rarity and demise.

3. EXTRATERRESTRIAL CAUSES FOR EXTINCTION

Allow us to relive the past to uncover the fate of species quickly snatched from Earth—interrupting gradual evolution described by Aristotle, von Humboldt, Darwin, Mayr, and a host of other biologists and paleoanthropologists.

Mass extinctions of species from extraterrestrial invasions differ from one another, the key event being uncovering the forces shaping the time-based boundary between the Cretacious and Tertiary periods.

An *overall* cooling Earth after the first five great extinctions eliminated vast number of species, survivors radiating into smaller regional ranges of modification and combat—making them more vulnerable to extinctions from still other climactic causes.

And yet, those that survived in the form of seeds, eggs, fungi, molds and mycenial polyps dependent on the Sun's radiation or chemosynthesis continued the evolutionary miracle of life on Earth—in diverse forms that even today are barely understood or appreciated by our species.

The father-and-son team Luis and Walter Alvarez wrote of their research findings in "Extraterrestrial Causes for the Cretaceous-Tertiary Extinction," in *Science*, June 1980.[9]

They explained that an asteroid six miles wide collided with Earth moving 45,000 miles per hour, its trajectory from a low angle relative to Earth hitting North America hardest with searing vapor and debris. It instantly exploded, releasing energy more than a million times the most powerful Hydrogen bomb ever tested, pulverizing debris spread around the globe, broiling the planet's surface.

It blasted the Yucatán Peninsula's rich sulfate aerosols, thereafter blocked out the Sun causing plunging temperatures, mass extinctions of forests, destruction of diverse plant communities, only millions of years later replaced by rapidly dispersing ferns.

Marine ecosystems were upended for half to several million years; non-avian dinosaurs completely annihilated; three-quarters of bird families destroyed; 80 percent of lizards and snakes gone forever, along with two-thirds of all mammalian families.

Sea plesiosaurs, momasaurers, belemnites, and ammonites disappeared. Among bivalves were heavy losses alongside brachiopods and bryozoans;

several groups of marine microorganisms were lost; and complete annihilation visited ninety-five percent of near-surface-loving foraminifera.

So it was following this so-called K-T extinction for millions of years, until life recovered its former level of diversity, with many surviving species shrinking.

Today, the resulting Chicxulub crater off the Yucatán Peninsula, is still buried under a half mile of sediment—the "Crater of Doom," Walter Alverez called it.

Throughout the world, ammonites did not survive the Cretacious period, having lived in shallow oceans for more than 300 million years. This was because their tiny eggs, a few hundredths of an inch across, floated near the water surface, drifting along currents, which possibly were so toxic their offspring *ammonitellae* could not survive.

Gathered by collectors today, their fossilized shells of many shapes appear in forms of evolved spiral shells divided into multiple chambers, exhibiting nine teeth—octopuses their closest living kin.

Remains of Other Mass Extinctions

Atmospheric concentrations of gases proved critical during other mass extinctions, too, their effects only partly known.

Overlapping the Denovian period (about 438-444 million years ago) and the Permian (about 245 million years ago), during the so-called Carboniferous Era (some 300 million years or so ago), oxygen concentrations in the air were about 35%—compared with 17-21% in our era.

As plants spread on land, their ability to photosynthesize drew carbon out of the atmosphere and converted it into two parts: organic matter incorporated, effectively buried, in the plants themselves; and plumes of oxygen released and added to atmospheric nitrogen and other gases, increasing overall pressure that effectively forced oxygen into millions of evolving insects through spiracle holes attached to their hollow-tube tracheas.

For reasons not yet fully understood, an *abrupt decline* in atmospheric oxygen content at the *end* of the Permian 252-million-years-ago led to the planet's greatest mass extinction of oxygen-conditioned life.

As the Sun reaching Earth was re-radiated to the atmosphere loaded with carbon dioxide, temperatures had soared; rains and fallout warming the seas some eighteen degrees; and as the seas became acidified, dissolving ocean oxygen, many organisms suffocated—an eyelash of 100,000 to 200,000 years collapsing the reefs they had built—eliminating some ninety percent of all species on Earth!

Possibly ocean heat incubated bacteria that produced hydrogen sulfide killing marine creatures, then released poisonous bubbles, bursting into the air killing almost everything else—the sulfate somewhat reducing bacteria changing the color of the oceans to a glassy purple, the aerated hydrogen-sulfate transforming the color of the sky to pale green.[10]

Other Forces Interplayed with Extraterrestrial Incursions

During the Permian, massive volcanic releases of atmospheric hydrogen sulfides and carbons likely were not timely-offset by weathering of mountains removing carbon dioxide from the air and washing it down into ocean basins to be buried in deep-sea sediments.

And accumulated atmospheric carbon dioxide stored solar heat, as today.

It is just as likely that *another* factor building into Permian mass mortality was long-term climate change, as the consequence of a rain of meteorite strikes and volcanic eruptions in north-central Siberian Steppes in the long moment of mass extinctions.

Major Extinctions; Fossils; Species Radiation and Competition

Together, the fossil record of the five major mass extinctions (including the Ordovician about 438-444 million years ago) reveal only mollusks, arthropods and fishes, whose remains quickly settled to the bottom of the seas and were silted over before they fully decomposed.

Using these fossils as *families of related species* are the best taxonomists can measure—each of these destructive spasms except the Permian eliminating roughly 12% of *all families*, the Permian destroying a staggering 54%.

Statistical methods have produced estimates that during the Permian era from 77-96% of all marine animal species were lost—an extremely close brush with total destruction!

Two highly distinctive and dominant groups—Trilobites and Placodermi fishes—met their end.

On land, too, mammal-like reptiles, distant ancestors of humanity, were devastated, only a few slipping through to reproduce. Adaptability of many insects and plants also meant their survival—reproductive fitness bringing them through later episodes of mass extinction.[11]

Test of Competition

Comparable destruction and modifications took place among many thousands of species in the reach of lesser extraterrestrial and earthly fulminations.

As each sister species forms that evolved from a common forebear often was challenged and radiated under altered geographic conditions and pressures, natural barriers sometimes appeared and hostile combative species evolved. These led to the competitive transformation of the sisters' anatomies and organs—these sisters reaching the moment of no longer appearing, performing, or their progeny mating, along the lines of the now separate species branches.

Speciation, or the descending appearance of distinct species, thereby emerged.

Still not all species survived the tests of competition, "tooth and claw" combat, or changed environments.

Darwin surmised that "the increase in every creature is constantly being checked by unperceived hostile agencies; and that these same unperceived agencies are amply sufficient to cause rarity, and final extinction."

And so it was. Those creatures that could not transform their structures; or lacked organic variability to change their proclivities and habits; or to use their inherited organs in new ways—or simply did not adapt to changed conditions—would indeed become rare or face extinction, severing a branch of the tree of life.

4. SPECIES GEOGRAPHIC DISTRIBUTION AND MODIFICATION[12]

In 1860, Alfred Wallace had laid out the framework dividing the world into six regions, using continental names and their features based on his vast collection of 125,660 specimen in 1860s of animals, and his delineation of what became known as Wallace's Line. These included:

- The Neotropica Region, covering Southern and Central America up to a line in Mexico, together with the West Indies
- The Neararctic Region, including North America, its Arctic island and Greenland
- The Palaearctic Region, including Europe and part of Asia, running across from Britain to Japan
- The Oriental Region, formed of the Indian Peninsula, the Far East with Southern China and Formosa, the Phillippine Island, and all Malaysian islands west of Celebes and Lombok
- The northern boundary of the Oriental Region, including the Himalaya and great mountain regions east of it and the desert areas north of them
- The Ethiopian Region of Africa south of the Sahara and reaching Madagascar
- The Australasian Region, taking in the islands east of Borneo, Java and Bali, with New Guinea and Australia, as well as New Zealand and the Pacific islands.

And Charles Elton added a seventh realm of life, the vast Eastern Pacific beyond the Australian continental arc, with its Milky Way of islands, including Hawaii, most of them with an exiguous number of species.[13]

Using other names, my own study of *Gaia* also covers the same designated regional realms of life that you can compare.[14]

And to determine species' geographic distribution and their modification, I ask you to join me as a bio-geologist to trace cooling, jeopardizing warm-weather species; and to compare cooling temperatures with the movement of landmasses and sea changes during continental drift.

Together we need to look to temperatures determined by longitudinal distance from the Equator to the northern and southern poles; and to examine isolines crawling from lower elevations up continental slopes and mountain sides.

Raising Pacific Anchor: Past Species Radiation, Modification and Extinction

So where should we raise anchor, geographically revealing past species' radiation, modifications, and extinctions; as well as human propensities for redistributing life's animals and plants on Earth?

Over the course of 3,200 years, coming across the Pacific archipelago, paddling and sailing their sea stallions, adventurers achieved deeds that easily equaled those of later European Vikings and colonists.

The Malayan-Polynesians explored and took the islands of the Pacific. Possibly about 1,200 B.C., some of the seafarers with advanced double-canoes and sails, paddles and hand tools moved to nearby islands from the Bismarck Archipelago that lays north of New Guinea; and they came with their dogs, chickens and pigs, taro, sweet potatoes, and yams.

Others also arrived from Malaysia, possibly even South America, to sweep Pacific Isles, islets and atolls—with the last Pacific landmass of New Zealand occupied around A.D. 1000. It is uncertain whether the DNA of all the Pacific invaders had a common origin, but through time, conquest, acculturation and exchange, they evolved common strains of language and cultural practices.

Periodic invasions of migratory Malayan-Polynesians broke up the social isolation and close-inbreeding of island populations that may have predated their arrival. The first wave of invaders often possessed superior tools and evolved genetic proclivities for adaptations to climate and endemic diseases.

Explorers stopped at various points in their journey; some likely reached South America, indigenous home of the essential food-plant kumara, also found today among agrarian peoples of the Pacific. Though proof is still lacking—given the great reach from the nearest South American islands of San Felix, San Ambrosio and Robinson Crusoe going west to Easter Island and then to Pitcairn—if some explorers made American landfall, they must have returned to their distant homelands, bringing the kamara with them. A genetic trail is in the asking: which sinew of humanity will lay claim to South America and plant its flag?

More certainly, other Polynesian explorers fanned out from the Marquesas between A.D. 350 and A.D. 600. For some 200 years, these adventurers, in outrigger canoes and speaking languages of the Austronesian family (Malaysian, Indonesian, Fijian, Hawaiian and Maori) carried live plants and seeds, the ceremonial Polynesian Viscosa herbal ground-weed, domesticated chickens, swine and goats.

Their quest for arable land might have been an outcome of their loss of traditional islands and atolls to rising, rolling seas during times of glacial ice melt in a warming global climate. Though Polynesians could not have known this, it remains a warning for our present.

Island Survival

Denied the possibility of agriculture where soil or volcanic pumice depth and water were adequate, Polynesians made use of what resources they found to make tools, hooks and other artifacts of stone. Lacking continental graphite, flints or other rocks, the sharp edges of volcanic *poh-hoi-hoi* were cut, ground, or polished into knives to cut food, spear points to hunt, and adzes to strip bark, make weapons, and clear trees for domesticating wild farm varieties.

Wild species were plentiful in some of the islands. In the malleable hands of farmers, bananas mutated. Their germ cells were not affected, but only those cells that shaped their corporal body—allowing their somatic mutation into over five hundred varieties (with still unnumbered, local island names) or species called sports blending into one another in unpredictable ways. Polynesians bringing the banana plants to Hawaii perhaps a thousand years ago unwittingly began an upward trophic pyramid of those creatures who lived on them—including sympatric snakes, minute insects, and banana moths that radiated into five entirely new species that could no longer mate with one another.

Island Biogeography

Island biogeography may have been determined by the original species that once walked, slithered, flew, swam or spread on the great Pangaean and Gondwana continents; but with these continents shattering into large pieces and bare slivers, whatever creatures and plants remained on each then radiated to the moment of human arrival—attacking and appropriating, destroying and domesticating both species and natural resources.

The first Polynesians on most islands hunted to extermination relatively helpless, flightless birds; but then had to seek food—especially protein—by gathering shellfish and crustaceans, as well as breeding seabirds, dogs, pigs and chickens They planted crops brought with them from outside Polynesia, or wild ones found on certain islands like the Hawaiian and New Zealand.

Most Pacific islands had had their own species of flightless rails, more than twelve species in the Hawaian archipelago—before humans invaded, bringing dogs, mice, rats and fleas, domesticated chickens, and goats gone feral. Lacking natural predators, all flightless rails had evolved on islands, counting New Guinea and New Zealand. But after human contact and rapacious hunting, perhaps 200 species went extinct on tropical Pacific isles, 60-odd living species quite flightless, still vulnerable to human reach.

Good soil and ample fresh water favored New Zealand. On other, more arid islands, strategies for food production ran the gamut of farming dry land crops varieties of yams, sweet potatoes, and taro; irrigated crops also included taro and coconut, banana, and breadfruit tree farming.

Yet slash and burn agricultural practices destroyed most standing forests on Pacific islands; and where rivers made taro irrigation possible, *corvée* or other obligatory labor systems were also used—as on Hawaii's Kauai, Oahu and Molokai.[15]

Caste and class partitioned social structures emerged both within islands and between them.

Crossfire of Species Invasions

As inter-island Pacific wars fulminated and subsided, occupying settlers developed varied cultivated crops.

Kahoolawe, Hawaii's mythic island-child of man and woman of Holani, became *he lopa*—"the farmer." And farm the new arrivals did—producing native popola plants, imported Maksli'i potatoes and taro; bananas, sugar cane, ape plant and ti; yams hoi, arrowroot (pia), and breadfruit; mountain apples (ohia), coconuts, and edible ferns. And seas were not only fished, but also farmed.

Also creating diverse plant diversity, later to be reduced by plantations, migrating birds and those driven off course by gales and the North Pacific's westerly currents brought seeds to the Hawaiian chain. Fallen trees and debris carrying plants and amphibians also floated from island to island over the course of millennium. Even the potato tuber, starch for new plants growing from its buds or eyes, must have floated or been brought by settlers to Pacific islands from the shores of tropical South America—to become a staple that islanders cultivated and interbred, enlarging their tubers.

Conifers and cycads, gentums and ginko evolved to produce 500 or so species of naked seeds (Gymnosperms) that migratory birds ate, spilled or defecated on many Pacific islands. The birds also spread herbaceous buttercups, roses, peas, violets, mints, and sunflowers, evolving into a riot of varieties, with their two or more seed leaves, net-veined leaves, four-or-five parted flowers, and wood fibers arranged in a ring around the pith.

Pondweeds, aroids, grasses, palms, orchids and lilies—Monocotyledons having one seed leaf, parallel-veined leaves, three-parted flowers, and wood-fiber scattered throughout the pith—radiated as well.

Among the emergent 140,000 or so of these flower-producing Angiosperm varieties on Pacific islands and atolls, insects and birds survived, adapted, radiated into new species, and altered their proboscis shapes to extract pollen from the corresponding contours of the flowers they lived on. Farmers planted some of them, too, for honey and sweetness, cultural ceremony, and gift-giving.

In the crosshairs of newly arriving plants, animals and humans, indigenous and endemic species suffered the fate of stress, competition for resources, rarity, and extinction. The progression depended on the isolation of Pacific islands from larger continents and island chains; their terrain accommodating new arrivals; and their ability to combat immigrants to propagate and radiate into new species.

Balancing the Pacific Ecological World

Before the so-called civilized Atlantic world crashed on the remote galaxy of island communities set apart by the seas, the number of human beings were regulated by indigenous customs, often harsh, but meeting desired tribal ends.

They balanced the Pacific ecological world with the presence of the small, but numerous black rat, derived from a Malayan form, brought by the Polynesians in their great migrations eastward and southwards.

These small *Rattus exulans*, much gentler in habits and less aggressive than the later-arriving larger western ship and Norway rats, were used for the sport of hunting-games and culinary delights; they did not harm bird island sanctuaries; and only later were exterminated on many Pacific islands by feral cats or the arrival of larger carnivors arriving on western ships— breaking the natural balance.[16]

So said Charles Elton in his classic 1958 study of the *Ecology of Invasions by Animals and Plants*, documenting both parallels and dissimilarities among alien invasive species on Easter Island, the South Atlantic's Tristan da Cunha group, the Hawaiian Islands, and other atolls and islets.[17]

There were thousands and thousands of small, remote islands that had been too far from evolved evolutionary centers of major continents to acquire more than a sprinkling of accidental immigrants before human arrivals began to make the process of species dispersal easier, faster . . . and more devastating.[18]

Examples of a handful of remaining islands with indigenous species had included: Palmya Island, a small atoll in the equatorial Pacific holding fast only fourteen species of native plants. And Midway Island, lying at the western extremity of the Hawaiian chain, with a tiny insect variety—six beetle species, and nine of flies.

Meaning the scale of dispersing island species was well underway by the late 1950s.

5. ISLANDS SHORN OF SPECIES; EVOLUTION FROM ALIEN INVADERS

Where do isolated species come from in the first instance?

We know species living on remote islands have evolved from aliens over millions of years.

And there are known instances when islands shorn of species have been reimbursed from flora and fauna of nearby islands.

The 1883 volcanic eruption on Krakatau Island destroyed two-thirds of its entire six-mile-long territory, once clothed in tropical rainforest, leaving a remnant of the southern volcano and two nearby neighboring islands, Panjang and Sertung. Surviving terrain was covered with hot-ash, thirty-foot deep, leaving neither plant, blade of grass, nor fly.

In part it was re-inhabited by epiphytic plants and birds, reptiles and insects crossing from the nearby, 25-mile, adjacent islands. And a so-called reassembly of a feeding chain was attempted by ecologists three years later, planting trees, to be followed by flying casuarinas migrating from Australia, and the introduction of a lizard monitor to keep the birds from undercutting the reproduction rate of ferns. Such reassemblies under human care usually amount to little more than the third stage of trophic gardening—after Asian Polynesians and European invaders.[19]

In recent years Krakatau again has blown its crater several times, belching sulfuric acid, lava and fiery boulders, burning grasses and ferns, threatening lizards, birds, and infesting insects, driving off intrepid volcanists.

Other Alien Invaders

By contrast, isolated, distant Easter Island once held fast only an original forest savannah of grass, five endemic small animals, and the single tree *Sophora toromiro,* now nearly extinct.

The island lay 2,280 miles from South America, the nearest vegetated Pacific reach from Ducie Island a thousand miles distant. Lore claims with some supporting evidence that the island people were divided by social classes, warring against one another for land to garden, for sustenance and housing—their battle destroying the last forests, burning their dead in mass graves, and, without remaining timbers to build ocean craft, preventing the escape of survivors.[20]

Much later, western adventurers brought earthworms, 44 kinds of insects, spiders and other invertebrates; lizards, birds and rats—non-native animals outnumbering in species natives 10-to-1, and far more so in total populations. Recent times brought South American wedding parties, flying in to celebrate at newly constructed hotels and inns, inadvertently bringing microbes, seeds and tiny animals with them—so that reconstructing past species diversity is now practically impossible.

Island isolation also beset the South Atlantic's three remote islands of the Tristan da Cunha group, even farther away from the nearest continent than Easter Island—2,900 miles from South Africa, 3,200 miles from Brazil, 4,500 from Cape Horn.

Fifty native and seventy alien species of flowering plants beset the three islands, besides nearly 300 of ferns, mosses and liverworts. Other than sea-birds, some five were native, some evolving differences on the separate islands.

Insect life includes twelve moths and butterflies, five natives, seven brought by humans. Of plant-lice of four kinds, two were brought by man.

Of twenty beetle species thought to be indigenous, two are predatory, the rest are herbivorous, being weevils, six of seven brought by our species. Besides vegetation running down the sides of an ancient crater on the upper part of Tristan island, live one species of small trees, tree ferns and heavy tussock grasses and liverwort.

The rest of the three islands are wild, except for the small human community living on a shelf of land three miles long, exploiting woods and seabirds, growing gardens and food-essential potato crops, alongside pasture—invaded by rats eating potatoes, a native moth lacking parasites and an introduced staphylind beetle, an European millipede, and still another on cultivated land.

Attempts to remove potato bugs and potato-eating rats today are aided by internet connections prescribing bioengineering techniques. If interested as an invasive, contact your travel agent.

Hawaii's Biological Order; Invasive Species & Survival[21]

Arguably the biological ordering of species in Hawaii is the most isolated and convoluted on the planet.

Its isolated archipelago of tropical mountains and rarified species lie, as the seabird flies, some 3,400 miles from Japan, extending some 2,484 miles from the nearest point in North America, and 2,800 miles from Fiji.

As the most geographically remote among the planet's archipelagoes, the islands also line up in a linear pattern.

Created by fulminating volcanoes, its territory is graced by the four great island of Hawaii, Oahu, Maui, and Kauai, each in size exceeding most other ocean island fastnesses, punctuated by various topographical elevations, mini-climates, and diverse habitats.

Each island in turn is separated by sufficient ocean water to hyphenate isolation promoting speciation, as well as remote from four others—Molokai, Lanai, Kahooawe, and Nihau—and more than a dozen northwesterly lying flyspecks.

Among their mini-climates, and diverse habitats, the warm, moist and mountainous island chain became a laboratory of radiated evolution and quickened genetic revolutions, framing the virtual splitting of original populations into many full-blown species.

The isolation of each island and islet, and the isolated niches on each in turn speeded such radiated evolution, creating a glorious order of surviving species, both endemic and newly-arrived. Together they are a bio-microcosm unknown on any other place on Earth.

Rosetta Stone of Hawaiian Life

The Rosetta Stone of Hawaii's island life existence turns on Nature's scripts that express a common condition: (1) Isolated small populations of living creatures and plants exist in particular niche environments; (2) that genetically evolved within their species group and radiated into many different species varieties to the point where they could no longer mate or germinate with their antecedents or distinct-and-separate species contemporaries; and (3) that formed independent linkages between themselves and with other genetically-variable, symbiotic or sympatric species—all these seemingly purposeful to the human eye, yet remaining without a preconceived design of the life-essential connections between them.

In Hawaii's many thousands of unique environments, the evolution of these plants and animals also became building blocks for an ecosystem composed of available foods, detritions, gases, biomass and soils, all interlocked with the cycles of sunlight and available water—together bringing about rapid species radiation and divergent evolution within species. And these impacted still other species in sustaining and balancing the environment and evolving interlocked food chains.

Biological inner-conversion of gases and nutrients thereby became a central factor balancing and quickening rapid plant transformation of carbon dioxide into oxygen, sustaining animal existence. And this natural assembly of species came on stream in thousands of unique environments.[22]

Let's try to illustrate how such complexities evolved.

Stages of Island Evolution

There have been three known stages in the evolution of animal, bird, insect and plant species in Hawaii: (1) the seeming pristine ancient, evidenced by excavated animal and bird bones; (2) the middle, viewed as the rapid adaptive radiation and the later trophic cascade of honey creepers, and their essential foods of insects and the pollens and saps of specialized plants; and (3) the modern invasion of alien humans, animals, birds and diseases, disjointing and undermining the middle period of radiation and initiating a series of rapid trophic cascades.

During first stage, the extinction rate for Hawaiian ancient birds may approach 80%, says Storrs Olson of the Smithsonian Institution's National Museum of Natural History.

This annihilation may have come in the aftermath of the 530-32 A.D. giant explosion of Krakatau; and another great explosion of the New Zealand Taupo Volcano about 186 A.D. that spewed upward some 24 cubic kilometers of rock and debris; and ejected deadly gases and ash so high into the atmosphere and stratosphere that the air currents and rotation of the planet carried the vapors and clouds throughout the Pacific. Sunlight was blocked out during plant growing seasons; entire forests and billions of insects died; and birds that survived on insect and plant life were destroyed.

During the middle stages, Hawaii's bird and animal species were isolated, confined and conditioned by the distance from other Pacific islands and mainland continents.

Bats and a single seal species were present. But there were no reptiles, amphibians or mammals that floated to Hawaii or settled on its shores. Some evolving birds were native; others came on flyways or were blown in by storms.

Ecologist Bill Bryson says millions of years of isolation had allowed Hawaii to evolve 8,800 unique species of animals and plants.

On the wing, bird colonizers, individuals or in small groups, sailing over the ocean, were the first to stumble on Hawaii, where they remained to live and breed.

This progression continued for millions of years, with a few species repeating the same process, even now. Then came evolution of new species, as many of these original colonizers changed through adaptation, filling unoccupied ecological niches in thse young islands, radiating into new species, and very likely fitting into vastly changed environments—signifying their epigenetic transformation.[23]

For endemic *Drepanidinae* descended from finch-like ancestors, for example, their own subspecies of honeycreepers rapidly adapted to changing environments, to feed, nest and reproduce—evolving into more than thirty separate species, seen in appearances (body structure or traits), specific living spaces (niche environments), unique calls (songs and warbles), and inherited marks (phylogenic markers) and proclivities, as well as physiological or habitual movements.

During the third stage, invasions by humans bringing in voracious alien species upset the balance among endemics species—setting off a downward death cycle.

At least one bird species brought by the early Polynesians still survives as an established alien species—the red junglefowl (*Gallus gallus*), ancestor of the domesticated chicken. Hawaiians call it *moa*, not to be confused with the extinct, flightless bird in New Zealand of the same name. How many other bird species may have been brought by the Polynesians and failed to become established is unknown.

Birds known to have been in Hawaii in the past, but which are no longer there, can be catalogued as follows: 35 or more species (sub-fossils, probably native residents) were extinct before Captain Cook's visit; 16 species (resident-natives) have become extinct since Captain Cook's visit; and about 150 species may have been introductions.

As already seen, Polynesians also brought dogs, chickens, small-bodied pigs, and the tropical rat *Ratus exulans*—that rapidly ran feral and consumed indigenous birds, their eggs, and chicks.

Eco-friendly Polynesians were not, fossil evidence suggesting a significant Hawaiian impact. For once the Polynesians arrived, they ate great numbers of birds. Unearthed in the archeological garbage pits of Hawaiians past are numerous bird bones, native birds dominating the earliest sites. In the later pits, the middens, there are fewer native bird bones and more and more Polynesian-introduced food sources such as dog, pig and chicken. These pits also show several extinct species of flightless land birds and numerous species of seabirds.[24]

The Hawaiians also cleared vast aeas of lowland forest for firewood gathering, for agricultural plantings of taro and sweet potatoes, and for grassland cultivation.

The loss of habitat drove certain endemic species and animals upland, to extinction, many others being forced into marginal areas.

And food needs to sustain a growing native population led to massive environmental destruction, as well as the killing of the *mamo, Drepanis pacificas,* for their yellow feathers festooning the crowning cape of *alii-ai-moku*, the chief political ruler.

New Plunder Topples Species Isolation

Human invasions ended species isolation and led to rapid destruction—continuing to our time.

In 1778 the Europeans had arrived with guns, plundering and carrying in their foreign animals and diseases.

Establishing his authority, the self-assured Captain Cook set loose ravenous goats and pigs. Captain George Vancouver followed with cattle and sheep—all going feral, destroying grasslands and woodlands. Then a holocaust was loosened:

- In 1778, from the Mexican west coast, anchoring at the island of Kauai, the British *Wellington* brought larvae of the *Culex pipiens fatigans* mosquito as the principal vector of avaian malaria infecting bird populations that dropped like stones. Later European colonists' poultry harboring a birdpox virus was also carried by these *Culex* mosquito species to endemic Hawaiian bird species, with darkened consequences.
- Without foresight, European seafarers, merchants and settlers burned forests, hunted everything moving, and carted in their chickens, goats and horses. Settlers destroyed native flora and fauna by turning native trees into timber, clearing and burning large lowland areas for planting sugar cane; later leaving no soil unturned, introducing pineapples. Even forests left uncleared were invaded by exotic weeds, undermining native flora.
- Nineteenth century arriving European ships also carried to landfall the black and Norway rat, the black varuety able to reach and raid the nest of even the most cautious bird. Still later arriving settlers brought cats and the terrible mongoose, both going feral too, added to axis deer introduced from India multiplying prolifically, destructively browsing forests.
- From arriving ships escaped once-caged, exotic birds carrying malaria and dropsy—bringing down most honeycreeper varieties; and ever new rounds of stowaway mosquitoes were brought to portside, again carrying avian pox and malarias—so dangerous that by the end of the 19th century whole communities of native birds vanished from the forests.
- Those islands at a higher topographical reach and those with still higher mountains traces acted as a last barrier to which endemic birds could escape the *culix*. Yet only a handful of survivors retreated to elevations where mosquitoes were absent or still relatively few. Altitudinal migration was not possible for the most spectacular of the honeycreepers, though. Necessity and habit set their evolutionary path and locked them into local environments and food resources—each going extinct in a sequence that, step-by-step, accompanied the invaders: the Hawaii *mamo* (*Drepanis pacifica*) last seen in 1899; the *black mamo* (*Drepanis funereal*) gone about 1907; similar missing *Ciridops ann*, a little red, black and gray finch-like honeycreeper, precise extinction date unknown.

The food resources for birds, some ten thousand insect varieties known in Hawaii today also evolved from some 400 or so immigrant varieties of insects originally reaching its shores.

Designating shifts in their own genetic inheritance, habitat and habits, certain insects changed their island dominance, feeding practices, and survival fitness, finding niches in Hawaii's many unique environments. Genetic isolation and drift, and environmental adaptation drove the fruit fly *Drosophilia* to radiate as 500 or so separate, competitive species that were reproductively incompatible. They specialized not only in the fruits, saps, and animal species they favored to feed and lay their larvae upon, but themselves also became the prey other specialized insects and birds made of them.[25]

Insects along with seeds, pollens and nectars of unnumbered plants became the feed for birds, particularly honeycreepers, whose ancestral species of a single pair or small flock of primitive colonists most likely arrived many thousands of years ago on the wing, or were blown to the islands by storm from North America or Eastern Asia.

As their population expanded, the honeycreepers exhibited an explosive adaptive radiation, so much so that they penetrated varied environmental niches and, in seeming concert, diversified their anatomies, feeding stations and behavior.[26]

Replacing Resident Species

In lowland forests, heavily impacted by one human invader after another, more alien insects, birds, and plants species became established in assembly, and reassembled as circles of feeding and protective chains of similar or sympatric species than anyplace else in the world. The foreigners, yes the aliens, humans and otherwise, displaced the resident, endemic species.

- Twenty-eight percent of the archipelago's insects, birds and plants imported were transformed into reproducing, non-native species.
- Each time an alien species succeeded in establishing its domain or niche, it triggered a reduction in the size of native species or it pushed them—to extinction.
- In the early 1890s, Hawaii was native home to about 70 avian species. By mid-summer 2005, 26 of these birds were extinct; 31 more were endangered; and of the U.S. list of threatened and endangered bird species, 290 flew Hawaiian.
- Birds known to have been in Hawaii in the past, but which are no longer there, can be summarized as follows: 35 or more (subfossils, probably native residents) were extinct before Captain Cook's visit in 1778; 16 species (native residents) have become extinct since Captain's visit; and about 150 species may have been introduced.

- Adding these to the 272 bird species counted in Hawaii by 2006, about 475 species of birds are known one time or another to have dropped their landing gear in what was once a large, lush aviary.[27]

The global picture of avian survival also remains bleak: Since the year 1600, worldwide 171 species and subspecies of birds have gone extinct, of which more than 90%, some 155, lived and died on islands.

Island isolation of bird species would be no more.

6. PACIFIC COUNTERPOINT—FOREVER ENDING SPECIES ISOLATION

"Looking to [species] geographical distribution," Darwin ventured, "if we admit that there has been during the long course of ages much migration from one part of the world to another, owing to former climatal and geographical changes and to the many occasions and unknown means of dispersal, then we can understand, on the theory of descent with modification, most of the great leading facts in Distribution."[28]

Species Distrubution

Darwin was atune to the migratory patterns that spread species throughout the globe. In the vast Pacific reach, Wallace would prove the master of species distribution and diversities. And Charles Elton would cap the consequences of human travel on plant and animal distribution.

Travel lust and urgent escape simultaneously led island warriors and chiefs to the seas following island crowding, exploding volcanoes, and the land submerging beneath rising seas. In their train came settlers, retainers and priests, seeds, animals, culture, and ideas.

Over seven or eight generations, ethnic and cultural makeovers would be the feature of the Hawaiian domain; it continued among the southern Pacific island group for another several generations up to the 15th century, when a Samoan group from Sawaii emigrated to New Zealand.

More changes followed. Through the 16th-19th centuries imperialists would recast the face of the entire Pacific—conquering with armored ships and soldiers; spreading European diseases, manufactured wares and methods of expropriation—destroying indigenous, island species of animals; and transferring the flora of one colonial territory to others to create new agricultural crops and gardens.

Trophic Cascades and Surviving Species

The cycle of downturn for certain interconnected and niche-confined island species thus turned on environments to which they could not escape or adopt to prevent starvation and maintain their fitness to reproduce.

In the case of Hawaiian honeycreepers unable to migrate to higher altitudes, their death set off a downward cascade: the plants they pollinated could not reproduce, left no trace of new plants that could absorb the Sun's rays, then decayed and emitted methane gases—in small ways leading to even cooler altitudinal midlands, where ever fewer honeycreepers could live—extinctions of a maelstrom.

The Dantesque scenario was often thus: Death of bird and other animal species, that initially undercut plant pollination and the spread of foliated varieties; eventually drying the soils; annihilating worms and other processors of biomass; cutting emissions of oxygen by living plants and methane by deadened ones; changing composition of atmospheric gases—together adding fuel to wildfires set by volcanic explosions and lightening.

Under the threat of fire, wild cattle, sheep, goats and horses, and the clearing of land for crops, luxuriant forest lines retreated from the coasts, and another one-quarter of their reach when Polynesians first arrived.

Ecological Vibrancy During Last Traces of the Ethocene

During the last traces of the Ethocene in the 1950s, Pacific Island ecospheres nonetheless retained their vibrancy.

Elton documented that:

- Over nine-tenth of the 1,729 species of flowering plants were found nowhere else outside the islands; 272 of them originally arriving to the islands by various routes.
- Some 3,722 of the 6,000 or so species of insects were endemic; 233 to 254 of them originally arriving to the islands by various stages.
- Of two large families of snails, the *Archatinellidae* with 215 species were unique to the islands, living in trees, one of them originally arriving to the islands from abroad; the *Amanstridae* with almost 294 species lived in trees and on the ground, 22-to-24 of them originally arriving to the islands by various paths.
- Endemic birds throughout the island chain numbered 77, with about 43 species and sub-species belonging to the *Drepaniidae*—a family within which more than a dozen ecological ways of life evolved as honey-suckers, wood-insect hunters, other insect-eaters, and those reliant on seeds, nuts, and fruits. Among them, about 14 originally arriving to the islands

by unknown routes—the *Drepaniida* coming from only one of these forms.

Global Web of Life Continues Still

Yet the procession of immigrants confronting endemic species continued over time, contining still.

Island biogeography was permanently being restructured. The worldwide linkages of transport, air and sea currents brought alien species into contact with once regional and island ecosystem plants and animals that were unable to move, resist or survive—leading to downward cascades of some species; further displacement by still others; or becoming interlocked with invading fauna and plants, together making up an entirely new biological web.

Newly-established insect species, for instance, destroyed both natives and other invading counter-pests, that arrived by tempestuous weather cycles, human carriers, or inadvertently in their train.

"It is quite an exchange and bazaar for species, a scrambling together of forms from the continents and islands of the world, a very rapid and efficient breaking down of Wallace's Realms and Wallace's *Island Life!*" Charles Elton wrote in 1958, of the shattering of regional isolation of unique forms of taxa.[29]

7. A NEW PANGAEA APPEARS

Long ago, species isolation had forever ended on the Hawaiian Islands and other Pacific islands and flyspecks.

Endowed by the new means of worldwide locomotion and communications, remaining species' environmental webs were soon blasted apart. In transformed ecosystems this, in turn, led to intense competition between species for fitness to survive and reproduce.

As human-driven warmer environments spread, many thousands of species sought to escape.

As each species had their own maximum and minimal heat tolerances, each sought geographies where it could survive and multiply—moving up to higher elevations or farther north or south from the Equator towards the polar caps.

A Few Examples

Not only animals, but trees as well can and do move, as a few examples help explain.

From 1984 to 2011, the march of mangrove forests more than doubled in size at the northern end of their Florida range, as temperatures had rarely dipped below 25°F, a vital threshold for the trees' reproductive fitness.

Mangroves quickly responded to the new inhospitable areas, floating seedlings soon taking root. At the most northern reach of the Atlantic seaboard at 30°F, mangroves enlarged by 170.7% to 3,324 acres in these 17 years. And geographically moving down the coast where 29ºF prevailed, mangroves enlarged 47.1% to 3,720 acres; reaching midlands Florida tolerating temperatures of 28ºF, expanding by 16.1% to 10,848 acres.

Colder ranges meanwhile led to Mangrove retreat, losing territory by 3.7% to 2,021 acres as temperatures ranged lower at 27ºF; and at 25-26ºF, by 4% to 18,793 acres.[30]

Even in colder regions, human unintended transport of invasive species can take advantage of climate alterations to destroy endemic and long-present forest species.[31]

Redraw your U.S. map of surviving species after the 2002 discovery of the invasion of the Asian Emerald Ash Beetle Borer in the U.S., for example. When the adult borer multiples its species, its larvae eat into the bark, burrow deep into the trunk insulating themselves from the winter cold; and as its larvae bores into the sap line of the ash tree, it cuts off the tree's essential supply of water and nutrients,

The beetles then emerge in the spring as adults, mating and laying eggs, leaving the next generation of larvae to tunnel through the tree's internal organic structure—able to kill an ash tree in as little as two years.

The death of groves of ash, then leave gaps in the leaf canopy that allows sunlight to reach parts of the forest floor that were previously shaded. Those gaps provide an opportunity for invasive honeysuckle bushes to grow unchecked. "In the worst case scenario," predicts Dr. Kathleen Knight, research ecologist at the U.S. Forest Service, "it becomes a dense, impenetrable thicket of shrubs in the understory." The thicket then prevents native plants from growing, and is likely to affect which kinds of animals can live there.

Proof of such destruction and the creation of a restructured environment is reported in a 2000 study in the journal *Biological Invasions,* listing 43 native insect species that rely on ash trees for food or breeding, these insect becoming the traditional food supply for birds, including woodpeckers.

Such a cascade restructures forest life among species. "You end up with a different ecosystem that different species prefer and where the old ones can't do as well," Dr. Knight reports. And some trees survive better than others by species and location—black and green ash are quickly lost; 60-to-70 percent of white ash are destroyed; white ash falls somewhere in between.[32] What is not in between is the old ecosystem—often lost forever.

Never-Ceasing Vectors of Invasive Species

Just as you and I often unknowingly carry our maladies, and diseases to others among our company, the vectors transporting or carrying particular invasive species, blight or microbe may be living animals or plants, air, water or shifting ground, or all of them.

A mighty 1,000-year-old oak can be felled by a fungus carried on a bay leaf tree or on an imported marigold flower. Asian hemlock wooly adelgid, the size of a poppy seed, making fluffy, white egg sacs to wrap themselves, fix to the undersides of hemlock branches, attaching a feeding tube to suck the sap out of the tree like a vampire bat draws blood—leaving the tree to slowly expire.

Working their way north from Virginia in 50 years, the adelgid kill-rate registered more than half in infected forests.

Fungal infections brought down entire forests of U.S, chestnuts more than 500 years ago. Hemlocks replaced them, today hemlocks largely being destroyed by the adelgid creating the ground for a new canopy.

Indeed, species are constantly embattled for survival, the most fit facing unperceived environmental and human challenges, surviving at the expense of others—this more than a metaphor, but constantly restructuring ecosystem on Earth.

NOTES

1. Charles Darwin, *Origin of Species*, 515-516.
2. Charles S. Elton, *The Ecology of Invasions by Animals and Plants* (Chicago and London: University of Chicago Press, 1958), 33-34.
3. Elton, *The Ecology,* 34.
4. Darwin, *Origin of Species*, 516
5. For mud footprints, see Jill Lawless,, "Scientists find 800,000-year-old footprint in UK," *Associated Press*, February7, 2014; reporting by Nick Ashton, of the team from the British Museum, London's Natural History Museum and Queen Mary College at the University of London, *PLOS ONE*.
6. Gould, *The Structure*, 1296.
7. Elton, *The Ecology,* 35-37
8. This and the following paragraphs are based on geologist Richard Smith's series on Australia, Nova, PBS, April 13, 17, and 24, 2013, 9-10 PM. See also Krooth, *Gaia*, 596-97.
9. Luis and Walter Alvarez, "Extraterrestrial Causes for the Cretaceous-Tertiary Extinction," *Science*, June 1980.
10. Kolbert, *The Sixth Extinction, passim*. During my own travels in both Costa Rica and Belize, I have seen similar chemical alchemy performed by microscopic bacteria living in water-filled, green-tinted, warm volcanic craters; as well as by microorganisms releasing bluish-purple steam from sulfuric-laced, boiling mud basins. Sun rays also oxidize species-released methane, creating a rainbow of lurid colors. The composition and colored-patina of their chemistries support Gaian findings—that transpiring plants and respirating tiny animals issue powerful gases that change environmental conditions, both making life possible for some species and destroying others lacking the reproductive fitness to survive under suddenly altered atmospheres—from blistering extraterrestrial incursions to seafloors disrupted by pelagic volcanic explosions, spewing hydrogen sulfides.

11. Wilson, *The Diversity of Life*, 29. See also Krooth, *Gaia*, 596.

12. This section compares Darwin, *Origin of Species*, 516, with Krooth, *Gaia*, Part V; Leakey and Lewin, *The Sixth Extinction*, 77-194. For Australia, see A. P. Elkins, *The Australian Aborigines*. For the Americas, see Reader's Digest, *The Mysteries of the Americas.*

13. Elton, *The Ecology* 36-37.

14. Krooth, *Gaia, passim.*

15. Krooth, *Gaia* ,Chapter 15

16. Krooth, *Gaia,* Chapter 15.

17. Elton, *The Ecology*, 77-93.

18. Elton, *The Ecology*, 77-93. See also Krooth, Gaia, 336.

19. See Ian Thortnon, Krakatou: *The Destruction and Reassembly of an Island Ecosystem* (Cambridge: Harvard University Press, 1996).

20. This and the following paragraphs are based on Krooth, *Gaia*, Chapter 15. See also Elton, *The Ecology*, 77-93.

21. Krooth, *Gaia*, Chapter 15.

22. Krooth, *Gaia*, 311-312.

23. Robert L. Pyle, *Birds of Hawaii* (Hawaii: Bishop Museum, 2006), *passim.*

24. This and the following paragraphs are based on Krooth, *Gaia*, 320. See also Rob Pacheo. "Nbird's Bone View of Hawaii," *Hawaiian Forests & Trails*, February 2006. And Quammen, *The Song of the Dodo*, 317-18, *et passim.*

25. Krooth, *Gaia*, 317, 326.

26. For avian radiation, see Wilson, *The Diversity of Life*. 94-9, 101-02. See also Pyle, *Birds of Hawaii, passim.*

27. Based on Krooth, *Gaia*, 326. See also the following: Richard Krooth, "Travels in Hawaii," *Notebooks*, 1971. Martha Beckwith, Hawaiian Mythology (Honolulu: University of Hawaiian Press, 1970), 168-73. Leakey and Lewin, *The Sixth Extnction*, 164-65. Jared Diamond, "Holistic Extinctions: A Rosetta Stone for Understanding Prehistoric Extinctions." In Paul S. Martin and Richasrd G. Klein, Eds., *Qiartenary Extinctions: Prehistoric Evolution* (Tucson: University of Arizona Press, 1984). And see Kate Wong, "Mystery Human,"*Scientific America* (March 2016), Vol. 314, no. 3, 29-37.

28. Charles Darwin, *Origin of Species*, 515-516.

29. Elton, *The Ecology*, 77-93.

30. Rachel Hartigan Shea, "Uprooted," *National Geographic*, June 2014.

31. Richard Krooth, "Travels on the U S. Eastern Seaboard and in the Midwest," Notebooks, 2004-14.

32. Dr. Kathleen Knight, research ecologist, U.S. Forest Service, 2014.

V

Destructions Past; Destructions Future

> Something far more dangerous [than exploitive relationships] is revealed [in state and military institutions]: the capacity of man to abandon his humanity, indeed, the inevitability that he does so, as he merges his unique personality into larger institutional structures. This is the fatal flaw nature has designed into us, and which in the long run gives our species only a modest chance to survive. It is ironic that the virtues of loyalty, discipline, and self-sacrifice that we value so highly in the individual are the very properties that create destructive organizational engines of war and blind men to malevolent systems of authority.—Stanley Milgram, *Obedience to Authority*

> Existence itself is being pushed by our species towards the vanishing point. War over resources—food, oil, water, land—and the fouling of the environment, pandemics, and technological nostrums hasten the day.—Richard Krooth, *Gaia and the Fate of Midas*

Humankind has not patiently, or most otherwise, accepted Nature's wonders. During the past 250 years or so, under the rule and guidance of constitutional states and those where lawlessness prevails, fossil fuels have been mobilized to drive machinery, aerating gases, debris and erffluents, many becoming long-lasting components of cloud formation, the atmosphere, and fall-out to Earth as bedrock.

Geographic extension of the industrial foundations for production and unnumbered commodities also has worsened environmental conditions, altering flows and circulation of air, water and soils. On practically all continents, organizational engines of war have brought further degradations of Nature. And the most recent Cyclops, nuclear power plants, have released irradiation and toxins on a global scale, undermining species ability to adapt and survive. Some cite as causes of environmental degradation the wealth accumulating drives of great enterprises, avariciously digging mines and extracting and burning fossil fuels and exploiting land and sea. Others view the profit-quest as secondary to using the bounty of Nature to secure human needs. Still others blame population expansion and human spatial designs as infringing on the natural domain. And considering all these forces imposing on the natural world, there is little doubt our species has developed a colossal blind spot that leaves little hope for securing unmolested future natural cycles.

Humans seems bent on walking over a proverbial cliff, when it comes to calibrating the directions in which their own actions and technological advances are taking them. As a species we either produce means of destruction with consequences our societies are unable to control; or our existing social organizations themselves overwhelm the existing means to sustain us. Either way entropy sets its heel, leading to local disorders and disasters, punctuated by social dystopias that rapidly spread to other regions and merge or otherwise consolidate into an overwhelming disclimax, jeopardizing the remains of natural environments and species reproductive fitness.

The following chapter elaborates these conundrums. Throughout, the question remains: Can we control the future pathway of the exponential growth in means of production jeopardizing natural cycles or threatening our species relationships?

DISCOVERIES, GEOGRAPHIC EXPANSION, TECHNOLOGICAL ADVANCES

The past of discoveries, geographic expansion, and technological advances bode ill for present and future human equities and species diversities. Led by its discovery of the New World, European expansion brought on environmental destruction and slavery. Destruction was advanced by intermediary conflicts and brigandage—between Spanish, Portuguese and British in the Americas and the West Indies and North America, establishing mercantile colonies. Then Britain changed policies from seeking the accumulation of wealth in the form of tribute paid in bullion to establishing colonial trading territories, from which British factories could draw raw materials and in turn manufacture them and then return them to colonial markets as finished goods.

Expansions past also reveal that geographic, biological, and ecological destruction led to excavating and domesticating forests and seas, creating new frontiers of colonial encumbrances, endangering the very roots of land and ocean plant life, bringing in alien species and releasing dangerous, enzootic life forms, along with rapid changes in climate and reduction of life diversities. Discoveries also released forces causing global warming that are so well known today that many species' survival and future reproduction remain at stake. And the appalling human lack of foresight advancing the technological sphere endangered and continues to endanger the biosphere in deep-seated ways, though often not immediately or completely understood.

TWO WARNINGS

There is no excuse for our species saying it did not know the deadly consequences its explorations, expansions, and technological advances were having; and the disasters that lay ahead for the interlocked web of life. Today technical expansion serving our species' needs continues the devastation of forests, the toxic impact on domesticated soils and oceans, and the blinded invasion of the deepest seas, jeopardizing species evolution.

Lester Brown, President of the Earth Policy Institute, author of *World on the Edge,* says of "the real threats to our security today, at the top of the list would be climate change, population growth, [fresh] water shortages, rising food prices and the number of failing states." As these forces remained fast unraveling, on April 3, 2012, Jean-Michel Cousteau warned that:

> Climate change is a reality. What we are responsible for and the consequences we'll have to face up to is that because our emissions of carbon dioxide are accelerating the process, things are happening much faster. One of the biggest impacts is the reduced protection of our coastlines due to corals dying and sea level rising. [Anticipate] increased storms and flooding; and millions being displaced around the world.

Al Gore adds that scientists can now create and control the conditions for, and channels of, species evolution by manipulating life's genetic codes of DNA and RNA—yet barely comprehending the possible path their evolving species might take—either beneficial to, or endangering, Earth and natural species. (See *The Future*: *Six Drivers of Global Change* (New York: Random House, 2013), Chapter 5, "The Reinvention of Life and Death.")

Using eugenic methods restructuring and interrupting natural selection and evolution, the diversity of species is already being placed in jeopardy. This "creative" design by self-styled scientists, hubris justifying their technical skills, at root turns on many factors, including the accumulation of wealth, laying the foundation for future risk to planetary life. These threats remains baited by our own hand, set to destroy practically every living species and so-called civilization.

Chapter Eight

Atmosphere and Deserts; Deluge and Extinctions

> For the first time since life emerged from the primeval soup, there are beings who understand how they have come to be what they are. To those who fear adding to the power of government and the scientific establishment, this seems more of a danger than a source of freedom. In a more distant future that we can still barely glimpse, it may turn out to be the prerequisite for a new kind of freedom.—Peter Singer[1]

> Nature is the best engineering teacher of all: it's been there for 3.8 billion years, constantly evolving brilliant solutions. Now there is a movement dedicated to learning from this infinite wisdom for human design needs. Can a leaf help us design a better solar cell? Biomimicry looks to nature for forms, patterns, systems, and strategies that we can study and design in accordance with, for maximum sustainability.—NextworldTV, April 7, 2012

Today the advance of technology and science is overwhelming Nature and the diversity of species. The complexities of unregulated scientific experiments and increasing government controls over the general population favor the accumulation of wealth at the expense of Nature and the web of life. What does this have to do with global warming?

DANGERS ANEW

"If global mean temperatures rise 2.3° to 4.1°F (1.3° to 2.3°C) above today's mean, then roughly 20-30% of the 1.7 million identified species could be facing a high risk of extinction. Some 400,000 species could vanish," warns Terry L. Root, co-recipient of the 2007 Noble Peace Prize. It's a dismal foretelling.

Should earthly temperatures continue to spiral out of control, for the first time in human existence it will place in jeopardy land-dwelling species, then flying creatures, next, the sea's unknown life varieties. Humans are turning on themselves, releasing to the atmosphere ever more heat, toxins, effluents, radiation, and deadly microbes. Life could expire as our species becomes lost in ignorance of the consequences of the very forces we have valued and loosened.

EARTH'S NATURAL CYCLES STILL IN CONTROL

The seeming exoskeleton of Earth is constantly in motion—driven by tectonic plates, upended sub-terrains, floes of molten rock, pounding seas, east-to-west movement on its axis, and atmospheric fulminations. The layered geologic strata and core of Earth fulminate and boil, pushing through *Tierra firma* with boulders, molten rocks, and lava from the planet's 30-mile inner reaches. Winds and seas driven by Earth's gyres along its axis ceaselessly carry soils, minerals, coastal sands and ephemera from one continent and island to another.

Skies become the ever-moving plane for transporting soils and dust, gases and moisture, whipping up atmospheric pressure points that radiate with water-droplets attached to minute particle matter readied for release as bursts from cloud cover as rain, snow, sleet and hail. Some precipitation descends to sea and land whenever their electrical charges signal lightening, thunder and moisture release—their relative weight giving way to forces of gravity, vulnerable to or able to overcome wind velocities and temperature pressure barriers.

The motions of the atmosphere set the pace and scale of surface evaporation; and, as foliage reaches an *apogée* of vitality and then decays, transpiration declines, then comes to a close, reflecting the rates at which plants release their life-giving moisture, along with oxygen, nitrogen and methane. So, too, land mass is transformed. Suffused with an assorted mix of soils and water, chemicals, nutrients, and gases, it releases them either on atmospheric schedules of seasons, or at irregular moments of wind velocities and precipitation, water flows and upheavals from volcanoes, as well as the cycled melt of glaciers, sea ice, and tundra. Species life forms depend on these changing conditions that are essential for fitness to sustain themselves, reproduce and evolve.

SPEEDING THE CYCLING OF EARTHLY SOILS, MINERALS AND HEAT

Gaia cycles these earthly soils and chemicals, essential nutrients, moisture and heat. The human-designed technological sphere hastens the process, also producing pollution and thermal outpours laced with sulfates, nitrous oxides, gases and debris particles. Together, soil and dust mixed with pollution-borne sulfates and greenhouse gases, condense water vapor and ice crystals, changing the atmospheric balance that absorbs the re-radiation of solar radiation reaching Earth. Hothouse Earth becomes not only a metaphor, but also a potential danger to planetary species life.

HOMO SAPIENS EMERGE; SHIFTING CLIMATES; SPECIES THERMAL LIMITS

Sometime between three million and two million years ago, perhaps on a primeval savannah in Africa, our species became recognizably human. Shifting climate had opened spreading grasslands, presaging changes in the physiology and organic structure of an earlier species of Australopithecine, the new descending lineage evolving longer legs, hands with an opposing thumb enabling tool-making, an enormous brain cavity, and thereby the means to one day rule the planet.[2] But for millennia the human population on Earth was too small to rule the environment.

Jump forward for a sharp remake of the environment 20,000 years ago, as Earth's gyres every thousand years or so added some 0.9° to 1.8° Fahrenheit (0.5° to 1° Centigrade) of heat and humans and other species faced melting ice sheets covering most of the Northern Hemisphere. The melt eventually drove up sea levels at least 100 meters (over 300 feet), forcing mobile animals and spreading foliage to move further Northward. Those that could not move to follow the spreading sources of food faced extinction, as geographers and paleoanthropologists have documented.[3]

The human species either did not know, or at first did not care, about the details of past cycles of population movements. But 250 years ago they themselves changed the environment by burning fossil fuels and littering the atmosphere with carbon dioxide, methane, nitrous oxides, and particle matter—some still circulating after more than a century, absorbing the Sun's radiant energy re-radiated from Earth. The air accumulated warmth, handed down to our time, though not yet approximating an earlier scale of warming.

Future temperatures are unknown, of course. But judging by the last 100 years, the twenty-first century likely will exceed the maximum thermal temperature limits most known land species can tolerate. If worldwide average temperatures *increase* by 11° Fahrenheit (6° Centigrade), the natural triage of life varieties of plants, animals and marine creatures will be jolted and largely undermined.

The so-called mid-latitude band of temperatures lying between 30-to-60 degrees North are critical to predicting species' future sustainability at their thermal maximums. This temperature band covers Russia, China, Europe, the United States and most of Canada. The average annual temperature in this regional area generally falls by 66% of a degree Centigrade with each degree of increasing longitude. But if there were five degrees Centigrade of warming in the 21st century, all species' survival would depend on withstanding or escaping heat by moving Northward.

Yet most of life can not move so quickly. For example, that would impossibly require larch, oak or cedar to seed new growth at an unsustainable rate; and it would make black and brown bear, deer and black squirrels, dependent on berries, nuts and foliage, take their fill in shorter springs and summers.

As heat spreads and dries soils, worms can no longer pry open corridors to reach and consume biomass of dead leaves, dried grasses, decaying foliage, wood chips and roots. Without these the worms die; and without worms grinding biomass to redistribute as loam pellates, the soil is not renewed. Dehydration of the land is thus an initial disaster that leads to a breakdown of life processes that sustain the feeding chain of worms, plants, invertebrates, and vertebrates on Earth's surface.

Leafcutters also depend on green foliage that would face dying tree roots and curled, yellowed leaves preventing them from storing their dens with molding food supplies. Still, heat escalation dehydrating Northern geographic forests and crops would advance more slowly than foliage in seared tropical zones impacting bananas, manioc, ten-foot island corn, cotton, rice and tobacco. But rising thermal temperatures are certain to reduce plant life everywhere.

Of course, not everyone agrees about the future environmental impact on Nature's forests or Gaia's weave of life forces or what role our species would play. But a few examples may illustrate the conundrums of global heat jeopardizing forests storing carbon dioxide and undermining species resources and diversity.

IMPERIAL JEOPARDY FOR AFRICA'S SOURCES OF AGRICULTURAL LAND AND WATER

During the 18th, 19th and 20th centuries, the imperial division of Africa concentrated on extracting resources and plantation output aiding industrialization and commerce in Europe and in the United States, though to a far lesser degree. By the start of the 21st century, many Western powers faced the domestic erosion of agricultural topsoil, the degradation of cropland, the lack of sufficient fresh water, and the inability to feed their expanding populations. This led to massive imports of food from Latin America, Asia and Pacific nations. For food, raw material and fuel imports, Europe had long established special relations with its former colonies in Africa. Japan, the U.S., and Middle East nations had concentrated on trade networks elsewhere, in Mexico, Southern Europe, South America, and Nordic regions.

Intensified competition to purchase, appropriate and re-divide African arable land, water, and other resources, along with control over an unfree and impoverished labor force, intensified in the early 21st century. Of the worldwide purchase of 80 million hectares of land by foreign nations and corporations from 2008-2009, two-thirds were in Africa, evicting farmers and taking away their pre-colonial tenure rights. Undemocratic government elites dealt directly with alien multinational corporations and nations seeking short-term profits at the expense of the evicted farmers, thereby jeopardizing national food production capability for domestic survival.[4] As the process now comes full circle in the early 21st century, local farmers deprived of land and water are likely to become laborers for foreign interests. Many of their governments are being forced to import expensive grains for survival and refined fuels for production, falling deeply into debt to imperial interests. The only source of their means of payment is the sale of their resources, their plantation mono-crops, and their populations' work efforts. It has become a sharp return to colonial systems of indebtedness to Europe and the "head-right" and *corvée* systems of subjugating labor.[5]

Ethiopia provides a classic example where the Nile waters originate, its rich loams beckoning foreign interests growing wheat, other grains, and bamboo, a grass that requires few nutrients, yet is the source for hardwood corporations making inexpensive furniture to supply consumer markets worldwide.

By 2012, some 8.2% of Ethiopia's agricultural land had been legally signed over for foreign use. As one citizen of the region of Gambella, Nyikaw Ochalla, explains:

> The foreign companies are arriving in large number, depriving people of land they have used for centuries. There is no consultation with the indigenous population. The deals are done secretly. The only thing the local people see is people coming with lots of tractors to invade their lands. All the land around my family village of Illia has been taken over and is being cleared. People now have to work for an Indian company. Their land has been compulsorily taken and they have been given no compensation. People cannot believe what is happening. Thousands of people will be affected and people will go hungry.[6]

The imperial ruination of Africa's remaining forests, cropland and fresh water is in sight for the 21st century. These natural sinks for carbon dioxide and methane likely will be destroyed, led by the destruction of forests and jungles, and especially the remaining big trees once festooning Earth.

BIG TREES STORE MOST FOREST CARBON

Dead or alive, big trees store most forest carbon, acting as a blotter or sink for CO_2 that otherwise would be stored in the sea and atmosphere. Giant old trees still dominate many forests worldwide, providing habitat for a wide range of organisms, from fungi to woodpeckers.

Eastern White Pines were once the most rapidly growing Northern forest tree and among North America's most important timber resources. From their clear-cutting, burning and girdling, farmland was extended.

In rural New England, much abandoned farmland then became pine forest. Clear-cutting did not result in the return of young pines, but rather the growth of a "weed forest" or "weed trees," largely scrub hardwoods such as grey birch, poplar, and wild cherry. After cutting, wind-borne seeds took hold, became tightly packed on the old forest floor, and grew before pine seeds had a chance to take root.

Rare indeed today are the truly old pine forests, aged a hundred or more years. Their aisles remain dark, mysterious and inviting. But no birds sing and little grows beneath the few remaining great trees. Year after year the needles and old limbs fall and, at last, the old trees themselves. Such ancient groves become sterile. They will not renew themselves, except, perhaps by fire.[7] And with death, fire and rot, they will release their stored carbon dioxide to the atmosphere again.

Globally, the axe, fire storm and giant plow have decimated forest cover despite the obvious fact that trees absorb carbon dioxide, the main greenhouse gas given off by human activities—meaning the death of a forest might shorten the life of those relying on the woodland's gaseous atmospheric and symbiotic life forms.

What, then, is left of old growth forest cover and growing stocks of commercial-size wood? In 1978, a U.S. report projected that by the year 2000, some 40% of forests would be gone in Latin America, Africa, Asia and Oceania. Dense human occupation of territory lasting over several hundred years already has led to the destruction of forests in North America, the Russian Federation, Japan, Australia and New Zealand, so whittling away their past heritage that forest declines superficially appeared less severe by 2000. Yet, overall, the worldwide fall in forests between 1975 and 2000 was devastating: 47% of all global growing stocks per capita; and 63% in less developed countries.

Crisis was at hand, insisted the *Global 2000 Report*: Deforestation was projected to continue until about 2020, when the total world forest area would stabilize at about 1.8 billion hectares, or 4.5 billion acres. Most of the loss was projected to occur in tropical forests of the developing world.

About 1.45 billion hectares of the industrialized nations' forests had already been stabilized (meaning that after being devastated, the remains would be protected or inaccessible), while about 0.37 billions of hectares in the Less Developed Countries was physically inaccessible or economically impractical to cut. By 2020 virtually all of the physical forests in the so-called less-developed-countries were expected to be cut.[8]

FOREST SURVIVAL; STORE OF CARBON DIOXIDE

Survival of tropical forests rests on a precarious balance of soils, land-forms, temperatures, rainfall patterns, nutrient distribution, and human destruction. Disturbed by extensive cutting in these forests, neither trees nor productive undergrowth of grasses will ever again reproduce. Even in less fragile tropical forests, extensive cutting destroys the great diversity of species.

Collateral damage of deforestation, especially in South Asia, the Amazon Basin, and Central Africa, reveals destabilized water flows leading to silting-up streams as well as reservoirs behind electronic dams (as with the Aswan Dam) and irrigation works. Cutting forests also causes depletion of ground water, intensified flooding during storms, and aggravated water shortages during dry periods.

Some geographies are more impacted than others due to population or agricultural concentrations. In South and Southeast Asia, one billion people live in heavily populated alluvial basins and valleys that depend for their water on forested mountain watersheds. Yet, forests were reduced an estimated one-half between 1975 and 2000, resulting in erosion, siltation, and erratic stream flows seriously impacting food production.[9]

Tree size determines the quantity of carbon dioxide that forests store, large trees slowing its release into the atmosphere, effectively securing Earth from becoming a hothouse, dehydrating jungle plants and trees. Already in the early 21st century, between 15 and 35 degrees North and South of the Equator, great swaths of the world's major deserts and semi-deserts have appeared.

Arid, these desert reaches are kept dry by atmospheric pressures that once dropped rain and dampened moisture near the Equator, but which now drive masses of hot, dry air down to Earth's surface. The descending air cells create high-pressure zones that usually tend to ward off rainfall, though periodic downpours sometimes create a momentary deluge on the desert floor.[10]

Compounding this cycle, near and far our species is tearing up more forests to extend food production, over-drawing on aquifers and water tables, making a desert of Earth. About 20% of all the planet's land area is now covered by deserts, mostly devoid of trees, bushes, and plant life—the largest being the Sahara, still expanding, covering nearly a third of Africa.[11]

The steps towards such a bleak closure are many, but certain it is that the death of big trees lines the pathway. "If a large tree dies, it can have an immediate effect, and takes a long time to replace. We should pay more attention to what's happening to them. Their fate is a harbinger of what's to come, perhaps," says University of Washington professor James A. Lutz.

Among their many other invaluable benefits, the oldest trees also have stored great quantities of carbon that otherwise would fill the skies. With University of Washington Researchers, Lutz surveyed a plot in California's Yosemite National Park, tallying above-ground biomass including live trees, snags, and downed woody debris.[12] They found that big trees (those with a diameter greater than three feet at chest height) accounted for only 1 percent of trees, but stored half of the area's biomass. They measured the diameter of every tree in the 63-acre plot of old-growth forest, the largest of its kind in North America. The researchers counted nearly 34,500 trees, only 489 of which were greater than one meter in diameter. The largest trees on site were sugar pines, followed by white firs. The champion sugar pine was nearly 7 feet diameter wide and 220 feet tall, probably 350 years old. Yet the site was selected because it was considered representative of average forests in the area and did not contain the most enormous individual trees.

Biomass is calculated by using species-specific equations worked out by past researchers measuring dead trees. Current models used by foresters do not work very well for the outsize giants, possibly because of the lack of ancient trees in some American forests and the difficulty in studying them. Models often assume, for example, that forests are made up of a relatively uniform mixture of different-sized trees. In reality, though, big old trees are special, and anything that affects them could have a disproportionate effect on the forests.[13]

And the loss of big trees by sheer age or clear-cutting in our time is also destroying one of Earth's great sinks for carbon dioxide. Along with the oceans and atmosphere, once forests with their giant sentinels having reached their *maximum absorptive capacity* storing carbon dioxide, there will be no large place to store additional carbons issuing from human activities and natural sources.

SPECIES' ENVIRONMENT AND BIOLOGICAL TEMPERATURE LIMITS

Almost all known species on Earth have biological temperature limits. As temperatures mount, the simple condition of species accustomed to sun-shaded climates where they have evolved and developed will find themselves forced to move towards cooler polar geographies and higher elevations, or seek underground shelter.

Plants usually can use their tiny pore stomata located on stems, or on new leaf growth, to absorb needed carbon dioxide; but as they also transpire water vapor through these same apertures, they can dehydrate when heat intensifies.

At the moment that heat destroys their roots, they drop their leaves and curl, releasing methane; and the danger of heat-dehydration also spreads to other plants. Even in a concentrated zone, plants *en masse* that get all needed carbon dioxide, need sustaining quantities of water; and water holds heat, meaning that any slight decreased in plant transpiration also may cause less atmospheric cooling, less precipitation, and thus further warming.[14]

PLANT AND ANIMAL ADAPTATIONS

Some plants and animals alternatively may adapt to both old and new changing environments.

But as metaphor of many species escaping higher temperatures, California's tit mouse can go no higher than the mountain tops; and even furry creatures once living at the edges of northern continents will not *suddenly* adapt to torrid temperatures by losing their coats or taking to the sea.[15]

One species moving into the territorial domain of other species also may impact survival resources to the moment when a fight for life between old and new claimants will turn on the Darwinian survival of the fittest—not just by tooth and claw as Herbert Spencer influenced Darwin to pen, but also by the competitive ability to develop new evolutionary structures for movement, new organs to respirate, and new systems to feed, digest nutrients, and eliminate wastes.[16] As a classic example, the lungfish evolved leg-like structures in water long before it moved onto land.[17] So too, the seven-layered-coat Musk Ox, handed down from 10,000-years-ago, was able to adapt to the 40-degrees below zero Alaskan temperatures, but was almost eliminated by human predation in the 19th century in what today is the Arctic National Wildlife Refuge.[18]

TIMING IS EVERYTHING; WITNESSES ABOUND

How species extinction or transformation will proceed is guesswork in part. Hope breeds stronger than species evolution perhaps. We still do not know if natural cycles will save Earth from human missteps and lack of foresight.

Clouds so far have borne witness, saving Earth by blocking some sunlight and refracting some heat into space. But under our species' reign of burning fossil fuels causing pollution and heat, natural cycling of particles into the atmosphere has grown larger, especially as cyclonic *dust storms* redistribute land mass between continents and islands.

As particle matters attach to water vapor in the clouds, winds often shift precipitating rain, snow and hail to distant terrain and seas below—causing changes in Earth's temperatures in different regions. China sends its dust storms over the Pacific to two dozen islands, including Hawaii, the western United States and Canada. Asian desert dust catches the jet stream invading Greenland and the Arctic Circle. The torrid soils and minute sulfate particles of Patagonia take flight to filter into Antarctica. Particle matter form tiny hummocks within glaciers and on icy seas, in communities of mosses and lichen, providing feeding stations in which insects and birds reproduce and develop innumerable varieties.

Wind shear also produces static electricity and heat rays across the face of dust storms as billions of particles of soil, rocks, sulfates and other chemical elements powerfully collide with one another and interlock with atmospheric moisture and evaporation from the Atlantic sea. Gathering as huge dust fronts, this mass annually moves across the Atlantic, intensifying heat, creating desert-like conditions in the sky, even refracting to space some sunlight that otherwise would be absorbed on Earth's surface.

The cooling effects are periodically modulated, though. Building up heated atmospheric islands, their cocktails of soils, sands and nutrients, seeds and water bombard one another, filtering and dropping alike on light-colored surface ice and desert sands.

Ice-covered land and glistening seas are blackened by dirt and carbon-laden soot, too. Blue waters are suddenly darkened by the fallout. Terrain, sea and ice that once re-radiated the sun's energy to space now absorb it, their warmth helping melt glaciers and permafrost, heating and opening land fissures—releasing methane gas that quickly combines with oxygen, producing carbon dioxide moving skyward.

Dumping their remaining loads on reaching landfall, the magnetic carpet of dust effectively redistributes Earth's mass, salts and heat, jettisoning them onto Brazil, Texas, Florida, its Keys, and barrier reefs; as well as on Iceland and dozens of islands and islets in the Arctic Circle. Here the collected debris accumulates in cuestas, making secure breeding hollows for plants, insects, bees and birds, as we've seen.

TRANSFERS OF ENERGY NOT YET OVER

The transfers of energy not yet over, Gaia's atmospheric cycling of soils and heat is offset, then intensified, as the enclosing human technological sphere dries terrain that winds then loosen, especially from African and Asian deserts, its swirls creating small wind-like twisters that grow into mini-cyclones, carrying the agglomerated dust mix across the seas.[19] Of Earth's circulation of an estimated two billion metric tons of dust each year, more than half comes from African deserts and dry-lands on a 6,400-kilometer ride on the Atlantic's western trade winds. Much of it is carried over the southern and eastern U.S. every summer, responsible for 75-to-80 percent of the dust that falls over Florida. Yearly an estimated 40 billion tons of dust loaded with life-sustaining minerals, including iron and phosphorous, are also dumped into the water-column of the sea, and onto the aerial carpet fertilizing Amazon's massive rain forest.

Gaia's circulatory system enlarges, too, as parts of North Africa blow from its desert and drying reach. Once in the air it begins to modulate climate by absorbing radiation from the sun. And as already seen, some radiant heat is re-radiated off of Earth, warming the atmosphere. Other radiation is refracted or absorbed by Earth depending on the chemical composition of the dust, its mineralogical content, the size of the dust particles, and the wavelength of light.

Certain propensities are known. For heat is absorbed by Earth itself when dust (especially if mingled with carbon soot) is imbued with long-wave radiation coming off the planet's surface; or, contrarily, heat is re-radiated back into space when the dust is imbued with short-wave radiation.

KEY EXAMPLE OF UNKNOWNS

Unknown exists almost everywhere climatologists turn, here offered one key example. Save for wet Tasmania, Australia makes up a great desert lying under a band of high pressure, 30 degrees longitude north to south. Dry air from the topside is constantly empowering the Sun to create a pressure cooker below—pushing downwards, dehydrating the land, wisps of cloud cover holding little moisture, though the sea winds of five surrounding oceans periodically offer small relief, or a rare deluge. But most of the time these seas are relatively cool, evaporation is modest, no rain is in sight, land dries, farms go broke, fires break out from a spark to grassland or timber, and indigenous animals survive if at all on poorly-watered, Sun-shimmering soils.

Only *El Niño* years draw sea waters eastward across the Pacific face, through the rampaging Coral Sea, offering momentary cloudbursts over Queensland, New South Wales and Victoria. So during the interval, spaced out by *El Niño* every two to eight years, Australia's big dry prevails.

Its climate has been worsened, too, by China, Japan, Britain and the U.S. mining just about every know Australian mineral and organic compound supporting their industries, loosening top soils and iron-, coal- and sulfur-laden particles and sands to the winds And when the winds finally come, carrying reddened compounds of dust, they roar, sweeping them away to Sydney, New Zealand, and westward across the Coral Sea to descend on dozens of islands.[20]

The jigsaw puzzle is made more complex still by clouds covering 60 percent of Earth's surface at any given moment, small changes in formations and propensities altering their reflection of light and heat back into space. One optimist argues the Anthroprocenic forces would never alter the planet's temperature: that a 5 percent increase in "short-wave cloud forcing" would cool the planet enough to compensate for all the increases in greenhouse gases that occurred between the years 1750 and 2000—that is, moments from the start of Britain's Industrial Revolution and its global spread.

This is contrary to what many scientists believe: That the overall impact of atmospheric dust storms pose a probable cooling of the planet, but not nearly enough to compensate during the present Era for the hothouse effects of greenhouse gases.[21] In sum, on their skyway journey, direct solar energy only in part is refracted into space from the particle-driven skies, while indirect solar radiation bounces off terrain and water into the atmospheric skein of dust, sweeping it westward—the dance of swelling heat once again searing planetary life.

In Australia, wildfires are a common feature of arid Australia's summers, with more than 100 homes razed in the southern state of Tasmania in January 2013 alone; and hundreds of blazes breaking out due to heat wave conditions in several states. The government's Climate Commission had warned that such extreme weather events are increasing in frequency and severity due to climate change, exacerbating the risk of fires. And the so-called Black Saturday firestorm in 2009 destroyed even more as heat in Sydney, hit 42 degrees Celsius or 108 Fahrenheit.[22] The drought, heat and fires from 2010-16 continued; and the fires next time will be fed from dried vegetation unable to absorb carbon dioxide, that then rises into the atmosphere as a heat dome over the seared ground below.

Australia is hardly alone.[23] The global long-term future aeration of carbons and greenhouse gases is destined to last for many generations, causing glacial and sea ice to melt, the level of oceans to rise, thousands of islands to be completely inundated, coastlines of major continents to be submerged under water, and species life to seek cooler temperatures by retreating to higher altitudes and the globe's polar regions.

NOTES

1. Peter Singer, *A Darwinian Left: Politics, Evolution and Cooperation* (New Haven and London: Yale University Press, 1999), 63. See also a comparison of the ideas of Jeanette Winterson, "Strange New World," *The New York Times Book Review*, September 20, 2009, a

review of Margaret Atwood, *The Year of the Flood* (New York: Nan A. Talese/Doubleday, 2009), with Stanley Milgram, *Obedience to Authority: An Experimental View* (New York: Perennial Classics, 2004).

2. Kate Wong, "First of Our Kind: Sensational fossils from South Africa spark debate over how we came to be human," *Scientific American.com*, April 2012, 31-39. See also Kate Wong "Mystery Humam," ScientificAmerican, Vol. 314 no. 3 (March 2016): 28-37.

3. Krooth, *Gaia, passim*.

4. Gore, *The Future*, 196-197.

5. Krooth, *Arms and Empire, passim. See also* Krooth, *Gaia, passim*. See herein Chapters 9-14.

6. Gore, *The Future*, 197, quoting the London *Guardian*.

7. Old Farmer's Almanac, 1976, 39.

8. The *Global 2000 Report to the President* (New York: Penguin Books, 1982), 26.

9. *The Global 2000 Report*, 36.

10. See herein Chapter 2.

11. Krooth, *Gaia*, 589-590.

12. Douglas M. Main, "Safeguarding Mammoth Trees, Champs of the Ecosystem," *The New York Times*, May 5, 2012.

13. Main, 'Safeguarding,' May 5, 2012. Also based on Richard Krooth, "Fifty Years on the Trail," *Notebooks*, 1965-2015.

14. See herein Chapter 2.

15. "Docent Lecture Tour," Oakland Museum, Oakland, California, May, 2012. Also see Alexander von Humboldt, *Cosmos*, Vols. I and II, *passim*.

16. Sydney Australia, "The Future is Wild," exhibit, Sydney Aquarium, 2011. See also Sean B. Carroll, *Species Most Beautiful, passim*.

17. Richard Attenborough, KQED, February 2012.

18. Alaska Geographic, "Arctic National Wildlife Refuge," Holand Hargrave, editor, KQED, June 8, 2013, 9-10 PM.

19. Krooth, "Travels in North Africa, Southern Europe, and the Atlantic," *Notebooks* 1962-66.

20. Krooth, "Into the Australian Cavil,"*Notebooks*, 2010-11.

21. See Jeffery Bartholet, "Swept from Africa to the Amazon," *Scientific American.com*, Vol. 306, No.2 (February 2012): 44-47.

22. Nerve-centre, at the heart of Australian-fires, January 8, 2013.

23. Matt McGrath, "CO_2 levels mark 'new era' in the world changing climate," *BBC*, October 24, 2016. In November 2016, five U.S. southern states that had lacked precipitation for four years witnessed massive forest and brush fires, with no surcease until the fuel burned out.

Chapter Nine

Dance of Pelagic Death—Part 1

By the year 2100, ocean acidity will be greater than at any moment over the last 20 million years. Invertebrates like snails and clams will be unable to form their calcium shells and skeletons—forcing them to adapt or die. So it is not only a question of oceans being stripped of some species, but as well of others evolving under entirely new pelagic conditions. Species extinction, evolution and development together will mark the march of oceanic life in a future sea that no human-constructed triage system can rehabilitate or guide.—"The Future is Wild," Sydney Aquarium, Exhibit, 2012

About 80 percent of the world marine fish stocks for which assessment information is available are fully exploited or overexploited. The average maximum size of fish caught declined by 22% since 1959 globally for all assessed communities. There is also an increasing trend of stock collapses over time, with 14% of assessed stocks collapsed in 2007.—Secretariat of the Convention on Biological Diversity, 2011

Though the sea rises and dilutes salt with glacier and permafrost melt, not a drop to drink remains more than a metaphor about its dense, salty brine. Acid rain, sleet and snow are showering down upon the sea—a warning that seawater is increasingly being infused with greenhouse gases, leading to down-welling affecting the feeding chain of marine ecosystems.

Sea currents also have changed the distribution of food in ocean water columns. And beset by solutions of carbon- dioxide, ocean acidity and species death is certain to wipe out many invertebrates.

The oceans cover 71 percent of the Earth's surface and 97.5 percent of all water on the planet. They have played a dominant role in regulating the planet's climate over at least the last 300 million years, today absorbing about one-third of the carbon dioxide produced by human activities, acting as a heat-sink offsetting rising atmospheric temperatures.

Yet the past reliance on the ocean as a store of CO_2 is ending. Marine ecosystems are being overburdened; and the seas face a fundamental and irreversible ecological transformation not seen in millions of years. As the seascape changes, so too will thousands of species face heat and stress, or expire. It is just as likely, though, that future, surviving species will evolve organic and cellular forms bred to new conditions, as well as change their metabolic tolerances.

ICE AGE CLOSING; ACIDIC SEA RISING

Most of Earth's water is brackish; the balance of 2.5% water is fresh; and 70% of this is still frozen in Antarctica and Greenland ice and snow. How this potential fresh water will supply the still-enlarging 7-11 billion people on Earth remains unknown, for existing technology requires more energy to remove salt and other minerals from seawater than is financially feasible. It also produces too little quantities of fresh water; and likely will require solar-energy evaporation techniques on a grand scale.

Most Darwinian sea life evolved to use the salt and calcium circulating outside their bodies to build their frail vertebrae or circulatory systems. By contrast, land animal species largely obtained their salt and calcium by food intake and using their internal circulation systems to process salt-based iodine, keeping their thyroid gland, the body's carburetor, functioning in balance. They ate of crops, land-animal protein, fish, and bivalve creatures.

COMMERCIAL FISHERIES AND POWERED VESSELS

North Sea fisheries went commercial as early as the 15th and 16th centuries. By the 17th century, European and other fishermen had already crossed the Atlantic to Newfoundland for cod, so plentiful that a day netting and salting them could fill a sailing ship. With Fulton's steam engine and Diesel's powered vessels, faster sea travel for netting cod and haddock invaded the North Sea as far as the Arctic Circle, able to freeze their catch in ice without salting, for delivery back to Europe.

By the late 20th century, a virtual explosion of global commercial fishing turned on three innovations: technologies first developed in Australia to locally freeze slaughtered mutton in the 1930s; giant vessels handed down from World War II sea carriers; and winches, cranes, and great purse-seines evolved from Norwegian mackerel and herring-ring netting equipment, redesigned for fishing schools of pelagic sardines, anchovies and tuna.

The technological sphere soon was trouncing sea species, the pace of fishing effectively overcoming their rate of reproduction. Alone, large-scale American vessels could encircle an entire tuna school and freeze up to 2,000 tons of fish. Other giant rigs from Russia, Japan, and Norway scoured the seas, taking a by-catch of marine animals and giant turtles, some dating back to 500 years before Columbus reached the Americas. Large-scale trawlers seeking penaeid shrimp and tropical lobster permanently damaged centuries-old benthic communities of interdependent species, killing those judged non-edible, throwing them into the abyss to make vessel-space for marketable varieties.

Despite regional bans in the South Pacific and so-called government action to restrict or control the number of fishing vessels operating in particular fisheries; or closing fisheries; or restricting the length of seasonal fishing—these have been undermined by privately-owned, profit-driven trawlers in the open seas supplying the growing world population dependent on fish protein.

SEASCAPE TRAGEDY

The mayhem continues today. Deep–swimming tuna, marlin and billfish are captured by so-called long-liners with many hundreds of hooks strung along 20 miles (30 kilometers) behind city-block-long vessels—virtual sea factories gutting, instantly freezing, storing, and bringing fish to markets.

Some fleets tens of kilometers in length know no limit on their catch of demersal fish on continental shelves or offshore tropical zones, targeting penaeid prawns and tropical spiny lobsters, already significantly over-harvested in many areas. Six of the 35 fish species in a commercial catch—pollack, mackerel, herring, sardine, cod and anchovy—make up more than half the total landings. And finfish of cod, hake and haddock in the early 1990s already made up 30 percent of the world catch.

But virtually no single country tracks the size of fish populations and the extent of natural, inter-annual variations in ocean characteristics such as sweeping current patterns, or upwelling and salinity affecting the survival of juvenile fry and resulting stock sizes.

Tragedy awaits seascape species, as fishing recklessnessly exceeds sustainable levels. Some 90-95 percent of global fish stocks are now fished at or beyond maximum sustainable levels—having reached 90-to-100 million metric tons a year by the late 1980s, rising steadily by 2004 to provide 10-to-20 percent of the world's animal protein.

There also seems to be no escape from the destructive nature of modern harvesting techniques for locating and catching fish. Witness the sharp decline of stocks of cod and haddock in the North Atlantic, herring in the North Sea, and salmon in the North Pacific. Not satisfied with undermining these stocks, commercial fishing enterprises have turned to less profitable species like squid and shellfish, switching their geographic focus from the Northern to the Southern Hemisphere.[1]

It will take state sea-policing and a cannon-shot across the bow of violators in the three, five, or ten mile limits of coastal waters to stop the brigandage. And without *tete-à-tete* government agreements, there is no way to stop over-fishing on the high seas.

BALANCING HUMAN FOOD NEEDS AGAINST SURVIVAL OF PELAGIC SPECIES

Can we put this seeming dilemma in perspective? The balance between food sources has frequently changed by the geographic and spatial range of human populations living along coastlines or river systems; or on savannah or open plains, or in forests or valleys. But the need for salt, protein and iodine has not changed.

As crop and grassland herding were degraded or limited by viable soils, fertilizers and access to water, there was a sharp turn in the early 1960s to commercial fisheries that weakened or destroyed one ocean species after another. By 2012, the growth in human populations and global per capita consumption of fish (from 22 pounds in the 1960s to almost 38 in 2012) placed 4.3 billion people on a fish diet for 15% of their animal protein.

This led directly to demand for fish outstripping supply, commercial fisheries accounting for about 90 percent of the world fish catch of which 75 percent was for human consumption (the other 25 percent for fish meal for animal feed); to overexploitation of the majority of world fisheries; and, to depletion of almost one-third of ocean fish resources—with stocks of large vertebrate fish like tuna, swordfish, marlin, halibut and flounder reduced by 90% since the 1960s.[2]

These losses have been framed by a larger picture. As the still-melting shards of the Ice Age come to a close, terrestrial heat exchange and atmospheric greenhouse gases impact temperatures, acidity, and oxygen levels—changing ocean food chains and major currents and altering global weather patterns.

Variations in air pressure, the evaporation-and-precipitation interchange with ocean H_2O gases, and sea currents began transforming both the down-welling and up-welling sweep of ocean food supplies, as well as their distribution and seasonal appearances. Also, though the increase in ocean acidity remains slight and seemingly not a present danger, it represents a major departure from the geo-chemical conditions that have existed in the seas for hundred of thousands, if not millions, of years.

Viewed though a dark, reverse telescope of prehistory, acidification was likely the force behind extinctions some 252 million years ago, when life in the sea and on land almost perished, taking 30 million years to evolve the relatively few species that survived.

GATEWAY TO THE NEXT GREAT EXTINCTION

Yet further evidence exists that we are well on our way to the next great extinction event, insists Ove Hoegh-Guldberg, the director of Australia's Global Change Institute at the University of Queensland, co-author of a critical *Science* journal report emphasizing:[3]

- The average temperature of the upper level of the oceans has increased more than one degree Fahrenheit over the past 100 years; and global ocean surface temperatures in January 2010 were the second warmest ever recorded for that month.
- Nutrient-poor "ocean deserts" in the Pacific and Atlantic oceans grew by 15 percent, or roughly 2.5 million square miles, from 1998 to 2006.
- Ocean oxygen concentrations have been dropping off the Northwest U.S. coast and the coast of southern Africa, where dead zones are regularly appearing.
- And there is paleo-ontological evidence that declining oxygen levels in the oceans played a major role in at least four or five past mass extinctions—so the possibility again presents itself.

Setting the stage to understand the rising carbon dioxide sequestered in sea water, we already know that chemical reactions make it much more acidic—3 percent more from the 1750 start of the Industrial Revolution until 2012. In our time of burning fossil fuels, about a third of all emissions have gone into the sea, effectively reducing atmospheric carbonate ion concentrations by 16 percent in the ocean. Yet the result has proven beneficial as well as damaging, because carbonate ions are required by sea creatures to construct their shelters, reefs and shells.

On the other hand, the survival of entire ocean ecosystems has been threatened by *rising* ocean acidity reducing the formation of phytoplankton feeding the sea web—sustaining zooplankton, coral communities, the entire food chain of the Antarctic, and sea urchins. Also, human consumption takes from oyster beds and wild salmon fisheries. From fish eggs to larvae to sea butterflies—as food for temperate regional herring, salmon, whales and seabirds—collapse of the sea web is threatening.

GEOCHEMICAL DEATH CYCLE

Geochemical conditions represent the foundation and root of fast-changing ocean acidification, with seas becoming more caustic more quickly than they have in the past. Major shifts in climate and mass extinctions in the distant past turned on atmospheric CO_2 increases absorbing solar radiation re-radiated from Earth—warming the planet, some of its atmospheric carbon dioxide entering the sea and making it more acidic. The best evidence shows that today the ocean is acidifying at least 10 times *faster* than it did 56 million years ago.

But there are different causal forces between past and present, as past increases in the atmosphere's carbon dioxide levels resulted from volcanoes and other natural causes, while today the spike in aerated CO_2 largely is due to human technological advances, and the extraction and burning of fossil fuels bringing on the inevitable consequences spreading heat to cooler land and seas that the Second Law of Thermodynamics foretells.

Many sea species now live at death's door as their evolutionary design took place eons ago in a low acidic environment, while today, high acidic conditions are pushing them beyond their physiological limits. "We know that life during past ocean acidification events was not wiped out—new species evolved [from earlier ones] to replace those that died off," says researcher Bärbel Hönisch, a paleo-oceanographer at Columbia University's Lamont-Doherty Earth Observatory. "But if industrial carbon emissions continue at the current pace, we may lose organisms we care about—coral reefs, oysters, salmon."

The scientific logic seems irrefutable. As the level of carbon dioxide in the atmosphere increases, oceans absorb some of it, which then turns into carbonic acid. As a measure of acidity (compared with alkalinity), the resulting pH drops, meaning the water has become more acidic. This dissolves the carbonates needed by many organisms like corals, oysters or tiny snails to form their shells.[4]

As carbon dioxide dissolves surface water to form a powerful acid, it further fans out towards the seabed, threatening tiny organisms facing their death-knell, and wasting the shells of barnacles, mussels, sea urchins, as well as other larger sea creatures.

In a review published in the journal *Science*, March 1, 2012, Hönisch and colleagues again found the closest modern parallel was about 56 millions ago during what is called the Paleocene-Eocene Thermal Maximum (PETM), when atmospheric carbon concentrations doubled, pushing up global temperatures. Extinctions in the deep sea accompanied this shift. (To judge the time-frame, recall that the PETM occurred about 9 million years after the dinosaurs went extinct.)

Ocean acidification also may have occurred when volcanoes pumped massive amounts of carbon dioxide into the air 252 million years ago, at the end of the Permian period, and again 201 million years ago, at the end of the Triassic period. Both are associated with mass extinctions, though there is much scientific disagreement about the scale of carbons that ended in the sea.

MERCURY INVADES THE SEA

Mercury also invades the sea from burning fossil fuels, especially those released from smokestacks of coal-fired power plants producing steam for electricity production, scaling high into the atmosphere, likely to fall into some distant sea.Then a descending concentration of the mercury among ocean species begins. So long as it remains at the sunlit surface of the ocean, mercury retains its elemental form. But when it is aided by near-surface plankton and Antarctic krill that absorb both radioactive particles and mercury, it descends to the abyssal depths, there to be transformed into a highly toxic form of methyl mercury—bacteria likely playing a significant bio-accumulating role.

Readily taken into living organisms, mercury further bio-accumulates in marine creatures like fish, shellfish, and shrimp, entering and remaining in their tissues. If krill eat bacteria containing mercury, they receive an initial dose; then the small fish that eat a hundred krill multiply the dose a hundred times and on up the marine food chain to the largest sea predators, with sharks and swordfish accumulating dangerous mercury levels. So eating these fish, humans, the largest land predator, bio-accumulate the very poison that was initially spewed out from smokestacks and other sources into the sea.

"[I]t is a remarkable fact," says ecologist Tim Flannery, "much methyl mercury has taken a journey through the three organs of Gaia—from its origin in a coal mine [coal's origin, its biomass, trees], to the atmosphere, and on to the ocean depths—before it comes up again through the food chain and lands on our plate. It's a perfect example of how everything is connected on planet Earth."[5]

Are there ways humans could stop the burning of black coal and other fossil fuels that release mercury and carbon dioxide to the atmosphere, ending in the sea, causing the future mass extinction . . . of us? Probably not without an overturn of political/social power, restructuring the use of all sources of energy on Earth.

EFFECTS ON SEA LIFE SO FAR

Though the patterns of ocean acidity are powerfully emerging, Hönisch and his colleagues are cautious, using typical academic equivocations: "The current rate of (mainly fossil fuel) carbon dioxide release stands out as capable of driving a combination and magnitude of ocean geochemical changes potentially unparalleled in at least the last 300 million years of Earth history, raising the possibility that we are entering an unknown territory of marine ecosystem change."[6]

How exactly have these changes affected sea life *so far*? Since the early 1980s through late 2010, the production of ocean phytoplankton, a crucial life form powering the ocean food chain, has become smaller and declined by six percent, with 70 percent of the decline found in northern seas. The reasons are many. As primary food producers using photosynthesis, phytoplankton convert sunlight and nutrients into chemical energy. These nutrients are concentrations of dissolved and suspended organic matter in seawater, absorbing light at depths down to 650 feet (200 meters), below which no light penetrates. Due to different wave-lengths, light penetrates to different depths, most primary phytoplankton production taking place only a few tens of meters below the surface—in the so-called euphoto zone, receiving 1% of sunlight penetration at a photosynthesis photic penetration range of 400 to 700 nanometers.[7]

Phytoplankton then becomes the source of food for almost all other marine organisms in a massive, pelagic feeding chain. But as ocean phytoplankton production is reduced due to sea warming, the chain of life is itself constrained.

WARMING THREATENS KRILL, CORNERSTONE OF ANTARCTICA'S ECOSYSTEM

The size of a pinkie, looking like a shrimp, one of the most abundant animal species in the sea, forming schools miles long and miles deep, thousands of krill crustaceans are packed into each cubic foot—sustaining the entire feeding chain of the Antarctic ecosystem. Sea warming is their enemy, carbon dioxide raining acids into the sea, preventing their larvae development and the hatching of their eggs—and presenting dangers to the entire Southern Ocean and Antarctica's life system as the food source for squid, whales, seals, penguins, sea birds, and unnumbered fish species.

This feeding chain covers a relatively small ocean geography, with currents arriving from the tip of Patagonia to the Antarctica land mass. With waters swirling West to East, vertical currents cause the Southern Ocean's deep water to continually rise, coming in contact with the atmosphere, absorbing heat and carbon dioxide into the ocean, then sinking again.

This affects the krill that also sink from the water surface into deeper water, where they normally hatch and swim back to the surface to eat algae, phytoplankton and zooplankton. Their bodies convert these into a form suitable for many larger animals dependent on krill for the largest part of their own diet.

We seem blind to the fact that krill growth and reproduction on their daily vertical migration provide food for predators on the surface at night (where the krill also feed) and in deeper waters during the day. Annually, over half of the krill population is consumed by squid, whales, penguins, seals and fish; the rest normally reproducing. And to feed farmed fish, in turn feeding a voracious human demand by a world population of more than seven billion, krill today are being sucked up by fishing trawlers from their swarms during ice-melt peaks in winter, in the fastest warming place on the planet.

The Antarctica Commission for the Conservation of Antarctic Marine Living Resources says the 300,000 tons of krill caught in 2014 were significantly lower than the 500,000 tons caught some 30 years ago, when former Soviet Union trawlers were the biggest krill fishers. And today krill are increasingly being taken to extract Omega-3 for human consumption in cereals, fruits and cosmetics.

Still the question remains: Could the krill population go into decline? Though the krill can adapt to changing conditions over their 12 larval stages before reaching adulthood, as they sink from the water's surface into deeper seas where carbon dioxide levels increase, they can't avoid the area of greater acidity and are poisoned.

Like other sea animals that can't build shells and exoskeletons in acidic seas, acidulation bodes ill for krill experiencing declining populations. Possibly their population will fall as much as 20 to 70 percent by 2100 if global warming, acid seas, and harvesting continue.

Here's the key to Antarctic krill survival at the bottom of the food chain: they make up an estimated biomass of around 379,000,000 tons, more than that of the planet's entire human population. So long as krill continue to reproduce more quickly than human and other species consumption, they will remain at the very cornerstone of the Antarctic ecosystem. Yet they might be undermined as the three interfaced parts of Gaia's functioning—mining the *land* for and burning Earth's organic fossil fuels; aerating carbon dioxide and radioactive mercury into the *atmosphere*; and, quickening precipitation and fallout descending to the abyssal *ocean* depths. These might be transformed into lethal acids and a highly toxic form of methyl mercury.

These connected stages are delivery systems reaching krill which absorb toxins and feed on bio-accumulated deadly bacteria. And these poisons directly or indirectly are transferred to the thousands of species consuming and surviving on krill, also ending on your menu of farmed, krill-fed salmon and other fish.[8]

PROJECTING WORLD FISHERIES

World fisheries are poised to lead in two divergent directions: either near-total depletion in the coming decades, or conservation that cleans up the sea and establishes fishing quotas in its commons. Reasons exist for optimism about the future of the world's fish stocks, despite their currently dire state, says the Prince of Wales, launching a report from his green think-tank, "Fisheries in Transition"—detailing 50 case studies of successfully managed fisheries in various parts of the world. His International Sustainability Unit (ISU) is aiming to build constructive dialogue between the fishing industry and those trying to secure Nature.

Still, declares the prince, speaking at Fishmongers' Hall in the City of London, the issue has been dogged by a "debilitating fatalism." So it is "critically urgent" that countries find better ways of dealing with over-fishing and other marine issues. He worries that, "Oceans are becoming more acidic; run-off from industrialized farming is causing waters to become too rich in nutrient, thus creating dead zones in the oceans; waters are getting warmer, and more and more square miles of them are being polluted with plastic. The human assault on the Earth's oceans is depressingly comprehensive." The report's overall conclusion is that, by regulating and allocating the catch in the common sea, fishermen would make more money for less effort, and the stock would be safeguarded against ruination.

"We recognize that there is no 'one size fits all' solution, every fishery is different," says Charlotte Cawthorne, International Sustainability Unit (ISU) Marine Program manager. She describes the ISU's work, in part, as compiling a "toolkit" of measures that can be deployed to make fisheries sustainable. Nevertheless, she insists, three things remain essential: scientific understanding of the ecosystem; funding for the transition; and sound management.[9]

POISONING THE PELAGIC WEB; EXPANDING FLOATING ISLANDS

Darwinian oceanic evolution and development are being interrupted by the dumping of human–made effluents, discarded medical equipment, toxic garbage, lethally irradiated materials, mercury, and acids—much accumulated on sea-bound, floating islands of plastics and debris that poison the pelagic web. Comparatively speaking, plastic waste falls into two broad categories: hard types that fracture, and soft that are literally softeners—both hazardous to organic life.

These are different from other types of industrial waste. The U.N. estimates that alone the amount of electronic waste is 85 billion pounds per year and growing. And over 90% of metals that end up at a recycling facility are in fact recycled because the process is much easier than dealing with plastics. By contrast, only 10% of plastics are recycled because they must be sorted into many different categories of density, color and type. So most plastic is, in fact, not recycled at all, much ending up on and in landfill and in the sea.[10]

As these elements enter the sea, they are carried by its currents and enter feeding webs jeopardizing ocean life.

Five great oceanic gyres circulate as Earth rotates on its axis, linked to thermohaline currents bringing up deep sea food supplies from, to, and through the web of sea life. Concentrating on the Pacific gyre, oceanographers have identified a large floating island made up of shards of plastic and a garbage heap of human waste, which supports a colony of feeding species, spelling future dangers for coastline communities.

From Asia to the Americas are the Pacific layered flows that discharge accumulated human wastes, especially non-degradable plastic filaments and those that fracture into minute particles. They are being pumped through the sea and concentrated at points where ocean feeding chains leave a trail of death of invertebrate populations, phytoplankton, zooplankton, crabs, clams, starfish—an endless list of every order.

Known to scientists as the *North Pacific Subtropical Gyre*, one Pacific garbage patch, twice the size of Texas, is composed of mulched up plastic and other detritions, riding upon and changing natural sea floes. On this island of sludge, the sea strider *Halobates sericeus* (related to pond striders seen in freshwater lakes) lays its eggs, as it also does on ocean floating objects like seashells, seabird feathers, tar lumps and pumice, giving insects a place to breed on the open ocean face. But the insects and animals that live on the large hard surfaces of the plastic patch also are sources of rot and disease, different from the animals that live on small soft surfaces or in the water column. In fact, the number of pieces of ocean plastic less than 0.2 inches (5 millimeters) in diameter has increased by about 100 times over the 40 years 1972-2012. These tiny plastic pieces give sea striders more room to lay their eggs, leading to high-density communities of their invertebrates.

By 2012, this plastic island had increased its hard surfaces into an ecosystem that historically had in residence very few separate species, although it was beneficial for the strider's main predator, crabs, increasing their own numbers. Yet the garbage patch community had very low biodiversity, potentially disrupting the wider oceanic food web, as the striders also eat massive quantities of tiny animals like zooplankton and fish eggs—favoring their voraciousness in the low-biodiversity rafting community at the expense any tiny animals captured from the high biodiversity of the water column.

The consequences are still evolving. Debris from the patch has impacted other marine life, including ingestion by fish and invertebrates at a rate of roughly 12,000-to-24,000 tons of plastics and garbage per year, according to the Scripps Institution of Oceanography. And the floating island also transports pollutants and alien species into new coastline zones.[11]

WASHING CLOTHES AND YOUR PLASTIC DRINKING GOURD

Sources building the garbage patch include poisons and debris coming from many sources, including washing clothes. Also, as the human population grows and people use more synthetic textiles, they mindlessly release contaminants, enlarging the discharge of micro-plastics affecting both habitats and animals.

Forensic evaluation of micro-plastics taken from sediments reveals that the proportions of polyester and acrylic fibers used in clothing resemble those found in habitats that receive sewage-discharges and effluents. And experiments sampling wastewater from domestic washing machines have demonstrated that a single garment can produce more than 1900 fibers per wash.[12]

Today micro-plastics contaminate shorelines at 18 known sites worldwide, representing six continents from the Poles to the Equator, with more materials found in densely human-populated areas. Though there is no clear relationship between the abundance of micro-plastics and the mean-size-distribution of natural particulates, it is certain that an important source of micro-plastic debris regularly appears in sewage contaminated by fibers that come from our species' simple act of washing clothes.

Indeed, plastic debris of more than a single millimeter makes up micro-plastics that are fast accumulating in marine habitats worldwide. And they become a potential pathway for the transfer of pollutants, monomers, and plastic additives to ingesting sea organisms, with potential dangerous consequences for their health and those in the feeding chain.

Melting plastics and shaping polymers to make throw-away water bottles also has endangered the feeding web. These bottles either do not decay or, if they do, leave their poly-filaments in the soil and sea for hundreds of years to come.

Filling these bottles with filtered water will supposedly rectify a small part of the problem of too little pure water on the planet for too many people. Yet, the fact that only 2 percent of all water on Earth is potable turns in part on the impact of urban life, farming and industrial discharges, with hoped-for methods of water purification still unable to support human need.

The crisis of inadequate potable water besets some 63% of the population in Brazil, Russia, India and China, countries that already live with no way to ameliorate severe water-distress. By 2030, an estimated 47% of the world's population will be living with scarce or unhealthy water.[13]

To rectify the lack, the tragedy of bottled water remains seized by and bound up with four distorting public relations myths.

Myth #1: Bottled Water is better for you than tap water. Wrong! Tests show that there are often higher levels of chemicals in bottled water.

Myth #2: Bottled water tastes better. Not so: consistency in taste tests shows tap water comes out ahead of bottled.

Myth #3: Bottled water comes from supremely clean springs in magical places. Not at all: One third of the bottled water in the U.S., including Dasani and Aquafina, is in fact filtered tap water.

Myth #4: If you recycle the bottle, really you haven't done that much harm. Sorry, even the recycled bottles end up in a landfill in India where they are "down cycled" to manufacture more useless products that pollute people and the environment every step of the way.

Worse of course is that the majority of empty water bottles end up buried in the ground, decomposing and releasing toxins long, long after the brief moments of thirst-quenching are enjoyed.[14] These toxins, laden with pcbs, then filter down to water tables, into rivers and lakes, and ultimately into the seas—poisoning species as they move, becoming especially dangerous to ocean life, ultimately jeopardizing human existence.

HUMAN NEED, PELAGIC DESTRUCTION

Most effluents and lethal byproducts are certain to end in destroying the seas. This is because of the multiplying of our own species population and its desires for necessities, amenities and pleasures, for clean water, and for the extraction, burning and transportation of energy drawn from fossil fuels.

The land may be the origin where humankind begins the extraction and firing of fuels to produce commodities and other things gratifying us as the seemingly dominant species. But the atmosphere is the conveyor belt, dumping the effluents we create into an acid sea. The ocean creatures that cannot resist the pollutants and poisons are diminished in size and viability and lose their reproductive fitness. And the few species that may survive on the seafloor might well prove out their dominance in centuries coming.

Perhaps our species will be swept from the planet, no paleoanthropologist left to record that we were once intelligent creatures who nonetheless lacked foresight and the will to save the planet for the diversity of life. Will it not then be apparent that we had forgotten: as individuals alone, we could do little to stop the dangers, but collectively, we might have stopped the dance of pelagic destruction and death?

NOTES

1. Pernetta, *Guide to Oceans*, 142-147.
2. Gore, *The Future*, 199-20. See also Pernetta. *Guide to Oceans*, 144.
3. Ove Hoegh-Guldberg, director of Australia's Global Change Institute, University of Queensland, co-author, *Science*.

4. See Julia Whitty, "The End of a Myth: Exploring a New Vision for Earth's Greatest Wilderness," *Onearth*, Spring 2012: 39-43. See also our Chapter 4.

5. Flannery, *Here on Earth*, 185-187.

6. See Bärbel Hönisch, a paleoceanographer at Columbia University's Lamont-Doherty Earth Observatory, *Science*, March 1, 2012. See also Wynne Parry, "Ocean acidification worst in 300 million years, study finds." *Christian Science Monitor*, March, 2012.

7. Pernetta, *Guide to Oceans*, 3-33, 142-47. To gauge the meaning of the euphoto zone, human eyesight captures light as a form of electromagnetic radiation waves of varying length produced by particles called phtons, which are reflected off the objects we see. But human vision captures only a narrow band of these wavelengths, extended from 380 nanometers, which we perceive as violet, to 780 nanometers, which we perceive as dark red. This visible light spectrum represents only a tiny portion of the full electromagntic wave spectrum. See Eric R. Kandel, *The Age of Insight: The Quest to Understand the Unconscious in Art, Mind, and Brain, from Vienna 1900 to the Present* (New York: Random House, 2012), 240).

8. Michelle Innis, "Warming Oceans May Threaten Krill, a Cornerstone of the Antarctic Ecosystem," *The New York Times*, Octoiber19, 2015. See also Wikipedia, "Krill." See also chapter 4.

9. Richard Black, "Prince Charles optimistic for fisheries' future," *BBC News: Science and Environment*, February 3, 2012. http://www.bbc.co.uk/news/science-environment-16875215. Garrett Hardin also wrote that the commons of sea or land could always be compromised by a single individual's or a nation's greed to disproportionately take from or overuse of the commons at the expense of others. See Garret Hardin and John Baden, *Managing the Commons* (San Francisco: Freeman, 1977)). See also Garret Hardin, "Political Requirements for Preserving Our Common Heritage," In *Wildlife and America* (Washington, D.C.: Council on Environmental Quality, 1979). And so far across the face of Earth it seems his view has prevailed.

10. Mike Biddle, a plastics engineer, set out to find a solution. He set up a lab in his garage in Pittsburg, California, and began experimenting with complex-plastics recycling, borrowing ideas from such industries as mining and grain processing. Since then, Biddle has developed a patented 30-step plastics recycling system that includes magnetically extracting metals, shredding the plastics, sorting them by polymer type and producing graded pellets to be reused in industry. This process takes less than one tenth of the energy required to make virgin plastic from crude oil. His company's recycling process is a breakthrough solution for closing the loop on plastics. He has new plants opening all over the world, and companies are eager to buy the recycled pellets. See Bibi California Farber, "Above Ground Mining," *Nextworld TV*. So, too, there may be a way to reconvert plastics into a limited variety of usable oils. The technology already exists to turn plastic into oil using a conversion machine, developed in Japan, that converts 1kg of plastic into 1 liter of oil. Classified as "light medium", it can be further processed for use in cars, motorbikes, boilers, stoves, generators. And the problem then will be the technique and cost of reprocessing. Inventor Akinori Ito of the Blest Corporation. See Bibi Farber, NextworldTV, March 16, 2012 See also Scripps Institution of Oceanography, San Diego, California.

11. Miriam Goldstein, research-graduate student at the University, Royal Society journal, *Biology Letters*, May 9, 2012. See also John Pernetta, *Guide to Oceans*, 3-33.

12. ACS Publications, Environmental Science and Technology, "Accumulation of Microplastic on Shorelines Woldwide: Sources and Sinks." http://pubs.acs.org/doi/abs/10.1021/es201811s.

13. Reports from the Paris-based Organization for Economic Cooperation and Development, at Krooth, *Gaia*, 633.

14. Bibi Farber, "The Story of Bottled Water," Next World TV.

Chapter Ten

Dance of Pelagic Death—Part 2

> The use of the sea and air is common to all; neither can a title to the ocean belong to any people or private persons, forasmuch as neither nature nor public use and custom permit any possession thereof.—Queen Elizabeth I to the Spanish Ambassador, 1580

> It's an old whale's tale of Moby Dick, maddened and harpooned, demolishing the whalers' entire ship.—Herman Melville, *Moby Dick*

Whalers once sailed the seas for the slow-maneuvering Right Whales to harvest, nary a thought that the sea giants might turn upon and obliterate them.[1] Today, though, there is a way to safely harpoon and destroy Earth's few remaining cetaceans, as well as to secure the whalers themselves.

RIGHT WHALES, MICROSCOPIC LIFE

There are big sea creatures with backbones and small ones that are spineless. Humankind has long posed a threat to ocean vertebrate, especially docile, huge-headed whales, which once were hunted to near extinction in the 19th and 20th centuries.

Because of their slowness, and buoyancy after being killed, the Right Whales were the "right" ones to hunt for their oil and baleen.[2] Once they numbered in the hundreds of thousands. Now, with only about 400 in existence, North Atlantic Right Whales have inevitably attracted attention when they venture close to shore. The vast majority of them bear the scars of ship strikes and entanglement with fishing gear—two leading causes of Right Whale mortality and a threat to their species' survival. In coming decades, changing planetary conditions also will affect the survival of these gentle giants of the sea.

How this plays out could be observed during a few days in Spring when the Cape Cod Bay witnesses one of the rarest scenes in the animal kingdom: dozens of surface-skimming North Atlantic Right Whales lumbering just a few hundred yards from shore. Yet this rite of spring was upended in 2012.

The critically endangered animals, which usually arrive in late March or early April to graze on shrimp-like plankton, began arriving slowly before Christmas as water temperatures hovered several degrees above normal. The distribution of ocean warmth is crucial to the development of phytoplankton as the food supply for Right Whales. Without sufficient food, their birth rate is affected. And thus they move to warmer venues both to survive and reproduce.

Water temperatures in and around Cape Cod Bay were more than 3.5 degrees above average during winter, 2012, increasing the sun-based development of phytoplankton. Before arriving early in Cape Cod waters as one of their main feeding grounds, Right Whales had had a difficult winter off Florida and Georgia, where they gave birth to fewer calves.

Female Right Whales normally venture as far south as Florida to give birth; and the past decade 2000-2010 witnessed an encouraging increase in calving numbers, with a yearly average of 20 and a high of 39 born in 2009. But during winter, 2012, aerial survey teams in Florida and Georgia counted only six new calves, including one that likely died, apparently from malnutrition. "It was a terrible year for Right Whale calves," said Tony LaCasse, a spokesman for the New England Aquarium, a leader in Right Whale research.[3]

Maybe the winter air and water were too cold off Florida and Georgia, fostering the development of too little food, leading to the falling whale birth rates. Or the disappointing numbers of Right Whale births in 2012 could be linked to changes in the animals' Florida and Georgia feeding grounds, brought on by water that was not only cooler, but also less salty because of melting Arctic sea ice.

"To me or you, 3.5 degrees [in Cape Cod waters] isn't a big difference, but in an ocean system it means different oceanography, different currents, and different biological processes," reported Charles "Stormy" Mayo, director of the Right Whale habitat studies program at the Provincetown Center for Coastal Studies. He suspected this could be driving changes in the warming distribution and timing of plankton blooms, in turn influencing the whales' "odd behavior" in arriving at Cape Cod months earlier.

How could the Right Whale, riding the waves and tides over thousands of miles away in Florida and Georgia, know Cape Cod waters would be warm months earlier, that sun-sustained plankton was in surplus, ready to feed them? "It's a confluence of remarkable things,' Charles Mayo said: "We've got extraordinarily rare animals, nearly extinct, acting very unusually."

Gaia's creatures well may know more than we do, giving pause to wonder why we, as an intelligent species, each possessed of 300 billion neurons, have allowed the oceans to be so polluted, over-fished, heated, and cleansed of whales, so much that it affects all of life's existence in the seas, on land, and in overhead skies.

DESTRUCTION OF WHALES AND OCEAN FISH

As modern fisheries management experts understand, the regulations of the International Whaling Commission (I.W.C.) have become a legitimate path for their domesticated killing of whales. In its early years, the I.W.C created scientific subcommittees to keep whaling methods out of the hands of scientists. In similar fashion, fishery biologists today are often isolated from exercising deciding-power over national whaling quotas, as well as from acquiring positions in important regulatory bodies like the International Commission for the Conservation of Atlantic Tunas.

In the past, so-called scientific investigation of whale anatomies, food sources, habits, and environments essential for reproduction led to their slaughter for human food, lamp oil, perfumes, canes, hoop skirts . . . and profit. Scientists during the late 19th century, for instance, closely followed the work of Britain's leading expert, Sidney F. Harmer, pursuing Antarctic expeditions that were designed to investigate the taxonomy, migratory patterns, and rates of reproduction of the great Southern Ocean whales. The scientists foresaw the need to keep the supply of whales available for the long term. So before whalers could wipe them off the face of the sea, they planned for a future "sustainable" whaling industry to try to avoid the loss of their future supply.

As the path of destruction became well-worn, Johan Hjort, director of conserving fisheries in Bergen, Norway, actually lobbied for closer collaboration with whalers to avoid the risk of a fall in over-all supply, to keep the whole stock abundant should a local decrease appear. The whalers did the rest, turning blubber into oil to light lamps and street lights, and making perfumes for ladies in waiting, ribs for canes, hoop-skirts, and umbrellas—taking down the slow-moving Right Whales from the Atlantic reach by the end of the 19th century.[4]

The whalers' quest then turned to a huge supply of Antarctic *rorquals*, made up of Blues, Fins and Humpbacks. A source of fabled wealth, these giant creatures fueled industries making soaps, fertilizers, glycerin for war, and margarine instead of butter to spread on the toast of the middling classes.

Scenes from the death-machine are so well-documented by Farley Mowat that later writers, describing the killing fields of the Arctic and Southern Ocean Antarctica, can only corroborate tales like those of the horrifying whaling stations reeking like charnel houses, where dying mother whales ejected their unborn fetuses to try to save them; where blubber-flensers made steps for their jack-boots up the flanks of whales, slicing apart the giant cetaceans as if they were steaks on a cutting board, to boil their blubber.[5]

Adventurer-journalist Peter Heller, in our time, describes the whalers' brutalities during the drawn-out process of butchering these intelligent animals:

> first shooting explosive harpoons [that] are supposed to kill them instantly—that never do. The harpoons hit them in the wrong place; the whales are too tough. So the whales just thrash around on the ends of these harpoons, and they start drowning in their own hemorrhage, crying out, their babies, if they are mothers, swimming helplessly around. Then whalers wheel the creatures into great ships, jabbing them with electric probes. They run thousands of volts through them to try to kill them, though that doesn't really work. And it takes 20 minutes for them to agonizingly expire."[6]

Now, too, Heller explains, the so-called research for killing whales is a subterfuge, a commercial operation—the whaling companies telling lies while the Japanese and Norwegians are violating international law. Only some protesters such as Captain Paul Watson and others in his Sea Shepherd Conservation Society, as well as Greenpeace, are actually confronting the whalers' harpooning and butchering. Save for Watson's eco-pirates and Greenpeace, there is no one defending the great cetaceans.[7]

Sometimes fin whales are killied merely for sport, as poignantly described by Farley Mowat. In one such instance, he followed a whale's last blowing, seeing bucket-size swellings erupting over her body from deep bullet holes. Slowly thrusting her entire head out of water at near oar's length distant from his skiff, she swam slowly past, one Cyclopean eye looking directly at him, and Mowat heard her last long, low, sonorous moan: "I will always believe that she had tried to span the chasm between our species," he wrote of the dance of death.[8]

There remain a million long-finned pilot whales, 200,000 short-finned pilots, 400 northern Right Whales, as well as unknown numbers of bowheads, blue whales, humpback whales and sperm whales—all of them more or less facing extinction.

DANCE OF PELAGIC DEATH

Dolphins are whales, however small they appear compared to the great cetaceans of the sea. Their anatomy enables them to use refractive sound, like a bat's radar, to enable them to "see" and sense objects near them—even the innermost anatomies of other organic species.

Their brains proportional to body size are larger than those of humans; and they learn our species' body-language signs while we have not been able to learn the precise meaning of their whistles, shrieks and moans. Likely they are more intelligent than we are, for they are eminently cooperative social creatures, even saving distressed sailors or swimmers, as seen embossed on ancient Greek coins—while we are relatively isolated land dwellers and war-makers bent on lawlessness, securing ourselves against others we fear and distrust.[9]

Sound travels more rapidly and farther in water than in air; and whales of all varieties communicate with one another at distances up to 100 kilometers, often regrouping for cooperative feeding and mating. Like bats, whales also can discern from upwards of 40-to-120,000 vibrations per second, allowing them to weave in and out, almost instantaneously changing directions.

But as tonal creatures, dolphins also are easily confused and beach themselves from ships' sonar sounds; or Japanese fisheries can "manage" them by banging on subsurface pipes to drive them into nets for slaughter.[10] The hunt is on as giant Japanese and Norwegian vessels take dolphins for their protein and for selling the females at upwards of $150,000 each to perform in sea-worlds. There they will display their wondrous antics, instructed by human body language and fed with high-energy stimulants to perform.

When dolphins are killed for the human palate, there are deadly dangers, for today their flesh is suffused with mercury—the second most contaminating source after irradiation poisoning by nuclear isotopes. And the sea where dolphins survive is itself experiencing an annual 1%-to-3% rise in mercury levels from human dumping and industrial and radioactive effusions.

As part of the food chain, mercury absorbed in fish obviously then poisons human carnivores. This happened in the 1950s at Miamata Bay, when human release of fish toxins attacked fetuses and adults alike. Industry had secretly dumped mercury in the bay, leaving local Japanese eating its fish, especially sushi-grade tuna, with double the level of mercury in their blood stream. This affected their brain neurons that then were unable to function—leading not only to memory loss, but also to suffering with severe autism, eyes that could not see, ears that could not hear, crippled bodies that could not move, and fingers that could not feel.[11]

Today the poisoning in Japan is not yet over. From September to March each year, Japanese fishing-corporations bloodily slaughter 23,000 dolphins that they drive into the Taiji Cove lagoon. The dolphin meat is then packaged and sold to unaware consumers, who do not know they are eating mercury-poisoned fish.

Until investigators using video cameras exposed the truth, mercury-rich, slaughtered dolphin meat had been part of Japan's obligatory children's school lunches, poisoning unknown millions of young people. Japan's authorities have consistently denied and covered up both the slaughter of the dolphins and the dangers of marketing their mercury-laden flesh.

Only the brave members of the Oceanic Preservation Society and its Japanese IMPT partners were determined to try to stop the dolphin slaughterhouse in the bay and protect against the feeding chain reaching the human table.[12] And yet, even now, mercury-tainted sushi remains part of the staple Japanese diet.

SEA LIFE'S PRECARIOUS BALANCE

Sea life is precariously balanced today, undermined by human sewage disposal, run-off of agricultural fertilizers, and the concentration of mercury and other poisons in the ocean bed. We are also increasing ocean acidity, reducing water nutrient supplies, and raising sea temperatures, leading to bursts of algal blooms—putrefactions invading the water column in a complicated dance of death.

Toxins moving up and down the feeding chain weaken some species more than others, altering the web of existence and intensifying frenzied feeding and extinction.

As poisons spread, each surviving species brandishes its fitness for combat and reproduction, some annihilating others, setting up new dangerous feeding and disposal linkages as entropies set their heel.

SPECIES INTERDEPENDENCE

Over millenniums, life in the seascape has evolved a storied interlock of species dependent on one another for survival and reproduction. Phytoplankton lies at the core. Without it the entire aquatic life system would be brought to its knees and collapse.

Phytoplankton is consumed by zooplankton and feeds the microcosm of krill and an estimated 10 million sea and avian species—including unnumbered fish varieties and sharks, as well as seals, cetaceans, sea lions, porpoises, dolphins, skates, rays, sea turtles, mollusks, and crustaceans.

The top predator remains the white shark (*Carcharodon carcharias*) that has swum the oceans for over eleven million years, an adult growing to 21.5 feet and weighing up to seven thousand pounds, its vast diet consuming all that swim.

Why worry if these apex predators are eliminated by human fisheries and no longer control the size, abundance and behavior of innumerable other species in the sea? Because much like other apex predators, the loss of white sharks balancing biodiversity might disrupt the entire ecosystem, moving them to a completely different feeding chain among a lesser set of competitive species.

White sharks already have become victims of human pollution, climate change, coastal development and commercial over-fishing using giant vessels that are city-blocks long, dragging nets and baskets from which no creature can escape. The loss of each, one day all, disrupts the ecosystem, initiating a new struggle for positioning in the sea.[13] Thinning out the sea's top predator has set loose a momentary uplift of less powerful life forms that once fed white sharks. As the surviving species multiply, extending their geographic reach in still unknown directions, they throw large creatures into combat with still other species for nutrients and living room.

And then arrives a new giant fighting to dominate the commons of the common seas. Humans.

COMMONS OF THE COMMON SEA

Recall Britain's Queen Elizabeth I's edict in 1580 about the ownerless, common seascape. It makes up 71% of Earth's surface, and this is where human overkill of species defies the Gaian order. Just as 100,000 sharks of all species are killed each year by the fishing industry scouring the oceans, the same fate awaits all giant ocean creatures migrating by instinct, intelligence and habit across the sea.

To satiate the human palate, today one-thousand-year-old turtles, long-lived swordfish and marlin, whales and tuna, especially the Bluefin, are being hunted. Extinction of any one or all is possible.

The demise of the Atlantic Bluefin is certain unless the catch is suddenly limited to a zero-quota and re-categorized by the International Commission for the Conservation of Atlantic Tuna from the status of human seafood to one of a protected form of wildlife.

This issue is particularly pressing as Japanese and Western whalers reduce populations of large tuna around the world. The most environmentally sensitive are Atlantic Bluefin. But the risk is fast spreading to other species, one subpopulation of Atlantic Bluefin already vanishing after heavy fishing by Japanese long-liners deploying hooks off Brazil during the early 1970's "Bluefin boom".

Humans are literally eating the last remains of the ocean reach, so any surviving Atlantic Bluefin stocks soon will be no more. Pacific Bluefin between California and Japan will follow suit, and then southern Bluefin plying seas around Australia.

In the Americas, the direct fishing pressure on Bluefin continues with the more encompassing problem of large quantities of their North American species caught accidentally as "by-catch," while the industry's long-liners deploy legions of fishermen using hooks in search of Yellow Fin tuna over the Gulf of Mexico—Bluefin's spawning ground. Yes, by law, nearly all Bluefin caught as by-catch must be dumped back into the sea. But by the time the day's hooks are put aside, the Bluefins are long dead.[14]

FILLING FISH CANS

Part of the tuna catch is processed and put in cans. This is a different kind of tuna than both the blue and yellow fins, and the methods of catching and marketing them also differ, as the fishing industry is large and complicated, and generally international. Asda is owned by world-girdling Walmart, for example; Princes is owned by Mitsubishi; other tuna ships at sea fly flags of Norway, Japan, and allegedly peaceful Costa Rica.

Fishing practices have become a technological venture that uses fish aggregating devices (FADs). These devices depend on a quirk of fish behavior. Something big in the middle of the open sea—such as a buoy, a big log—causes many species to congregate around it.

One theory is that small species see it as a shelter from predation, though bigger carnivores come to see hovering objects as floating restaurants. Whatever the reason, one can now purchase sophisticated buoys equipped with sonar and GPS. They can be dispatched into oceans to send fishing vessels a continuous stream of data indicating where the fish are, roughly what volume of fish is gathered around, and when the moment is ideal for capturing them in a purse-seine-net.

Technology has moved in to help catch the main ingredient of canned tuna—Skipjack (*Katsuwonus pelamis*), a small and fast-swimming, fecund and gregarious species that is generally thought not to be significantly depleted, despite the huge fishing effort deployed to capture them. As Skipjack are often found in close proximity to other tunas such as Yellow Fin and Bigeye (*Thunnus obesus*), when FADs are the tool of choice, young Bigeye and Yellow-Fin are caught in the same nets that snare Skipjack.

The Bigeye is listed as vulnerable on the IUCN *Red List of Threatened Species*. But FADs also lure sharks, many species of which are also on the Red List—and these too are killed in the purse-seine-nets, whose mechanical *modus operandi* is to close like a purse around an entire shoal of fish.[15]

The alternative is to fish for Skipjack with bait, line and hook, giving work to small-time fishermen. If the Pacific Islands were to provide all the tuna they currently do, but only through pole and line fishing to feed the fish-hungry world, they would first have to catch their bait, such as sprat and anchovy[16] —and that as well would seriously deplete stocks of those species.

KRILL AND FISH-FARMING ENTROPY

A last effort to feed a fish-craving human population is farming fish, as one would garden soil with nutrients. As most farmed fish are literally fed great quantities of netted smaller *wild* fish like krill, the resulting processed fish-meal is the equivalent of raising a pound of land-fed beef on six bushels of wheat. Salmon in farmed areas, confined by nets, are being fed at a ratio of five pounds of ground wild fish for each pound of farmed salmon sent to market. And wild larger fish thereby are being starved of the smaller wild fish in the open ocean feeding chain.

In 2013, the crisis bordered on entropy as fishmeal, pet food and agricultural fertilizer supplies turned on an over-catch of krill in the Antarctic Ocean, breaking up the feeding chain where tiny planktonic krill is a main food source of baleen whales. An unfolding drama and dilemma thus appears as humankind, the top predator with its voracious appetite for fish species, mobilizes fishing vessels, sonar, GPS, FADs and purse-seine-nets to capture wild, swimming creatures, severing their evolutionary web of oceanic life in one moment of a pelagic cascade overwhelming Gaian life cycles.

SEA MOMENTS[17]

Oceans have many recesses and food webs supporting life diversities, many not yet known. In those darkened reaches where pollution reigns, species diversity declines. Where the human reach is relatively slight, species diversities exchange their nutrients, excrescences, and very selves as foods and sources of sustenance in this web of existence. Nature's realm cycles these essences; evolutionary diversity flowers; and Gaia's creatures evolve and revel.

COMPARISONS ARE MANY.

Though the much-abused Eastern coast of the United States has seen its species diversity plummet, Australia's 74 Whitsunday Islands around the Great Barrier Reef rejoice with life. Here, find a vast, blessed underwater wonderland, home to over 1,500 fish species and hundreds of coral, sponges and mollusks. The reef stretches some 2,300 kilometers along the East Coast of Queensland—from just below the Tropic of Capricorn to the edge of Papua New Guinea—its massive size, bigger than England, Scotland and Wales combined, providing sustenance to the profusion of life among more than 2,900 individual reefs.[18] Yet some of these reefs and sea creatures are now at risk.

As the sea is deprived of its largest fish, what's in store for other sea species, including oceanic birds that feed on both pelagic life and interconnected species on the surface of the land? Will the human trawlers and gigantic ocean fish factories hollow out the sea of life, starving Poseidon's bounty?

Sea moments can powerfully and suddenly change the links with air and soil, driving species in directions hitherto unknown, smashing, submerging or lifting entire atolls, islands or continental shelves with their plants, animals, even fishes. The entrenched forces of Nature operate with both precision and terror, tearing apart what once appeared stable and synchronous.

HERE'S A GENERAL PORTRAIT

Oceanic heat distribution and marine life turn on the sea's vertical and horizontal movements shaped by solar impact, winds, and unpredictable tremors and fulminations on land. The Sun's periodic intensity, its emissions spread, warming the tropical Atlantic Ocean, causes surface evaporation which then rains down hither and yon, intensifying sea salinity and density.

As some of this warmed water flows North, passing between the coasts of Iceland and Britain, it releases heat to the atmosphere, the predominating westerly winds carrying the air across Western Europe, warming the winter season. As the sea continually gives off its heat to the atmosphere, the surface water temperature drops close to the freezing point, setting off a relay of unforeseen forces: the density of the water further increases; the denser surface water then sinks, mixes with, and slips under the less-dense underlying bottom water—spreading on the ocean floor and flowing South as a force of the warming, salty thermohaline circulation cycle.

The remaining cold surface sea then carries water-soluble methane, carbon dioxide, and toxins of the industrial age, its dense mass moving to the Caribbean and Southern seas, poisoning fishes and every living strata of salt water creatures.

POISONING THE ATMOSPHERE ALONG WITH SEA AND SOIL

Aerated pollution converges with the putrefaction of sea and terrain. It is not only atmospheric moisture that stretches along the North Atlantic seaboard. Aerosols like industrial sulfur are pumped into Northern air currents and drift Southward, reflecting sunlight and momentarily if only partially cooling land surface temperature. Then they remain suspended in the atmosphere absorbing more heat, while fall-out poisons living plants and animals.

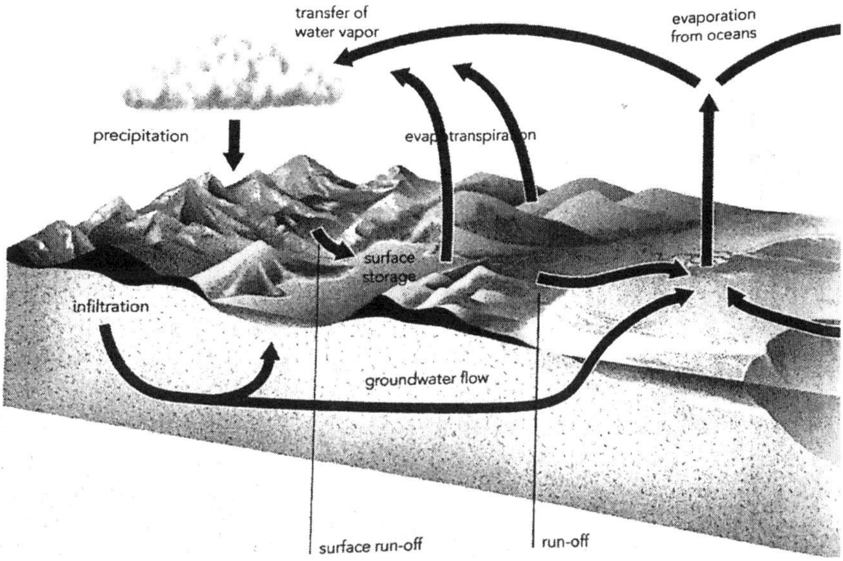

Figure 10.1. Hydrological Cycle.

GAIA'S UMBRELLA: SOIL, ATMOSPHERE AND SEA

There is no separating von Humboldt's massive aerial ocean from the sea and soil. Each and all rise and fall as one, sustaining or destroying life's forces under Gaia's umbrella. Infused with greenhouse gases that absorb solar radiation directly or which are re-radiated from Earth's surface, by 2009, overall atmospheric concentrations of carbon dioxide had already reached 386.8 parts per million of air molecules, up 38% from pre-industrial times. And methane was recorded at 1,803 parts per million, up 158% from the pre-industrial era.

This flow of heat and pollutants converged in Gaia's three great organs—soil, atmosphere, and sea. Combined with the overall 2010 rise of global temperature by 0.8 degrees Centigrade (1.4 degrees Fahrenheit) above pre-industrial levels, it meant the ice-caps would continue to melt, while ocean levels would rise and seas evaporate, leading to atmospheric fall-out—together floating ever more pollutants to the Caribbean and beyond.[19]

By 2015, the carbon dioxide count was over 400 parts per million of air molecules—with an expected rise to 450 by 2050. With warming land, warming seas and warming atmosphere floating more pollutants southward, as well as continuing to melt glaciers, the result was more rising sea levels, submerging islands and continental shorelines, pushing up the temperature of terrain and over-exposing and poisoning species life. And of all continents, Australia suffered worst.

AUSTRALIA'S THERMOHALINE CIRCULATION: LITTLE KNOWN, MUCH IN CHARGE

Strong evidence linking Australia's weather events to global warming comes from Australian and other researchers at the American National Oceanic and Atmospheric Administration. The Australian Government advisor, Professor Ross Garnaut, points to the powerful energy in both the atmosphere and the seas around the Australian continent. The black, brackish ocean deep is whipsawed by the energy of wind motion, ocean currents, and the unfolding of Australia's dry climate, now converging with global, human-caused, greenhouse gas emissions, raising temperatures to searing scales.

A moment of recent global heat has been reached affecting extended periods of Australia's daily weather patterns. The year 2013 continuing into 2016 and early 2017, for example, brought the most torrid months in Australian *modern* history—reaching 111°F. in January 2014.[20]

Such land-based heat was also increased in 2015-16, brought from *La Niña* and periodic wind storms into the Australian atmosphere, raising sea temperatures by hot and salty thermohaline circulating currents. A rough map of thermohaline circulation shows how relative temperatures lock Australia into an environmental vice during particularly hot years associated with *El Niño*'s blast of eastern-driven sea surface waters.

Cold and salty deep sea currents usually sweep south, then northeast of both Australia and New Zealand, bringing on an upwelling, transferring heat from air to ocean. At the other extreme during *El Niño* years, warm, shallow ocean currents sweep downward to the northern reach of parts of Indonesia and Australia as even more atmospheric heat is transferred from sea to sky, releasing moisture to the atmosphere. Thermal logic then prevails: a heated ocean evaporating to the atmosphere leads to sweeping downpours and cyclones.[21]

So, expect the worst for Australia if global warming during our Anthropocentric Era continues while we burn more fossil fuels, filling the skies with greenhouse gases heated by the Sun, re-radiated from the surfaces of sea and soil—with aerated particle matter holding droplets of water and nitrous oxides, periodically raining down poisons.

I saw first-hand the deluge beginning in the Queensland, Australian during the summer of 2010-11. A devastating downpour was nearing—a cyclone hit, then another, Gaia reeling out her wrath, tracing out humans boiling the Planet. And this was hardly the last of it.

Cyclonic futures are certain to follow.[22] Sci-fi it may seem, but the latest heat waves signal what's in store. For Australia, with all its natural beauty and species, with its generous people, is fated one day to suffer under a Last and Lasting Heat and Sea Wave—the entire continent submerged, disappearing with most of its wondrous forms of life.

NOTES

1. Herman Melville, *Moby Dick or The White Whale* (New York: The New American Library, Signet Classics, 1961, first published in 1851. For a discussing of Right and other whales, see also Krooth, *Mexico, Nafta and the Hardships of Progress* (Jefferson, North Carolina: Mc Farlnd, 1995), *passim*.
2. Krooth, *Mexico, passim.*
3. Peter Brannon, 'Scientists worry that warming seas may be harming the endangered right whale," *Washington Post*, April 30, 2012.
4. D. Graham Burnett, The Sounding of the Whale: Science and Cetaceans in the Twentieth Century (Chicago: The University of Chicago Press. 2011), passim.
5. Burnett, The Sounding of the Whale, passim.
6. Peter Heller, *The Whale Warriors: The Battle at the Bottom of the World to Save the Planet's Largest Mammals* (New York: Free Press, 2011).
7. Brita Belli, "Defender of the Seas," *E-The Environmental Magazine*, January/February 2012, 18-25.
8. Farley Mowat, *Bay of Spirits: A Love Story* (Toronto, Ontario: McClelland & Stewart Ltd, 2006), 353-354.
9. Elizabeth Rosenthal, "More Guns = More Killing," January 6, 2013. See also Krooth, *Arms & Empire, passim*.
10. Richard Krooth, "Travels in Florida and the Gulf of Mexico: Interview with Captain 'Nemo', "Ten Thousand Islands, National Refuge: *Notebooks*. 2005-07, 2011-2012.
11. See *Look Magazine* pictorial displays, 1950s. See also Richard Krooth, Morris Edelson, Hiroshi Fukurai, *Nuclear Tsunami*: *The Japanese Government and America's Role in the Fukushima Disaster* (Lanham, Maryland: Lexington Books, 2014), *passim*.
12. See Oceanic Preservation Society and IMPT Partners, "The Cove," Director Louie Psihoyos, Producer Fisher Stevens, 2009, 96 minute DVD (www.Lions Gate, DVD.com), explaining how humans are the biggest threat to dolphins, as well as their only hope of being saved.
13. See Terry Tempest Williams, "One Patriot," chapter in *Rachel Carson: Legacy and Challenge*, edited by Lisa H. Sideris and Kathleen Dean Moore (Albany New York: State University of New York Press, 2008), 31-32.
14. Paul Greenberg, "Tuna's End," *The New York Times Magazine*, June 27, 2010, 33.
15. http://en.wikipedia.org/wiki/Purse_seine#Purse_seine.
16. Richard Black, "Green tuna goes FAD-fere, *Earthwatch*, BBC, 9 March 2011.
17. Gore, *The Future*, 200-201, *passim*. See also Michelle Innis, "Warming Oceans May Threaten Kriil, a Cornerstone of the Antarctic Ecosystem,' *The New York Times*, October`19, 2015;. And see Wikipedia, "Krill."
18. Lonely Planet, "Whitsunday Islands," 23 October 2010. http://lonelyplanet.com/australia/whitsunday-islands/travel-tips-and-articles/76132.
19. Louise Gray, "Global warming has slowed because of pollution," 26 November 2010, Telegraph Media Group Ltd. 2010. http://www.telegraph.co.uk/ earth/environment/climaatechange/8159991/Global-warming-has-slowed-because-of-pollution.html.
20. Justin Gillis,"Scientists Trace Extreme Heat in Australia to Climate Change," *The New York Times*, September 29, 2014.
21. Richard Krooth, "Travels in Australia," *Notebooks*, 2011-12.
22. Matt McGrath, "CO_2 levels mark 'new era' in the world's changing climate," *BBC*, October 24, 2016.

VI

Forces Driving Global Warming

As is true of our present crisis, the Paleocene-Eocene thermal Maximum began . . . with the burning of fossil fuels. At the time [65 million to 34 million years ago] the supercontinent Pangaea was in the final stages of breaking up, and the Earth's crust was ripping apart, forming the north-eastern Atlantic Ocean. As a result, huge volumes of molten rock and intense heat rose up through the landmass that encompasses Europe and Greenland, baking carbon-rich sediments and perhaps even some coal and oil near the surface. The baking sediments, in turn, released large doses of two strong greenhouse gases, carbon dioxide and methane. Judging by the enormous volume of the eruptions, the volcanoes probably accounted for an initial buildup of greenhouse gases on the order of a few hundred petagrams [a thousand million grams] of carbon, enough to raise global temperatures to a couple of degrees.
——Lee R. Kump, "The Last Great Global Warming"

Chapter Eleven

Thermal Maximums and Entropy

> Our greatest climatologist, James Hansen, has given us eight more years to reverse the flow of carbon into the atmosphere, or else soar past the red lines that guard, among other things, the stability of the great ice shelves above Greenland and the West Antarctic.—Bill McKibben, 2008 [1]

Earth's biosphere is drawing ever nearer to a possible future collapse. The convergences of regional environmental disturbances and global disruptions are multiplying, leading to overlapping entropies and a possible worldwide disclimax.

The Anthropocene, moreover, is reaching towards its outer limits of destruction of Nature, to be followed by a long, dismal Eremocene that biologist E. O. Wilson describes as an Era of Loneliness for humankind and desperation, rarity and annihilation of most surviving species. Some say the cataclysm will take hundreds, even thousands of years, with periodic signs of impending disaster and death likely appearing for only certain landscapes and the largest species and their phylum.

But what if a disclimax makes an earlier appearance in still unknown ways from forces directly causing global warming, or from quickly advancing technologies endangering the biosphere? Could rapidly changed environmental conditions then undermine species survival on a grand scale? Or will a cascade of species be set in motion by other now unperceivable forces blocking evolution and development of existing species life? However remote these possibilities may seem today, they might impact evolving species, limiting the survival of their progeny.

1. CLOCK RUNNING OUT OF TIME

The clock is running out of time for reducing global heat levels and rebalancing the conditions and resources supporting life on Earth. By most calculations there already is more than sufficient heat caused by re-radiation of the Sun's energy reaching Earth to the cloud cover of aerated carbon dioxide, as well as other greenhouse gases, aerosol evaporation and plant transpiration, along with wind-carried particle matter. Together these have led to a surge in global warming, the massive melt of glacial ice, the rising of seas, the inundation of landmass, disappearing landscapes, and the reduction in species diversity.

Over Eons Past and Expected Futures

Over eons past, long before our species took charge of other life varieties on the planet, extremes of bitter cold and lethal heat periodically endangered species diversity. These extremes were relayed by the link of global hydrological systems to atmospheric and territorial conditions in a great cosmic loop that affected, and still affects, the evolution and development of species life at different elevations, subterrains, longitudes and meridians.

In 1786, the scholar Georges Cuviera taught at the Parisian Museum of Natural History that extinctions were caused by planetary catastrophes. Later, during his moment in the 19th century, Scottish geologist Charles Lyell argued that most extinctions happened gradually over millennia, not in cataclysmic spasms—as Darwin later agreed in his study of evolution of species and the extinction of some organic forms.

In contrast, at Chixalub, Mexico, a University of California father and son team proved in 1980 that an asteroid slamming into Earth had caused mass extinction of dinosaurs and nearly all else. Though 99.9 percent of all species ever inhabiting Earth are now extinct, the known Big Five incursions from space together account for only four percent of all extinctions that have taken place over the last 600 million years of Earth history. Recovery of surviving species through evolutionary speciation into entirely new organic and structural forms took many millions of years after each extraterrestrial invasion, as we have shown in previous and following chapters. The rest of earthly extinctions took place naturally on the planet itself and in the seas, either violently or gradually, leaving diminished living species forms that then evolved into other organic forms, as Darwin taught.

Gradual Versus Immediate Extinctions

Can we separate gradual extinctions from those caused by extraterrestrial invasions, volcanic eruptions, and other sudden unperceived causes? During the Paleocene-Eocene, we know atmospheric heat spiked when pelagic volcanic explosions released greenhouse gases. As heat reached beneath and destabilized cold sea floors, it caused an explosion of frozen methane-hydrate (called clathrate). Some of the thawed hydrate bubbled to the sea surface, releasing methane and carbon into the atmosphere. And as the methane trapped heat more effectively than aerated carbon dioxide, it quickly combined with oxygen to form still additional carbon dioxide.

So long as elevated atmospheric concentrations of released methane continued, they also absorbed re-radiated solar heat reaching Earth's surface, raising atmospheric temperatures and extending Earth's drought conditions—drying, baking and, by lightening strikes or volcanic eruptions, burning existing land-based biomass, releasing even more carbons in a surge of burning forests, peat and coal. A critical mass of torrid heat thereby was spread to Earth's cooler regions.

Every round of warming bred still more warming, melting polar ice caps and permafrost, uncovering and burning dead plant matter, releasing surges of methane gas. Rising oceans meanwhile became vast storage vaults for methane and carbon. When these accumulated gases were released, they turned into carbonic acid, acidifying the sea and killing 30 to 50 percent of living species of microscopic *foraminifera* thriving on the seafloor and within its sediments.[2]

Most then-existing land-based species migrated to escape the lethal heat if they could. But in the race to survive, only some developed organic and metabolic structures to adapt to and successfully reproduce in what became new, emergent, bio-geographical environments.

The Race for Life

Darwin had judged that, "The inhabitants of the world at each successive period in its history have beaten their predecessors in the race for life, and are, in so far, higher in the scale, and their structure has generally become more specialised . . . . "

Darwin also thought we could trace their evolutionary forbears:

> Extinct and ancient animals resemble to a certain extent the embryos of the more recent animals belonging to the same classes . . . . The succession of the same types of structure within the same area during the later geographical periods ceases to be mysterious, and is intelligible on the principle of inheritance . . . . [A]ll the chief laws of paleontology proclaim that species have been produced by ordinary generation: old forms having been supplanted by new and improved forms of life, the products of Variation and the Survival of the Fittest.[3]

Exactly when and how this happened may remain partially unknown. During the Ethocene of an equable climate lasting some 200,000 years, overall global warming of some five degrees centigrade sustained surviving species.[4] But thereafter, species survived or not depending on challenges and changes in both the natural flows of energy and human influence. Human inventions alone altered environments, jeopardized life diversities, and placed the entire planet at risk.

Detailed accounts of these forces on the trajectory of species evolution, development, and entropy have been explored and elaborated on by many paleoanthropologists, geologists and biologists examining environmental disruptions and the possibility of a disclimax on a global scale. To these we turn.

The Great Aerial Ocean and Its Gases Mutating Climate

The place many studies begin with is the atmosphere—Alexander von Humboldt's concept of the great aerial ocean that allows the Sun's radiation to pass through nitrogen and oxygen layers to the surface of the planet, incubating life.[5] Without this passage, the Sun's radiant energy would be blocked out by the opaqueness of methane, carbon dioxide, other carbon-bearing gases, and vaporized water attached to particle matter. These make up so-called "dark matter" that directly or indirectly traps infrared light and heat—creating the greenhouse syndrome.

Thereby does the atmosphere's energy reach Earth to warm its environs and life—to be re-radiated back to dark matter, which is the vapor cloud cover, and to some gases directly able to trap the Sun's infrared heat in the atmosphere itself.[6]

Analyzing these gases in the atmosphere during the middle of the 19th century, John Tyndall, the British physicist and professor of natural philosophy at the London Royal Institution, explained that:

> As a dam built across a river causes a local deepening of the stream, so our atmosphere, thrown as a barrier across the terrestrial rays, produces a local heightening of the temperature at Earth's surface. Without the atmosphere, you would assuredly destroy every plant capable of being destroyed by a freeze in temperature . . . . The atmosphere admits the entrance of solar heat, but checks its exit; the result is a tendency to accumulate heat at the surface of the planet.[7]

Tyndall elaborated that variations in the balance of the constituent gases in the atmosphere not only "must produce a change in climate" but also "may have produced all the mutations of climate which the researches of geologists reveal." These included swings from Ice Age glaciers gouging out and plowing terrain, to torrid desert fastnesses dehydrating and obliterating plant and animal species.[8]

Climate is changed and "mutated," so to speak, by its variegated atmospheric combination of gases. So, as key gases absorbing infrared light and heat, how much carbon dioxide and methane, for example, would be just right to satisfy a "Goldilocks Climate"—neither too much, nor too little heat to sustain life?

Swedish chemist Svante Arrhenius took up Tyndall's challenging theory of the interlocking processing of the Ice Ages, advancing and retreating glaciers, and "some points in geological climatology"—concentrating "on the influence of the absorption [of gases] by the atmosphere upon the climate." He calculated that cutting atmospheric carbon in half would lower the world's temperature by four or five degrees centigrade, putting Earth in a frigid state.

And on the contrary, he believed that it would take some 3,000 years for carbon dioxide to double in the atmosphere to increase temperatures by five to six degrees centigrade. This, he thought, would prevent a new Ice Age and bring more equable, improved climates, particularly in the colder regions of Earth, speeding more abundant crop production as carbon dioxide stimulated plant growth for the benefit of rapidly-increasing humankind.[9]

In 1938, Arrenius's theory was borne out by the studies of steam-engineer Guy Stewart Callendar, revealing that carbon dioxide was so increasing in the atmosphere that it would lead to global warming and climate change, albeit benefiting agriculture and delaying the return of deadly glaciers.[10] But the so-called Callendar Effect was only a temporary rediscovery of the idea of global warming—leading us to look to Roger Revelle, the theoretical grandfather of the greenhouse effect.

The carbon dioxide cycle causing climate modification, Revelle elaborated, was offset in part by an ocean storage sink that contained 60% more carbon dioxide than the atmosphere, thereby absorbing most of the carbon dioxide that largely came from people burning fossil fuel. It was a neat little theory that, because the sea was the carbon vault, human activity that released carbon would have little if any effect on climate. The problem was it was wrong, as Revelle later found. For he discovered "the ocean is stratified with a lid of warm water on the cold, and the mixing between them is limited"—constraining the ocean's ability to receive and absorb carbon dioxide.

By his own research, Revelle faced the contradiction that the ocean would still absorb most of the carbon, so there would not be carbon-caused global warming; yet the sea would only *temporarily* absorb most of the carbon due to the chemistry of its own water and the lack of interchange between warmed and cooler levels. So the carbon dioxide inevitably would seep back into the atmosphere, re-absorbing the Sun's radiant energy striking Earth's surface.

Moreover, Revelle surmised, the build-up of atmospheric carbon dioxide might "become significant during future decades if industrial fuel combustion continues to rise exponentially. Human beings are now carrying out a large scale geo-physical experiment of a kind that could not have happened in the past nor be reproduced in the future."

Requiring Proof

Revelle's theoretical focus required proof. This his *protégé*, chemist-geologist, Charles David Keeling, pulled together in the form of a diagram from atop the meteorological observatory in Hawaii's Mauna Lao—known as the Keeling Curve.

With it, and the data drawn from ice cores preserving ancient atmospheric-trapped tiny air bubbles extracted in Greenland and at the Russian Vostok station in Antarctica, a steady increase in atmospheric carbon dioxide levels were discovered on the following scale: [11]

- Pre-industrial age: average 275 to 280 concentration parts per million of air molecules
- 1959: average concentration 316 parts per million
- 1970: 325 parts per million
- 1990: 354 parts per million
- 2000: 350 parts per million
- 2010: 395 parts per million;
- 2015; 400 plus parts per million and for
- 2050: estimated 708 parts per million.

Keeling judged that in 30 years from 1969, "if present trends are any sign, mankind's world . . . will be in greater immediate danger than it is today."

Speculation and Facts About Global Warming

Keeling's theory set off wild speculation about the effects of global warming or a new Ice Age. As these alternatives rocketed through political circles from Washington to Moscow, precise proof was still lacking.

Delay set in. Earth with only a single atmosphere had to await computer programs to express by mathematical algorithms the physical laws governing the dynamics of gases, heat and moisture in the atmosphere itself—the patterns and flows of air globally providing climate models, the most prescient predicting that a doubling of atmospheric carbon dioxide would increase global temperatures by three to four degrees. [12]

When the original thinker, Revelle, was politically "exiled" from becoming chancellor of the new San Diego University of California campus, he became a Harvard professor of Natural Sciences, lecturing in Autumn 1968 on "Pops and Rocks"—Human Populations and Natural Resources. Here he expounded on his theories of environmental connections, on which his students, among them Albert Gore, cut their teeth:

> By bringing fossil fuels to the surface and burning them, human beings are simply returning the carbon and oxygen to their original state. Within a few short generations we are consuming materials that were formed and concentrated over geologic eras. There was probably never more carbon dioxide in the air at any time in the past billion years than today . . . . We can think of the increase in atmospheric carbon dioxide as a gigantic, unintentional experiment being conducted by human beings all over the world, that may give us greater insight into the processes determining climate.

There also were countervailing forces regulating possible increases or decreases in global temperatures, Revelle said. For: (1) Burning fossil fuels over the next few generations would add vast amounts of additional carbon dioxide to the atmosphere, likely increasing temperatures with "significant effect on the earth's climate." And (2) high temperatures would increase evaporation of water and thereby increase cloudiness, "which in turn will reduce the amount of increasing solar energy, and tend to lower the temperature."[13] Indeed, this had happened before—in historical time.

Limited Carbon Dioxide: Geographic Exposure During the Little Ice Age[14]

Looking backwards to the Middle Ages, uncertainty still surrounds the precise causes of the so-called Little Ice Age, evidence of its first record appearing around 1300, then extending through to the mid-1800s. It was the coldest interval between warming periods blanketing the Northern Hemisphere for one thousand years or so. As glaciers descended from the Alps to engulf villages, famished and insecure populations suffered extreme cold, crop failures, and deficiencies in nutrients; sickness and periodic plagues followed.

The climate had become colder and wetter in most of Europe in the early 14th century, and many crops failed or rotted on the stem, starving populations. The rotting of the crops caused the release of methane and other gases into the polluted air, further weakening all oxygen-breathing species. The cold was to last four and a half centuries, though the English industrial surge was not much impeded after 1750.

But what underlying forces caused the 14th century climate change, further causing crop failures that jeopardized biodiversity? One theory posits that as carbon dioxide levels rose in the atmosphere blocking sunlight reaching the troposphere, temperatures fell, jeopardizing the natural feeding chain of plants and animals. Or did these forces cause temperatures in the troposphere to rise?

The Question Seem Unresolved.

There were no photographers on the scene. But picturing the bitter cold between 1525-1569, Pieter Bruegal the Elder's six-portrait series of *The Seasons* captures the essence—the futility of Alpine huntsmen striving against Nature's throw of wintry dice. His hunters heading home have shot no quarry, their dogs with downcast eyes marching behind them in the snow. High above, ravens descend on the town's snow-capped roofs; cold and ice hang heavy on the trees. Outliers are feeding their last deadwood-fires. The returning burgher hunters are cold, heavily bundled, wrapped in thick coats and tight-fitted, brimmed hats. And as night descends, we surmise the skaters on the ice below may have no supper. There was no wild game hanging from the hunters' ropes and pouches slung over their shoulders.

The worsening effects of the Little Ice Age lasted half a century and more. Perhaps there was a sharp drop in solar radiation reaching the European continent. Modern proof remains sketchy. Isotopes of carbon in tree rings and beryllium in ice cores clearly show a drop-off in solar radiation during much of the period. And sunspot observations that began around 1610 show a near-absence of reported flares between 1645 and 1715.

Recent studies have played down the relative importance of this solar impact bringing on the Little Ice Age. But heat and perhaps a vast sulfuric, carbon and particle cloud cover over Europe had issued from volcanoes that erupted more frequently after 1500 than during the medieval warm period that preceded 1300. So no one knows *exactly* what forces impacted global temperature during this pre-industrial age.

Unlike the medieval warm period, the Little Ice Age appears to have been strongest over the continents of the Northern Hemisphere; and with little evidence of impact from the tropics and Southern Hemisphere, both phenomena were likely regionalized, as distinguished from today's global-scale of warming.[15]

2. CAUSES OF VARYING HEAT AND COLD ON EARTH

Changing conditions and temperatures on land, in the atmosphere and at sea are not uniform. And their variances are forever changing and affecting climate. Indeed, the causes of varying heat and cold on Earth after the Middle Ages come from many sources, posing questions not yet completely answered about environmental conditions even today.

Do Volcanic Eruptions Cause Global Temperature Variations?

A look at volcanism may help explain both regional and global periods of extreme cold, escalating warming, and species vulnerabilities. If we jump forward to the 1815 eruption of Indonesia's volcano Tambora, we view one of the most violent blasts ever recorded on Earth.

The burst led to a world-girdling atmospheric cloud blocking out the Sun's irradiant warmth reaching Earth, disastrous cold across much of the globe, the next year without a summer, bringing on crop failures in Northern Europe as well as snows in North America's Vermont as late as early June.

Still, the possible connections remain uncertain, as when volcanoes erupt, they emit a mixture of gases and particles into the air, having a dual, contrary effect. Some of the ash and gases, such as sulfur dioxide, have a cooling effect, because they (and the substances they catalyze) reflect sunlight away from the Earth. Others, such as carbon dioxide, admit some sunlight to the Earth's surface, but then block its re-radiated exit to space, causing warming by adding to the greenhouse effect.

The idea that volcanoes affect climate is hardly new. Benjamin Franklin, in May of 1784, wrote of a constant "dry fog" in 1783, which so dissipated the Sun's rays that, "when collected in the focus of a burning [candle-lit] glass, they would scarce kindle brown paper." Franklin correctly attributed the dry fog to a large Icelandic volcano that erupted in 1783; and he argued that the consequent reduced solar heating was the cause of unusual cold in 1783-84. He even discussed the concept of the climate feedback loop of heat and cold that would result in increased snow cover.

The cooling influence on the atmosphere is particularly marked in the case of large eruptions able to blast Sun-blocking particles all the way up to the stratosphere—as happened at the Philippines' Mount Pinatubo in 1991, which caused a significant dip in global temperatures in the following year or two, a process discerned elsewhere by examining the average global temperature change after multiple eruptions.[16] After being dormant for more than four centuries, a series of powerful volcanic steam explosions over two months led to the June 15th Mount Pinatubo blast, with enough force to expel molten lava at six hundred miles per hour, flooding an area of 250 square miles, requiring the evacuation of 200,000 people. Within a few hours, a plume of sulfur dioxide and ash penetrated the stratosphere, reaching an altitude of 21 miles. Three weeks later its aerosol cloud encircled Earth.

Remaining for nearly two years, its load of twenty million tons of sulfur dioxide mixed with droplets of water acted as a gaseous mirror reflecting solar rays skyward, reducing by more than 10% the amount of sunlight reaching the surface of Earth—also cutting global temperatures by nearly as much as three-quarters of a degree Celsius, offsetting a temperature rise that had been caused by the previous century of heavy industrialization—and that had made the 20th century the hottest in at least a thousand years.[17]

The effects of the eruption were felt worldwide. It brought vast quantities of minerals and metals down to the surface environment. It injected great magnitudes of aerosol into the stratosphere—more than any eruption since that of Krakatoa in 1883.

Expelling sulfur dioxide gas into the stratosphere, the aerosols formed a global layer of tiny droplets of sulfuric acid haze over the following months. Global temperatures dropped by about 0.5°C (0.9°F), and ozone was depleted substantially, if temporarily, then increased. Such droplets, or aerosols, remained suspended in the upper atmosphere for several years following the eruption.

Mount Pinatubo's aerosol cloud, which circled the globe within a few weeks of its appearance, led scientists to speculate that incoming sunlight might be scattered. Ellsworth G. Dutton, a meteorologist with the National Oceanic and Atmospheric Administration (NOAA) Climate Monitoring and Diagnostics Laboratory in Boulder, Colorado, traced the effects of Pinatubo's cloud with ground-based instruments that directly measured the strength of sunlight. Dutton said his results showed a 20-to-30 percent decline in the amount of solar radiation that reached the ground without being scattered or reflected, and a 2-to-4 percent decline in total solar radiation.[18]

Other Forces

General experience reveals that the cooling influence of an individual volcano will dominate for the period immediately after an eruption, but the warming impact will last much longer. So the significance of each blast depends on the time frame, a year or two, possibly even a century. And a very large volcanic eruption, say in 2016, might significantly reduce temperatures in 2017, but slightly warm them thereafter.

The literature is mixed, though. Both underwater and land-based volcanoes are estimated to release in total around 100–300 million tons of carbon dioxide each year. That's a large quantity, but comprises only around 1% of the amount that humans release from burning fossil fuel alone. As the British Geological Survey puts it: "The contribution to the present day atmospheric carbon dioxide loading from volcanic emissions is relatively insignificant."[19]

And yet volcanic eruptions, if large enough, can blast gas and dust into the Earth's *lower stratosphere*. Strong winds at these altitudes, about 15-25 kilometers per hour, rapidly blow the volcanic material around the world. Over a period of weeks, the main gas emitted by the volcanoes, sulfur dioxide, combines with oxygen and water to form sulfuric acid gas, that then condenses into fine droplets or "aerosols" that form a Benjamin Franklin haze. These are different than a *tropospheric haze* that originates mainly from the sulfur in fossil fuels, reducing visibility in the United States, Europe and other industrialized regions.

Historic evidence of the volcanic *lower stratospheric* variety lies with records of composite global *surface temperature change* near the periods of the five volcanoes producing the greatest optical depths (how far you can see in front of yourself) since 1880: Krakatau (1883), Santa Maria (1902), Agung (1963), El Chichon (1982) and Pinatubo (1991).

The similarity of these blasts, stratospheric impacts, and moderating surface temperature reveals that the volcanic haze scatters some of the incoming sunlight back to space, thus reducing solar heating of the Earth's surface. This change in heating Earth is an example of atmospheric alterations that change surface conditions by lowering temperatures, *forcefully* transforming the troposphere. An opposite effect underlies increasing greenhouse gases that are directly and indirectly heated by solar radiation and re-radiation from Earth's surface to cloud cover, pushing up Earth's thermometer and particle fallout.

Pressure Cooker Earth

To prevent Earth from becoming a pressure cooker, scientists and governments have been working on individual scenarios that seem more like science fiction than a way to offer safe passage for the planet and its species. One must still guess at what the future will hold to provide ways to offset future global warming.

For its part, the Intergovernmental Panel on Climate Change already has developed an end of the 21st century predictive range of Earth's average temperature rise varying from either 1.1-2.9 degrees Celsius to 2.4-6.4 degrees Celsius (equal to 4.3 to 11.5 degrees Fahrenheit)—the latter far higher than at any time in recorded history. The great fear of many climate scientists is that such warming could melt Arctic permafrost which stretches for thousands of miles across Alaska, Canada and Siberia—holding twice as much carbon dioxide beneath the tundra as there is in Earth's atmosphere—its melt threatening to release enormous stores of methane, nearly 32-times more potent than carbon dioxide, a so-called "game over" scenario for life on Earth.

To offset such a possibility, serious scientists and their corporate colleagues have proposed ejecting particles into the atmosphere that would block the Sun's radiant energy from reaching Earth's surface, causing the so-called smallest unwanted side effects. It's a roll-of-the-dice that God would never play, trying to mimic the action of volcanoes like Pinatubo. Pumping particles of sulfur dioxide, or similar reflective chemicals into the stratosphere, would surely risk a global disaster.

Indeed, if great quantities of sulfur dioxide were artificially pumped into the stratosphere, planetary destruction of life would be on Earth's agenda. As already shown, sulfur dioxide combines with oxygen and water, producing sulfuric acid, which most assuredly destroys the ozone layer protecting all surface animals from deadly cancers.

Less toxic compounds or synthetic particles have *never* been tried; and scientists, who care not for approval from nations with democratic forums, are spending great sums of money, intellectual energy and schemes, planning to experiment by raising the reflectivity (*albedo*) of Earth to block the Sun's rays reaching the planet, which would otherwise be re-radiated and absorbed by the cloud cover of greenhouse gases. Looking to the future, it may one day lead to scorching the Earth.

Some say it would be less dangerous to reduce the Sun's impact by seeding clouds or spreading giant mirrors in the desert. Yet, there is a significant possibility that, if tried, these schemes would flounce out of control in trying to lower the temperature of Earth. They would not halt ocean acidification destroying coral reefs and wiping out enormous numbers of aquatic species. They would not remove carbon dioxide from the atmosphere the way trees perform by "capturing" carbon dioxide in their leaves, metabolizing it in their branch systems, and storing it in their roots. They would not store it underground or in the ocean beds. Even recycling carbon waste in an expensive form of industrial photosynthesis would take at least thirty years to develop, using low-temperature heat from factories worldwide.

Experimental Non-Cooperation

Humankind is almost unilaterally polluting the Earth's atmosphere on a mine-by-mine, factory-by-factory, and nation-by-nation basis. Yet among national or productive units, there is no overall cooperative plan to stop polluting the ground, atmosphere or sea.

Lacking a full awareness of a clear and present danger to which only some governments respond, scientists and corporations just might take it on themselves to experiment by replicating a Mt. Pinotubo solution, says geo-engineering expert Ken Caldeira:

> If it becomes necessary to ring the planet with sulfates, why would you do that all at once? If the total amount of climate change that occurs could be neutralized by one Mt Pinatubo, then doesn't it make sense to add one per cent [of sulfates] this year, two per cent next year, and three per cent the year after that? Ramp it up slowly, throughout the century, and that way we can monitor what is happening. If we see something at one per cent that seems dangerous, we can easily dial it back. But who is going to do that when we don't have a visible crisis? Which politician in which country?[20]

Volcanic Background Noise, Scientific Chatter

As we have seen, the Mt. Pinatubo eruption produced the greatest volcanic aerosol haze in the twentieth century. But from just a single volcano eruption, it may be hard to identify a solitary climate "signal" from among the large amounts of weather and *unforced chaotic fluctuations* of the atmosphere and ocean—that is, background climate noise.

To study the effect of volcanoes and other exogenous, *forceful events* on climate, a Goddard Institute for Space Studies (GISS) "Pinatubo" research team pulled together a group of scientists, students and educators for an initial look at the average climate response after the five largest volcanoes this century. They found that there was a small cooling, about 1/4° Centigrade (1/2° Fahrenheit), which peaked 1-2 years after the eruption, confirming that volcanoes do indeed cause a *small global cooling*. A detailed test of climate models soon will be possible for the Pinatubo eruption, based on global satellite measurements of the volcanic aerosols and the resulting climate change.[21]

But until then, all the talk and preparations for putting up a sulfate buffer around Earth seem like madness. Journalist Michael Specter says:

> We are polluting earth unilaterally. Whether it's life-taking decisions, like wars, or something like a trade embargo, the world is about people taking action, not agreeing to take action. And, frankly, the [about-to-be submerged, sea-level island nation of] Maldives could say, "Fuck you all—we want to stay alive"—throwing up an inexpensive atmospheric sulfate shield surrounding the planet and its seven seas. Would you blame them? Wouldn't any reasonable country do the same?[22]

If any among the scientists, corporations, or nations ring the planet with sulfates or their equivalent, put on your gas mask and lead jumpsuit the next time you step outside your bunker.

3. GREENHOUSE GAS EMISSIONS AND THE IMPACT OF WATER VAPOR[23]

During the Anthropocene, the greatest global warming dangers are beyond sulfate aerosol tinkering, awaiting a future still filled with uncertainties. What is needed is to stop burning fossil fuels from aerating greenhouse gases to the atmosphere. Only then will the Sun's radiation reaching Earth, re-radiated to the cloud cover, reduce excess heat locked into the planet. Somehow humankind will have to cut its dependence on fossil fuels and develop technologies to burn energy-laden hydrogen that will leave behind no trace but water.

Greenhouse gas emissions levels are today *both* constantly and perilously rising. There is no going backwards to an ideal, Utopian balance.

Measuring contemporary energy consumption by a single three-dimensional unit of a cubic mile of oil, it would fill a pool a mile long, a mile wide and a mile deep. Translating that into a yearly power-pack for the world, today it would take three cubic miles' worth of fossil fuel, amounting to a trillion gallons of gas. Fast forward: within fifty years of economic advances, the demand for energy could easily be nine cubic miles of oil, adding by weight three times the carbon dioxide pumped into the atmosphere.

The danger is clear. Once in the atmosphere, concentrated carbon dioxide appears with water vapor, methane, nitrous oxide, fluorinated gases, and a host of other minor emissions. The one that causes the most warming overall is water vapor from natural forces and through human activity affecting its scale. The scientific evidence suggests that the warming caused by man-made emissions of carbon dioxide and other greenhouse gases boosts the rate of evaporation, increasing the amount of water vapor in the air.

Thus, human release of aerated carbon dioxide re-absorbs solar warming reaching and re-radiating from Earth. And the ground-level rising temperature thereby boosts evaporation, amplifying the warming. Indeed, many scientists believe that moisture evaporation is the most important positive feedback loop in the global climate system, as the water vapor interlocks with carbon dioxide and other greenhouse gases, and attaches to particle matter, later returning to Earth as precipitation.

Unlike carbon dioxide, which can persist in the air for centuries, water vapor cycles through the atmosphere quickly, evaporating from the oceans, lakes, rivers and land surfaces before coming down as fog, rain, hail, sleet or snow. Since the rates of evaporation and transpiration from plants—as well as from animal sweat, human perspiration, excretion and exhalation—increase with Earth's rising temperature, the amount of water vapor in the air at any one time (and, in turn, the amount of absorbed solar warming re-radiated from Earth) is *directly* related to the amount of other greenhouse gases locking heat in the atmosphere.

Countdown on Gas Releases

To best understand the greenhouse gases that humans *directly* emit into the atmosphere in significant quantities, an overall picture reveals:

- *Carbon dioxide* (CO_2) accounts for around three-quarters of the warming impact of current human greenhouse-gas emissions. The key source of CO_2 is the burning of fossil fuels such as coal, oil and gas, while deforestation (offset by carbon stored in *dead* trees) eliminates a sink for CO_2.

- *Methane* (CH_4). Accounts for around 14% of the impact of current human greenhouse gas emissions. Key sources include agriculture (especially aeration from livestock and rice fields); fossil fuel cracking/extraction; and, the decay of wetlands and organic waste landfill sites. Methane doesn't persist in the atmosphere as long as CO_2 (as it quickly combines with oxygen to produce carbon dioxide), though its warming effect can be much more potent (between 23 and 500 times for each gram of gas released).[24]
- *Nitrous oxide* (N_2O): Accounts for around 8% of the warming impact of current human greenhouse gas emissions. Key sources include aeration from agriculture (especially nitrogen-fertilized soils and livestock waste) and residues of industrial processes. The warming impact of nitrous oxide absorbing re-radiated astral radiation reaching Earth's surface is even more potent per gram than methane.
- *Fluorinated gases* ("F-gases") account for around 1% of the warming impact of current human greenhouse gas emissions. Key sources are industrial processes and consumer use of industrial refrigerants and aerosols. The warming impact of F-gases is even more potent per gram than nitrous oxide, as much as 10,000 times!

Human activity also changes the planet's temperature in other ways. Examples include aerosol vapor trails from planes; soot aerated from fires; and fluorinated gases cutting holes in the tropospheric ozone at the highest layer of the atmosphere.

Created *indirectly* by local pollution, all these tend to increase the absorption of solar radiation. Yet, *aerosol particles* produced by some vehicles, fires and industrial processes tend to refract sunlight away from the Earth, temporarily reducing greenhouse warming.[25]

These particles are lethal. They filter down to Earth's surface and enter the lungs of animals; or cause growth-disruption and poisoning of certain plant life, as well as the chain of species that feed on cellulose and pollens—descending through the food web to animals that consume other animals.[26]

Sources of Aerated Greeenhouse Gases; Comparative Heat Sources

The premise of this book is simple: The overlay of gases entering the atmosphere act as a gaseous cloud cover absorbing re-radiated sunlight reaching Earth, fast raising global temperatures and posing threats to life forms and their diversity. For example, total global gas emissions were significantly higher in 2012 than they were in 2005, when the World Resource Institute broke down global, Anthropogenic greenhouse gas (GHG) emissions into the following sectors:[27]

Industry 14.7%

Industrial Processes 4.5%
Transportation 14.5%
Energy – Electricity & Heat 24.9%
Other Fuel Consumption 6.0%
Agriculture 5.6%
Land Use Change 12.2%
Waste 3.2%
Fugitative Emissions 4%

These categories proved illusive, though, as emissions from the three main categories of the *technological sphere* were industrially-related uses (67%); agricultural/land-change uses (26%); and, following the second law of thermodynamics, spreading heat to cooler regions, lost energy and waste (7%). Comparing past periods of greenhouse gas emissions with the present releases reveals the past may have caused much less scorching heat than the present Anthropocene.

This is because today's enlarged Anthropocentric carbon releases to the atmosphere capture the Sun's radiant energy re-radiated from Earth's surfaces, heating the planet's atmosphere—accelerating climate change. As we have already documented, the vicious circle of the technological sphere's burning of fossil fuels worldwide has loaded the atmosphere with carbon dioxide that combines with methane, nitrous oxide, particle matter and moisture—which together are able to store solar heat for as long as a century or more!

Historic Record

The historical record reveals that the increased pace of impacting such greenhouse effusions into the atmosphere has come from human industry and farming over the course of the 200 years, 1750 to 1950. Together they *raised* atmospheric concentration levels well over 100 parts per million of air molecules—a rise of nearly 14,000 times *faster* than during the previous 610,000 years, during which the long-term *average* change was limited to 22 parts per million, larger fluctuations varying during the transitions between glaciation and interglacial conditions.[28]

More recently, inescapably storing more astral heat, the 2002-to-2005 growth rate of carbon dioxide emissions rose by more than 2.5% a year (amounting quantitatively to some 7.9 billion tons), having risen from less than 1% (6.8 billion tons) during the *entire* 1990s.[29] These aerated carbons acted as hitching posts that collected vast amounts of Earth's re-radiated solar heat, hermetically sealing them into global temperatures.

- By the year 2011, fifteen of the past sixteen years had ranked as the warmest ever *recorded* on Earth. The world's major scientific research groups all independently determined that 2001-2010 was the hottest decade on *human record.*

- This was followed by the 2015 spike of 400 molecules per millon level of carbon dioxide in the atm*osphere, above the average for the ten years 2005 to 2015.* And this was caused by drought conditions in tropical regions produced by *El Niño;* so, less vegetation absorbed less carbon dioxide; and drier conditions caused extra emissions from fires.[30]

The Intergovernmental Panel on Climate Change (IPCC), a U.N. scientific body convened to assess scientific data relating to climate change, stated in its 2007 assessment of the evidence, that "warming of the climate system is unequivocal." The same report—drawing on the full range of published science papers on the subject—pointed to a rise of about three-quarters of a degree Celsius in the past century, with much of that warming taking place over the past few decades.[31]

Of course, the fact that the world is warming doesn't mean that it is getting hotter in a uniform way. The long-term rise in temperature is affected by shorter-term factors such as changes in solar activity and regional cycles like the oscillation between *El Niño* and *La Niña* conditions in the Pacific. And some regions—such as the Arctic—are warming significantly faster than others. Taken as a whole, though, and averaged over decades to remove year-to-year variability, the warming trend is clear and undeniable.

Which Fossil Fuels Release the Most Carbon Dioxide in the Short and Long Terms?

The shift in the geographic displacement and movement of wild and domesticated plants and animals reveals the effects of the increase in the burning of fossil fuels. Could we stop the pressure for these migrations?

One needs to ask which human-burned fuels release the most carbon in which geographies. Why, though, should anyone care to sort out the worst, less-worse or tolerable worldwide use of fossil fuels, if there is so little we can do about the consequent surge in global heat? Only you can decide if the following strikes a chord in your judgment and changes the way you act to secure the planet that sustains you, the other seven billion humans like you, and at least 8.7 billion diverse species.

A few facts about the risks of burning fossil fuels may change your mind about what to do. The more carbon-rich a particular fuel, the more danger it presents to the future of animals that breathe oxygen released from photosynthesizing plants, which in turn absorb carbon dioxide. (Several fuels when burned also release impurities like sulfur and mercury, nitrous oxide and particle matter, all potent environmental pollutants.)

Ranking Carbon Releases to the Atmosphere

Ranking carbon releases from burning different fossil fuels elevates our understanding of the dangers to the atmosphere and earthly life. In 2002, for example, the burning of fossil fuel released a total of 21 billion tons of carbon dioxide into the atmosphere. Of this, coal contributed 41 percent, oil 39 per cent, and gas 20 per cent. These percentages, however, did not reflect the *relative tonnages* of carbon burned, as some fuels contain more carbon than others.

We need to eliminate the burning of all fossil fuels and keep them in the ground. To better understand the impact on Earth, here is a description revealing the effect of burning each of the major fossil fuels releasing carbon dioxide.

- *Hydrogen* as a source of energy produces more heat when burned than carbon; and in doing so produces only water.
- *Oil* as a fossil fuel is structured with two hydrogen atoms for every one of carbon. So burning oil releases less CO_2 per unit than coal.
- And the best *black coal* is almost pure carbon. Each ton burned creates 3.7 tons of CO_2—the most carbon and dirtiest released to the atmosphere among the fossil fuels.[32]

Carbon Dioxide Equivolents; Global-Warming Potential

To better describe the dangers of the *atmospheric impact* on species life when all these gases are glommed together, we need to study their shape as a single number called CO_2e or carbon dioxide equivalent, a scientific standard unit for measuring carbon footprints. Using the standard of CO_2e, it is possible to express the different impact of each greenhouse gas in terms of the *amount of* CO_2 *that would create an equivalent amount of warming.*

It follows that global warming potential (GWP) is a relative measure of how much heat a given greenhouse gas traps in the atmosphere. Global warming potential compares the amount of heat trapped by *a certain mass of a particular gas* to the amount of heat trapped by a similar mass of carbon dioxide.

A GWP is calculated over a specific time interval, commonly 20, 100 or 500 years. GWP then is expressed as a factor of carbon dioxide (whose GWP is standardized to 1). For example, the 20-year GWP of methane is 72, which means that if the *same mass* of methane and carbon dioxide were introduced into the atmosphere, that methane would trap 72 times more heat than the carbon dioxide over its next 20 years.

Atmospheric Gases—Single and Agglomerated

Such calculations have proven to be important in negotiations between the leading industrialized nations on how to reduce atmospheric greenhouse gases. What this means for single or agglomerated atmospheric gases is increasingly becoming critical in judging global heat levels and their consequences for the thermal maximum endurance of life varieties.

In 2009, for example, the UK directly released 474 million tons of CO_2. But if you include its emissions of methane, nitrous oxide and F-gases, the country's *total* emissions work out at 566 million tons of CO_2e. In other words, those additional gases added the equivalent of 92 million extra tons of CO_2.

In making this calculation, though, how can we make one gas equivalent to another? Quite easily: Standard ratios are used to convert the various gases into equivalent amounts of CO_2. Again, these ratios are based on the so-called global warming potential of each gas, which describes its *total warming impact relative to* CO_2 *over a set period* – say a hundred years. Over this time frame, according to the standard data of one ton of various gases equals One:

- methane scores 25: meaning that one ton of methane will cause the same amount of warming as 25 tons of CO_2
- nitrous oxide comes in at 298: meaning that one ton of nitrous oxide will cause the same amount of warming as 298 tons of CO_2 and
- some of the super-potent F-gases score more than 10,000: meaning that one ton of super-potent F-gases will cause the same amount of warming as 10,000 tons of CO_2.

The *length of time* that each gas remains in the atmosphere *vis-à-vis* the others also is critical. This is because the global warming potential of each is a relative measure of how much heat a particular greenhouse gas traps in the atmosphere over a given time. Their ratios change significantly if a shorter or longer period than a century is used as the standard. For instance, a ton of CO_2 emissions may warm the planet gently over many centuries. But a ton of methane emissions creates a powerful warming rise over a much shorter period, as methane combines with oxygen to form CO_2.

Relatively speaking, then, the impact of aerated methane from its sources, such as seabed effusions, agricultural waste, cattle-and-pig-raising remains, landfill sites and wetlands, depends on the specific time it remains in the atmosphere—decades or even centuries.

Over a period of 20 years, again, atmospheric methane's global warming potential rises to 72—which means that if the *same mass* of methane and carbon dioxide were introduced into the atmosphere at the same moment, methane would trap 72 times more heat than the carbon dioxide over the next 20 years. But as atmospheric methane dissipates over a vastly extended period of 500 years, its *overall* global warming potential falls to just 7.6.

4. FUTURE GLOBAL WARMING POTENTIAL; CORPORATE PROFITABILITY; SPECIES LONGEVITY

By any throw of the dice, all this is an arcane science that we need to master if we are to understand how greenhouse gases affect global heat and species thermal maximum and longevity—now and into possible futures. During the summer of 2012, the U.S. was hit with as much drought as the Dust Bowl in the 1930s. Colorado burned in extensive firestorms. In 2011, Atlanta, Georgia, and Madison, Wisconsin, recorded their hottest days in history, some in sequenced humidity. Dallas, Texas, suffered a full 40 days end-to-end at above 100 degrees, continuing even hotter in 2012. In 2016, parts of California, like Paso Robles, reached 109 degrees.

Reaching the critical point of lethal heat, known energy alternatives and corporate investment choices in using them have turned, not on cooling the environment, but on economic scales of corporate profitability at the expense of species life—with underlying commercial scenarios such as these:

- Splitting—fracturing—shale (pumping in chemicals and water that together poison ground water and alter subterranean geologic strata), is supposedly less invasive to atmospheric conditions than burning fossil fuels from coal mining and oil, as well as natural gas extraction.
- With these and other comparative costs, analysts insist that water, wind and solar generation of electrical current can't compete economically. Even government subsidies and intervention with these alternatives might well remain too little and too late. Innovative breakthroughs may never come.[33]
- Though known alternative technologies already exist, such as burning hydrogen, wind farms, tapping geothermal power, and building hydroelectrical dams, they are usually not emphasized as major power sources.

A New Bio-geological Age[34]

A new biogeological age is now upon us. Humans have dominated Nature and its species during the Anthropocene, reaching the present moment of disappearing landscapes and the decline species, signifying the onset of a new age that biologist E. O. Wilson calls the Eremocene.[35]

Humankind and other domesticated animals have by and large altered physical matter and evaporated gases, bringing on torrid heat and new environmental conditions, cycling the decaying remains of Earth, causing species' rarity and demise. Global warming has shaped and limited the ability of species to survive, evolve, and reproduce—also leading to a possible disclimax of Earth's interlocking spheres:

- The Land Mass: Scorching heat scouring and drying the land, causing rapid evaporation and plant transpiration of moisture to the atmosphere.
- The Great Aerial Ocean: Accumulating atmospheric heat able to overwhelm the thermal maximums most species on Earth can endure before facing rarity and annihilation.
- The Seas: Receiving carbon-carrying atmospheric precipitation, acidifying the seas, destroying phytoplankton, zooplankton and the feeding chain of creatures unable to form their carbonate shells.
- And the warming Seas: evaporating water molecules and once-frozen seabed methane (clathrate) to the atmosphere, accelerating their combination with oxygen to form still more carbon dioxide added to the air.

Anthropocentic Speed and Tranformation

This cycling between Gaia's three major spheres has been speeded and transformed during the Anthropocene. As the Sun's radiation striking the planet has been re-radiated from Earth's land and seas to the atmosphere, a thermal point has been reached, changing the several-million-year-old life-sustaining relationship of plants, invertebrates and vertebrates on a disproportionate scale.

Intense heat has now increasingly caused vascular plants to dehydrate and die. Many invertebrates without plant food, sufficient water, nutrients and trace minerals have expired and become rare. And vertebrates, losing their food supply, have faced the same end.

Alongside the ensuing cascade of land species, entropy also has beset the oceans. Already under attack from acidifying seascapes, the Sun alone has not sufficiently supplied phytoplankton with the ability to photosynthesize enough food to feed the ocean web of life. And too few minerals and rock-weathered carbons are flowing into the sea.

As thermal conditions on Earth have escalated, insects and other land-dwelling arthropods have expired, ringing the death-knell for amphibians, reptiles, birds and mammals. And without sufficient moisture in desert-like conditions, flowering plants, the physical structure of most forests, and other terrestrial habitats of the world are facing collapse.

The Future Promises No Relief

The future condition of the environment is not promising. The land surface and its biomass will begin to rot, as dead vegetation piles up and closes channels of the nutrient cycle, so that other complex forms of vegetation will die off, with remnants of land vertebrates slowly starving.

Free-living fungi will continue to thrive for a while, as their population multiplies by stupendous proportions. But, as most species perish, they too precipitously will collapse.

If these events should come to pass, Earth then would return to its approximate condition in early Paleozoic times—covered by mats of recumbent, wind-pollinated vegetation, perhaps punctuated by clumps of small trees and bushes, with few signs of animal life.[36]

Past Records; Present Fears; Thermal Maximums; Human-Caused and Robotic Heat

Past records of climate change also may act as trail markers for the future. Over periods ranging from decades to millions of years, environmental change has periodically warmed the planet and left lasting records of alterations in the distribution of weather patterns and the planet's thermal temperatures. Climate change has also been caused by biotic processes and variations in solar radiation received by Earth, as well as by volcanic eruptions.

As we know, humans continue to warm the planet, causing climate change, too. We have poured greenhouse gases and particle debris into the atmosphere, putting things we value and love in danger—from coral reefs to the Jersey Shores, from homes threatened by wildfire and rising seas, to farms endangered by drought and potential hunger on a global scale.[37] Though today more and more people are aware of the dangers, most among us have acted as if all is well, going about supporting technologies regulating and overcoming the biological domain—as one form of life after another loses its vibrancy and reaches its thermal limits.

Even looking to the future of a robot-directed world, James Lovelock, founder of the Gaian hypothesis, contends that Earth is a single, self-regulating organism—and foresees that robots will "start losing organic life" that they no longer depend on for their own existence. While humans remain locked into practices of exploitation of Nature and burning fossil fuels, programmed robots can destroy organic life without cutting the essences that they themselves require. For their own existence, robots can accommodate infinitely greater climate or other changes than humans can. Though initially programmed by our progeny, robots potentially might escape our grasp and even increase global temperatures to points destroying many species including ours—and still maintain a safe platform on which to exist.

They might mathematically *calibrate* that Gaian cycles not be changed too much, at least not beyond the point where they lose control over maintaining their power sources of hydrogen, electricity and nuclides.[38] And what if they turned against Mother Nature's symbiosis overseeing the scales of heat that regulate most life on Earth? Could humankind then revanch, policing the creation of dangerous robots or pulling the plug on their energy sources?

NOTES

1. Bill McKibben, "Meltdown: A Global Warming Travelogue," *What Matters* (New York: Sterling Publishing, 2008), 8.

2. Lee Kump, "The Last Great Global Warming," *Scientific American*, Vol. 305 No. 1 (July 2011): 57-59. See also Tim Flannery, *Weathermakers, passim*. For parameters of temperatures altitude and longitude, see Alexander von Humbolt, *Cosmos*, Vol. I., *passim*.

3. Darwin. *Origin*, 394.

4. Sean Carroll, *Endless Forms, passim*. See also Richard Dawkins, *The Greatest Show on Earth: The Evidence for Evolution* (New York: The Free Press, 2009}, *passim*.

5. Krooth, *Gaia, passim*. See also Flannery, *Weathermakers, passim*.

6. Krooth, *Gaia, passim*; Flannery, *Weathermakers, passim*; see also John Tyndall's three central studies, *Heat Considered as a Mode of Motion*, *The Forms of Water*, and *The Glaciers of the Alps* (London: British Library copies, 1853, 1872, 1873).

7. Quoted by Daniel Yergen, *The Quest*, 427; see also John Tyndall's three central studies, *Heat Considered as a Mode of Motion*, *The Forms of Water*, and *The Glaciers of the Alps* (London: British Library copies, 1853, 1872, 1873).

8. Yergen, *The Quest*, 428; John Tyndall, *The Glaciers of the Alps* (London: British Library copy, 1873).

9. Yergen, *The Quest*, 429-430, citing Svante Arrhenius, "On the Influence of Carbonic Acid in the Air Upon the Temperature of the Ground," *The London, Edinburgh and Dublin Philosophical Magazine and Journal of Science*, April 1896, 237-76 (for "absorption of the atmosphere"). See Julia Uppenbrink, "Arrhenius and Global Warming," *Science* 272, no. 5265 (1996): 1122. For a more complete, original account see Svante August Arrhenius, *Worlds in the Making: The Evolution of the Universe* (Stockholm, 1903, 1905, 1914), trans. *H. Borns (New York: Harper & Brothers, 1908);* and *Theories of Chemistry* (Stockholm, 1903, 1905, 1914).

10. Yergen, *The Quest*, 430-431. Citing G. S. Calendar, "Can Carbon Dioxide Influence Climate?" *Weather* 4 (1949): 310-314. And James Roger Fleming, *Historical Perspectives on Climate Change* (New York: Oxford University Press, 1988): 113, 115. And Spencer Weart, "The Discovery of Global Warming" and "The Carbon Dioxide Greenhouse Effect," at http://www.aip.org/history/climate/co2.htm.

11. Yergen, *The Quest*, 435. Citing Roger Rainger, "Patronage and Science: Roger Revelle, the U.S. Navy, and Oceanography at the Scripps Institution," *Earth Science History* 19:1 (2000): 58-59; And Gustaf, *Oral History*, Scripps Institution Of Oceanography Library, April 11, 2006. And R. Revelle and H. Suess, "Carbon Dioxide Exchange Between Atmosphere and Ocean and the Question of an Increase of Atmospheric CO_2 During the Past Decades." *Tellus* 9 no.1, 1957. And Spencer Weart, "Roger Revelle's Discovery," *The Discovery of Global Warming*, http:/www.aip.org/history/climate/Revelle.htm.

12. Yergen, *The Quest*, 440-442.

13. Yergin, *The Quest*, 451-52, citing Roger R. Revelle, *Lecture Notes*, Mc6 Box 55, Folder "Natural Sciences 118," Scripps Institution of Oceanography Archives.

14. For information on the Little Ice Age, see Krooth, *Gaia*, 170, footnotes 422-425. See also Edward O. Wilson, *The Diversity of Life*, 31. And Rudiman, 'How Did Humans First Alter Global Climate?" 46-52. See also Fagan, *The Little Ice Age, passim*; and Steve Mithen, *After the Ice: A Global Human History, 20,000 to 50,000 BC* (Cambridge. Mass., 2003), *passim*.

15. See Robe http://www.guardian.co.uk/profile/robert-henson rt Henson, *The Rough Guide to Climate Change* (2009).

16. Mount Pinatubo is now an active strato-volcano that before 1991, had no recorded historical eruptions. It was heavily eroded, inconspicuous and obscured from view, located on

the island of Luzon, near the tripoint of the Philippine provinces of Zambales, Tarlac, and Pampanga. It is located in the Tri-Cabusilan Mountain range separating the west coast of Luzon from the central plains, and is 42 km (26 mi) west of the dormant and solitary Mount Arayat. (http://en.wikipedia.org/wiki/Stratovolcano. "Mt. Pinatubo's cloud shades global climate," *The Free Library*, by Farlax, http://www.thefreelibrary.com /Science+News-p210.

17. Michael Specter, "The Climate Fixers," *The New Yorker*, May 14, 2012, 96.

18. "Mt. Pinatubo's cloud shades global climate," *The Free Library*, by Farlax, http:// www.thefreelibrary.com/ Science+News-p210; http://en.wikipedia.org/wiki/ Mount_ Pinatubo.

19. *See the British Geological Survey and the U.S. Geological Survey, 2001-2014.*

20. Specter, "The Climate Fixers," May 14, 2012, 96-103.

21. Hansen, J., et al. 1996. "A Pinatubo climate modeling investigation." In *The Mount Pinatubo Eruption: Effects on the Atmosphere and Climate*, G. Fiocco, D. Fua, and G. Visconti, Eds.. NATO ASI Series Vol. I 42, 233-272. Springer-Verlag. Heidelberg, Germany. Contact inquiries about this research: Dr. James Hansen, Affiliation: National Aeronautics and Space Administration, NASA Goddard Institute for Space Studies, 2880 Broadway, New York, NY 10025 USA.

22. Specter, "The Climate Fixers," May 14, 2012, 103.

23. This section is based on Specter, "The Climate Fixers," and Richard Krooth, "Public Speaking Advocacy Tours," 2001-2014.

24. See on line http://www.ipcc.ch/ publications and_ data/ar4/syr/en/figure-spm-3.html.

25. See on line http://www.ipcc.ch/ publications_and_ data/ar4/syr/en/figure-spm-3.html.

26. See Liz Perera, "Climate Change May Be Hazardous to Your Health," *Catalyst*, Fall 2011: 7-9; Kate Sheppard, "The Ex Files," *Mother Jones*, May/June: 47.

27. See World Resource Institute, 2012; and on line http://www.wri.org/chart/world-greenhouse-gas-emissions-2005.

28. Krooth, *Gaia*, 628-29.

29. "Carbon emissions show sharp rise," BBC *News*, November 17, 2006. Citing Dr. Mike Rapauch, Co-chair, Global Carbon Project; and Professor Corinne Le Quere of the British Antarctic Survey.

30. Union of Concerned Scientists, September 14, 2011. http://us.mc306.mail.yahoo. com/ mc/ welcome?ymv. And World Meteorlogical Organisation, *2016 Greenhouse Gas Bulletin. See* Matt McGrath, "CO2 levels mark 'new era' in the world's changing climate," *BBC*, October 24, 2016.

31. Contribution of Working Groups I, II and III to the Fourth Assessment Report of the Intergovernmental Panel on Climate Change Core Writing Team, Pachauri, R.K. and Reisinger, A. (Eds.) IPCC, Geneva, Switzerland: 104.

32. Tim Flannery, *The Weather Makers*, 70-71.

33. David Leonhardt, "There's Still Hope for the Planet," *The New York Times*, July 22, 2012.

34. See Wikipedia, http://en.wikipedia.org/wiki/Atmospheric_lifetime; "Global Warming Potential," http://en.wik_Global warming potential#citenote-1; and Global-warming-potential#cite_note-1. The global warming potential of each particular greenhouse gas depends directly on its infrared spectrum. And the use of infrared spectroscopy to study anthropogenic greenhouse gases has become centrally important in the effort to understand the impact of human activities on global climate change and species survival.

Using these means reveals that preserving present day species diversities faces limits in regions of soaring regional and global heat.

Global warming potential depends on the interlock of:

- the *location* of a gas that absorbs given light wave-lengths, designating its infrared spectrum
- the absorption of infrared spectral radiation by a *given species*,
- thus the potential *atmospheric lifetime* of the species.

Thus, a high global warming potential of a particular gas correlates with a given species' infrared absorption over its thermal atmospheric lifetime, that also shapes the conditions essential for that species' existence.

That we remember this may explain why the anthropogenic release of greenhouse gases interrupt flows of Gaia's energy cycle, species' potential existence, and our own collective longevity on Earth.

35. See Krooth, *Gaia*, 549-550, citing E. O. Wilson. See also Smithsonian, "Genome: Unlocking Life's Code," *Exhibit*, September 2013. Human genetic telemers set a limit on the renewal of our body's organic structure. Other creatures seem to be able to regenerate aging organs or structures. Naked mole rats, for example, live 30 years, seem to feel no pain, and don't get cancer, because their cells remove damaged proteins more efficiently than do human cells. What humans cells lack may one day be provided by genomic research, applying similar processes by human medications.

36. Krooth, *Gaia*, 549-550, citing E. O. Wilson. See also Wikipedia, "Global Warming Potential," http://en.wik_Global warming potential; and http:www.guardian.co.uk/environment/2011/apr/27/co2e-global-warming-potential. Scientists actively work to understand past and future climate by using observations and theoretical models. A climate record—extending deep into the Earth's past—has been assembled, and continues to be built up, based on geological evidence from borehole temperature profiles, cores removed from deep accumulations of ice, floral and faunal records, glacial and periglacial processes, stable-isotope and other analyses of sediment layers, and records of past sea levels. More recent data are provided by the instrumental record. General circulation models, based on the physical sciences, are often used in theoretical approaches to match past climate data, make future projections, and link causes and effects in climate change.

37. On Julian Simon and Paul Ehrlich's famous wager over the Earth's carrying capacity, see Paul Sabin. "Betting on the Apocalypse," *The New York Times*, September 7, 2013.

38. Decca Aitkenhead, "James Lovelock: Before the end of this century, robots will have taken over," *Guardian.com*, September 30, 2016. The elaboration of this paradigm can be found at Richard Krooth and Hiroshi Fukurai, *Lethal Robots and the Death of Nature* (Berkeley, California:. Harvest Publishers, 2017.) Copyright © Richard Krooth and Hiroshi Fukurai, 2017.

Chapter Twelve

Gaia Cycles Anthropocentric Heat

An ARKStorm [is] a superstorm that could lead to a flood that would devastate the West Coast of the United States. The name comes from "atmospheric river" (AR): a vast air current that transports water vapor around the earth. These currents can be more than 300 miles wide and transport the equivalent of 15 miles, the average water flow of Mississippi at its mouth. If an atmospheric river hits a mountain range, it can turn into an ARKstorm, which could drop 10 feet of rain in a month, blast winds of 125 m.p.h., displace 1.5 million people, and damage a quarter of California's buildings.—Alex Prud'homme, "California's Next Nightmare" [1]

One of the strongest storms to hit Western Alaska in nearly 40 years battered coastal communities with snow and hurricane-force winds, knocking out power, ripping up roofs and forcing some residents to board up their windows and seek higher ground. "This is a storm of epic proportions," said meteorologist Jeff Osiensky with the National Weather Service. The highest wind gusts recorded—89 mph—were at Wales at the western tip of the Seward Peninsula, which forms the U.S. side of the Bering Strait, said Bob Fischer, lead forecaster for the weather service in Fairbanks.—Dan Joling, Associated Press, "Alaska Battered by Storm," 2011

You can't, as the climate-change deniers love to say, blame any particular hurricane on global warming. They're born, as they always have been, when a tropical wave launches off the African coast and heads out into the open ocean. But when that ocean is hot—and at the moment sea surface temperatures off the Northeast are five degrees higher than normal—a storm like Sandy can lurch North longer and stronger, drawing huge quantities of moisture into its clouds, and then dumping them ashore.—Bill McKibben, " 'Frankenstorm' is Just Right for Hurricane Sandy," *The Daily Beast*, October 30, 2012

A committee of the American Association for the Advancement of Science, the world's largest general scientific society, released a stark warning March 18, 2014, on global warming—that the effects of human emissions of heat-trapping gases are already being felt, that the ultimate consequences could be dire, and that the window to do something about it is closing: "The evidence is overwhelming: Levels of greenhouse gases in the atmosphere are rising. Temperatures are going up. Springs are arriving earlier. Ice sheets are melting. Sea level is rising. The patterns of rainfall and drought are changing. Heat waves are getting worse, as is extreme precipitation. The oceans are acidifying." More pointed still were the March 18, 2014 summaries online by Climate Central:[2]

- Concentrations of carbon dioxide traced on the Keeling Curve show a steady rise in carbon dioxide, as well as the seasonal variations in the curve.
- In Northern Hemisphere spring, plants burst to life and suck carbon dioxide out of the air until they die off in the fall. Atmospheric carbon dioxide usually peaks in May. As levels continued to rise in April or May 2014, it was the first time the monthly atmospheric carbon dioxide average was above 400 parts per million. Estimates for when the atmosphere last contained this much carbon dioxide range from 800,000 years ago all the way to 15 million years—the Sun and Anthropocentric heat potentially causing vast changes in climate, contributing to heating the planet's surface and ocean temperatures, which in turn melts ice and raises ocean levels.
- When carbon dioxide was last at these levels, polar ice melted and flooded the oceans, raising levels up to 130 feet higher than today's levels. Oceans have already risen 8 inches in the past century and may rise another 3 feet by the end of this century with grave dangers and costs to coastal communities.
- Prehistoric temperatures were also up to 11°F above today's temperatures. In the next century, temperatures are projected to increase by up to 8.6°F if emissions aren't reduced. The big difference between the current carbon dioxide levels and the last time they were this high is how fast they've increased in recent times, with the future rate showing little sign of slowing.

GAIA'S BIOLOGICAL SPHERE

As our species has thrust its technologies into the biological sphere, there is overwhelming evidence that human activities are driving rapid changes in atmosphere, seas, terrains and subterrains on a scale similar to the five past cataclysmic events destroying almost all earthly life—undermining the essential conditions for the existence of most species, threatening the life forms of others.

Humans have now reversed Nature's essential motion, leading towards disorder of the conditions essential for existing species. Britain's most elevated scientific institution, called the Statigraphy Commission of the Geographic Society of London, has unanimously concluded that:

> The combination of extinctions, global species migrations and widespread replacement of natural vegetation with agricultural monocultures is producing a distinctive contemporary biostatigraphic signal.
>
> These effects are permanent as future evolution will take place from surviving (and frequently anthropogenically related) stocks.... The Earth has entered a statigraphic interval without close parallel in the last several million years.

Global torridity is elevated by burning of fossil and nuclear fuels; human destruction of forests; the turn to monocultures and the relocation of species on the world map—together leading to extended species rarity and annihilation, bringing on future evolution from surviving species.

Corroborating evidence of continued emissions of greenhouse gases, according to the November 2, 2014, UN's Intergovernmental Panel on Climate Change, will lead to "long-lasting changes in all components of the climate system, increasing the likelihood of severe, pervasive and irreversible impacts." Indeed, as the atmosphere and oceans absorb ever-greater amounts of thermal energy within a blanket of heat-trapping gases, abrupt and irreversible changes are almost certain.

These consequences will accompany unforeseen, discordant links between land, sea and atmosphere, causes and effects difficult to tease apart, but nonetheless reflecting the complex global impact of warming climates. Examples follow.

JET STREAMS, PRECIPITATION, AND ATMOSPHERIC RIVER STORMS

Violent storms are the hallmark of the crossing point between intense jet streams bringing down precipitation and Atmospheric River Storms (ARK storms): Vast air currents that transport water vapor around the Earth, capable of dumping an average water flow equal to the Mississippi River at its mouth.

Rooted in torrid heat fostering climate change, an ARK storm is massive, while a jet air stream high above Earth's surface controls daily weather patterns. But now a point of global heat is being reached in which the two become interwoven.

"The base state of our climate has changed dramatically over the last 10 years, and that change is starting to alter the jet stream's behavior," says Jeff Masters, cofounder of the Weather Underground.

> Atmospheric levels of heat-trapping carbon dioxide, for example, have increased more than 40 percent, primarily because of the burning of coal, oil and natural gas. The extent of summer sea ice in the Arctic is down nearly 50 percent since 1900, affecting heat flow in the atmosphere and ocean. Solar energy reflecting off the earth's surface has changed significantly because we have modified more than half of the planet's landscape with crops, pastures and cities. Massive clouds of sunlight reflecting, and sun-absorbing soot and pollution, belch forth from power plants, vehicles, buildings and industry. A huge ozone hole disrupts upper-level winds over the Antarctic.

Such warnings are a wake-up call. A torrid Earth will produce uncontrollable massive storms, threatening worldwide terrain, inland waterways, farms, rural areas and cities.

Scientific debate over climate as a whole, and the jet stream in particular, centers on whether or not a tipping point has been crossed into a new long-term state brought on by the Arctic's fast change in the jet stream, driving and contorting it into extreme weather positions—bringing greater droughts, floods, heat waves and deep freezes in many parts of the world. Many scientists agree that a rapidly warming Arctic might interfere with the jet stream—the river of air high above us flowing eastward around the northern hemisphere, bringing daily weather.

Sometimes, the jet stream flows relatively directly, from West to East; but other times, it makes long, wavy loops, with Arctic warming also making the jet stream more wavy and loopy on average, bringing dramatic weather consequences. The atmospheric physics behind the idea, as von Humboldt long ago explained, is based on relative latitudinal and altitudinal heat differences.

These differences of changes in longitudinal and topographical isolines, driving heat-varying isotherms, set the stage for unpredictable climate changes. Warm air expands, and naturally there is much more warm air at the Equator than at the Polar regions. The atmosphere thus is thicker at the Equator, and the jet stream's motion is driven by the relative decline in atmospheric thickness as one moves in a pole-ward direction – so the atmospheric river effectively flows "downhill," moving North.

Regional warming also changes relative heat differences. As the Arctic is warming faster than the mid-latitude regions, the heat difference in atmospheric thickness should fall as you move in a pole-ward direction. This should reduce the difference in thickness moving through middling latitudes, slowing the jet stream, leading to more loops and turns—with weather of all types getting stuck in place for longer, harsher weather conditions, their intensity causing more destruction. But not always.

CLIMATE CHANGE: CASE EXAMPLES ARE MULTIPLYING

Winter starts with storms in the San Francisco Bay Area, according to Mike Pechner, the owner of Golden West Meteorology, a private weather forecasting service. The reason for a change in winter weather, Pechner and other meteorologists proclaim, is a shift in the jet stream—again, the current of air high above the Earth that tends to drive storms.

In a normal winter, the jet stream moves in over California and brings rain, wind and some snow. In 2011, however, the jet stream was diverted North by a ridge of high atmospheric pressure. The result was a near-record blizzard in Alaska and unusually dry conditions along the West Coast.

In this classic case, only 3.37 inches of rain had fallen in San Francisco in 2011 from July-to-December, 32 percent of normal. The last significant rain had been on Thanksgiving weekend. December, usually a rainy month by the Bay, produced only 0.14 of an inch of rain in San Francisco, making it the driest December in 22 years, and the third driest since records began in 1849. The seemingly last measurable rainfall in San Francisco had occurred on December 30, and that was only 0.02 of an inch.[3]

Then came a deluge of nearly a foot at the end of March, 2012.

It was a great show of unpredictability, for a few days or a month hanging on a single branch of Nature, though only an example of a single shift of weather, not an indicator of long-term climate change as judged by the components of an ARK storm that devastates mountains and terrain, crops and pastures, river networks and entire cities. And yet the rapid advance of climate change may make the two indistinguishable.

GAIA'S THREE OVERLAPPING SPHERES—ATMOSPHERE, SEA AND LAND

Climate change and precipitation themselves turn on the planet's three overlapping spheres: air, land and oceans. These spheres' web of energy and gyres are regulated as an unseen entirety. They cannot be segregated into isolated departments like the Great Aerial Ocean, the Seven Seas, and Continental Territories and Island Fragments. The circle of their energy flows encompasses all of them, each one dependent on sustaining, and drawing on, the others.[4]

Their natural gyres of terrain, atmosphere and sea also reveal near-instantaneous feedback loops on a material scale. One event on Earth breeds another, as with a tsunami raising sea levels, rapidly spreading from one continent to another, changing temperatures, aerating mists and fogs, bringing down moisture.

How to define and predict Earth's present ecological trajectory and future condition turns on both natural occurrences as well as cumulative human activities, causing momentary disorder in planetary energy flows. Clearly the planet balances, rebalances and reverses entire Ages of frigid temperatures and torrid conditions. It shapes and alters terrain, atmosphere and water flows. Sometimes its tectonics transform subterrains and seafloors, and bring on devastating earthquakes, volcanic eruptions, hurricanes, and tsunamis, as well as altering the scale and direction of ice floes.[5]

From beneath sub-sea floors, volcanoes periodically burst, forcing molten rock and sulfurs through all barriers, their surge turning salty ocean into the quality of vinegar or lemon juice, killing most nearby creatures whose remains are swept into the vast organic water columns on which surviving species feed. Sub-sea volcanic lava also sears and explodes through frozen methane (clathrate) just beneath the sea floor, destroying marine creatures. Great bubbles of heated methane push upward to the ocean surface, and combine with oxygen as carbon dioxide, rising into the skies to trap and absorb the Sun's radiation either directly or as refracted from the planet surface.[6]

Ocean floor earthquakes sometimes dislodge tectonic plates, one butting against or sliding over another, setting off tidal waves or giant tsunamis speeding 450 m.p.h. across the ocean, making landfall hither and yon. These tempests tear up coastlines and up-end shorelines in a circle of devastation, drowning trees, foliage of every variety, and animals, including humans and their placid communities, along with fishing fleets, entire farms, cattle keeps, and whole living cities. Ground to bits, these once vibrant lives and membranes become the sands, clays, and stones of time.[7]

NEW SPRINGS

Decades later, perhaps even hundreds of years will pass before new springs bring surviving seeds to flower, insects and birds to plunder them, hawks and raptors to take of smaller aviaries, and invading animals to renew the dance of survival and reproduction.[8]

If not a volcano or tsunami, perhaps the Great Aerial Ocean between space and Earth will release downpours for renewal or destruction. A vapor-wave of condensed and speeding water may suddenly descend to Earth in great volumes. Or fueled by rapid changes in relative air pressure and gradations of temperatures, the rush of aerial water might overflow land itself, overwhelming elevation barriers, changing sea depths and currents, as well as the material or chemical composition of rivers, lakes and seas. Expect these forces to inundate or otherwise transform the habitat and conditions essential for the fitness, reproduction and momentary survival of millions of different species—humans not exempt.[9]

Then comes the sweep of Moon- and Sun-driven tides, the changing gyre of Earth's revolving planet, the wind-swept currents and changing levels of the sea setting in motion revolutions of twisting water funnels. What future awaits Earth will surprise Utopians as well as scientific skeptics and religious fanatics expecting an Apocalyptic end.

AERIAL TSUNAMI; SUPERSTORM; MEGAFLOOD

The end of life on Earth is not near, but other possibilities seem endless. The forces set loose by polar gyre and centrifugal motion around the equator, and the speed of warm and salty thermohaline ocean currents regulating the evaporation of seas, are quickened today by the Anthropogenic release of carbon dioxide, other greenhouse gases, and particle matter storing and refracting the radiant heat of the Sun. Plant transpiration, animal sweat, human perspiration, urine and dehydrating feces: together these aerate massive spirals as cloud formation. Wetlands with their decaying biomass evaporate, carrying vast stores of methane, which become sky cover to seal in the Sun's radiation directly reaching and re-radiated from Earth.

NEW LIFE

Land-locked water sources' evaporation and precipitation renew themselves, supporting life, and then again aerate methane, carbon dioxide and other greenhouse gases. In the U.S. and Canada alone, freshwater marshes rebuild the source of abundant plant and animal foods, sustaining varied birdlife nesting in dense vegetation. Canada has more glacier-melt lakes than the rest of the world combined. These, with other rivers, streams, brooks, ponds, and reservoirs, liven insect incubation and the feeding of fish, the diving of kingfishers, kookaburras and raptors, and the bears' taking of berries, salmon and smaller fare.

Renewed by precipitation, too, Southern wooded swamps and Florida salt water mangroves provide habitat for amphibians, giant snakes and fish, billions of insects, wild seeds and flowering plants, water mammals like beavers, weasels, and alligators, large bears, deer and fox, as well as large mountain cougar varieties, wolves and coyotes.[10]

EVAPORATION, TRANSPIRATION AND PRECIPITATION

Once evaporation and plant-transpiration load the atmospheric gases with water-vapor, precipitation descends. Forests again transpire moisture to the air, which disgorges water vapor as haze, rain, snow and sleet sustaining ground-level life. Forests also release inestimable flows of oxygen to creatures inhaling their essence and breathing out their own carbon dioxide, sustaining more forests still—Siberia alone supporting 25% of the world's forests, the Amazon at least as much. More water still evaporates to form cloud cover from the seven seas; Africa's lakes and rivers; Russia's Lake Baikal; the Amazon and Mississippi; America's other immense rivers; and, the Great Lakes, the largest inland seas in the world.[11]

Oceans and atmospheres exchange not only moisture, but also heat and motion. The higher the density of surface water compared with that of air, the slower is the rate of evaporation of sea surface by comparison with the speed of the atmospheric absorption of heat released from the sea, to again bring thunderstorms and be reimbursed for precipitation worldwide.

As the ocean surface absorbs the Sun's radiant energy during the summer and releases it over winter, heating the atmosphere in the Northern hemisphere, it also speeds evaporation and turbulent air movements that in turn drive waves and currents. This symbiosis of atmosphere and sea is clearly manifest in regions where the temperatures of the sea surface waters are high, as seen off the coast of Africa or throughout the Caribbean—bringing on powerful tropical storms and hurricanes.

OVERLAPPING NATURAL FORCES

Sea levels themselves are affected by these overlapping forces, including the volume of water locked in deep ocean basins; surface layer thermal expansion; and the changing cycles during Ages of planetary Ice and Heat. From water vapor, snowfall, rain, and sleet to the immeasurable volumes of water alternately locked into ice caps on land and the release of water from icebergs into the ocean basins, no one has an accurate measuring gauge. During the Pleistocene Era, for example, the sea rose by up to 400 feet (130 meters)![12]

Could glacial ice caps fully melt again is a question scientists ponder, but seemingly can do little to stop. Ninety percent of the world's ice covers Antarctica, locking up some 70% of all fresh water in the world. Similar to deserts, with average yearly total rainfall of two inches, the driest places on Earth, with an absolute humidity lower than the Gobi desert, all but 0.45% of Antarctica is a block of ice. Antarctica may fracture and fall into the sea, floating away as icebergs. But it will take more than a camp-size burner to reduce it to sea melt; and only if planet Mercury's daytime temperatures of 430 degrees Centigrade (806 degrees Fahrenheit) were to recur on Planet Earth would it melt, raise sea levels by some 300-400 feet, and put all islands and continental fringes suddenly awash in sea.

Few doubt that the final close of the remains of the present Ice Age glaciers and permafrost could inundate all coastal areas, islands, and most human-constructed barriers throughout the world. Berms initially built in Asia to hold back canal and rice field waters would be destroyed. Dikes in the Netherlands designed to protect reclaimed lower-than-sea-level lands would be hundreds of feet under the sea. California's levees of earth, clay, sand, rock and cement, once meant to prevent waterways from overflowing, would prove poor barriers to keep out the raging storms of invading Pacific walls of water. The unrelenting pressure would undermine the foundations of the old levees, drowning cities and their still-blind populations. And, yet, these possibilities buiding great storms of rock and waves remain.

HOW THE GREAT WAVE MIGHT GATHER ITS FORCE

Precisely how the Great Wave would gather its strength remains unknown. Though relative air-water temperatures and evapo-transpiration from land and wetlands, lakes, rivers and seas spell innundation of land, the aerial ocean ceaselessly is also loaded with billions of tons of water vapor that is charged electrically by air convection and heat radiation. This vapor is driven by winds criss-crossing around Earth's circumference, its clouds of moisture ready to plunge into regional seas and to deluge distant terrains.

Imagine a visual waterspout as a tropical tornado passing over the sea, sucking upwards a great column of water, rising as high as 300 feet (100 meters). The sea surface is directly drawn up into the atmosphere—into a column around a core created by falling cold air, in turn creating a low-pressure area into which warm air spirals upwards with strong winds that take up the surface water. As the water is vaporized, it is ready to download wherever prevailing winds and relative temperatures carry its charge.[13]

Imagine too that the imbalance of both sea and atmospheric temperatures regularize this water vaporization process, so that waterspouts expand in frequency, the diameter of each one enlarging, drawing up millions, perhaps billions, of gallons of water as vapor to be carried by atmospheric winds to distant locations like continental Australia—sudden rains pouring from the skies disrupting agriculture, communities, lives. Some say such a deluge could never occur; but check Australia's 2011 carbon dioxide reading at over 385 parts per million of air molecules, causing violent downpours along with the cold/hot flux of its surrounding five oceans. Waterspouts are often unseen, unrecorded and thus seemingly rare. But happen they often do under these conditions, along with the massive flooding I witnessed in New South Wales during 2011.

It is difficult to predict the precise interlock of these forces on an ocean basin scale, especially in the Pacific tropics during the years when *El Niño*

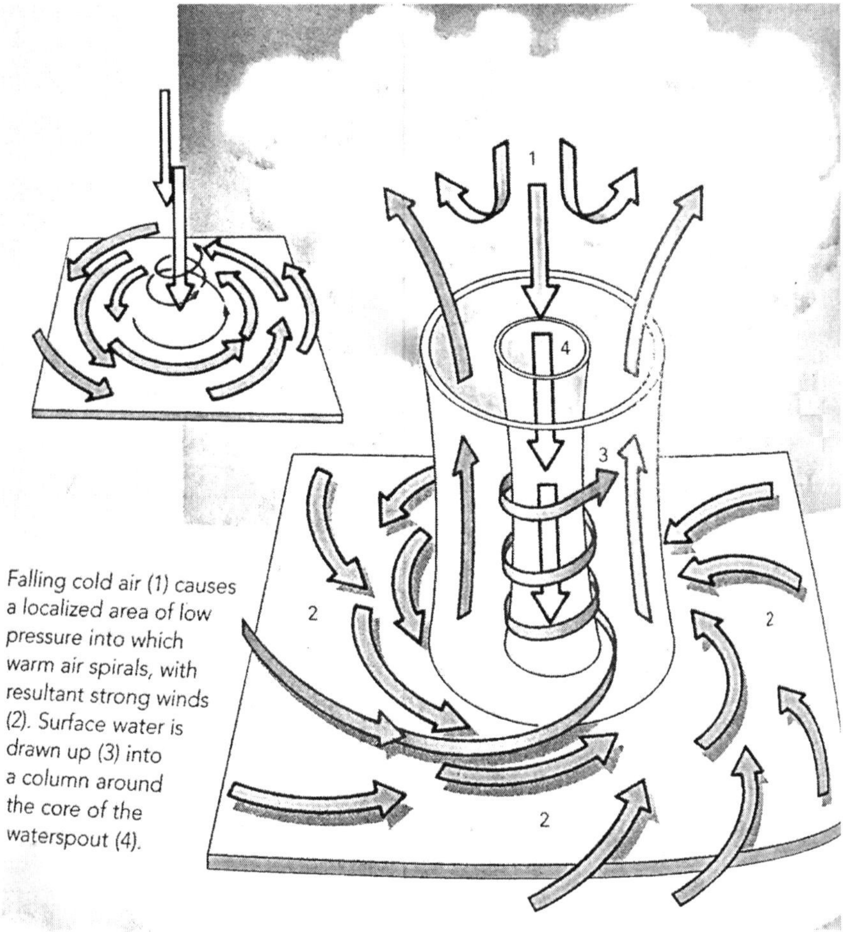

Figure 12.1. Origin of Ocean Waterspout.

changes Southern Pacific currents, whipsawing tropical weather patterns and sea level surface warming, and causing an unpredictable climate. Such air movements—weakening westerly winds, driving eastward the ocean gyre of the Southern Hemisphere—are splayed across the globe from the Japanese Kuroshio Sea currents to wind patterns in the South Atlantic; from alternating rain-and-drought in Southeast Asia to the India monsoons; to unpredictable Australia's "big dry" seasons with sandstorms and raging fires, to flash inland floods linked to evaporation from the warming, salty, ocean thermohaline.

Figure 12.2. Waterspout.

Part of Australia (except wet-weather Tasmania) is composed of a great desert lying under a band of high pressure, 30 degrees longitude North to South. Dry air from the upper atmosphere is constantly creating a pressure cooker below—pushing downwards, dehydrating the land, wisps of cloud cover holding little moisture as the sea winds of the five surrounding oceans

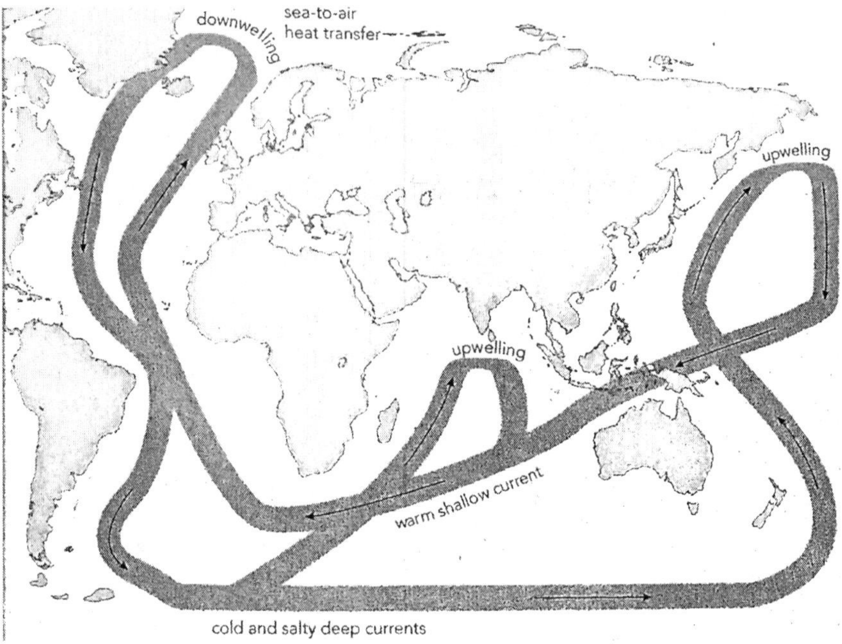

Figure 12.3. Map of Thermohaline.

periodically offer only brief relief—or a rare deluge. But most of the time these seas are relatively cool, evaporation is modest, no rain is in sight, land dries, farmers go broke, fires break out from a spark to grassland or timber, and indigenous animals barely survive on the searing soils.

The unfolding of Australia's arid climate has converged with human-caused greenhouse gas emissions to raise temperatures to scorching levels—in various regions reaching 111 degrees Fahrenheit in January 2014, and periodically, as during all 12 months in 2013.

Such thermal readings have provided strong evidence to scientists, Australia's five professional and scholarly groups issuing reports in 2014, tracing such extreme heat to climate change, linking Australia's daily weather patterns to long-term global warming caused or worsened by human greenhouse emissions.

Is there no respite? Only *El Niño* years draw warm seawaters eastward across the Pacific face, offering momentary and powerful cloud bursts over Queensland, New South Wales, and Victoria. So although *El Niño* surges up every two to eight years, Australia's big dry, unbearable heat, hot winds, and fires prevail.

GAIA'S INTERDEPENDENCE

"When we try to pick out anything by itself," John Muir once philosophized, "we find it hitched to everything else in the universe." Ancient Incas believed *Quadrilla* markers (indicators of seasons) were the hitching posts between Sun and Earth. Ancient Egyptians saw the Chariot of the Sun bringing night to day. In our moment, some scientists believe Earth is caught in the crossfire of many forces—organic-molecules carried in comet ice and rock; amino-acid frozen in meteorites; lunar cycles stirring tides; the gyre of polar-equatorial forces; winds wildly shifting in directions and velocities; powerful sea currents shaping the atmospheric destiny, swamping islands and continental coasts.

Blinded by ignorance, humankind continues to draw on groundwater and natural gas (fracking), which dries and compacts sediments, lowering land surfaces relative to seascapes, causing floods. Mined coal and minerals pile up toxic effluents that run off with rains and snow into surface and ground water. Add drilling and fracturing shale, which poisons aquifers; add again refined oil and gas sucked from deep-sea, inland, coastal and near-shore areas—fouling and absorbing Sun-radiation heating the atmosphere. Or focus on spills and explosions causing periodic blow-outs of oil wells, firestorms, and the spewing of crude oil into oceans, causing appalling, seemingly irreversible destruction of sea and coastal life.[14]

More devastating still is the loading of greenhouse gases and particle matters into the atmosphere that absorbs and stores heat radiation re-radiated from Earth's surface, modified only by rays reflected back into space. And the dangers are magnified, writes Mike Ives in the *New York Times,* September 27, 2016, to include warning from the World Health Organization that 9 out of 10 people on Earth are breathing air with concentrations of particles causing three million deaths a year—mostly from cardiovascular, pulmonary and other noncommunable diseases.

As the particle-infused cloud-cover becomes a heat-blanket that takes up and holds ever more water-vapor from sea and soil evaporation and plant transpiration, a moment of chaos is reached when atmospheric internal heat-meets-cold and the density—and sheer weight—of the water vapor suddenly will be expunged from the aerial ocean—making up an ARK river, or 2016 incidents of "water-bombs" drowning entire European cities.

THE FRAGILE PRESENT FORETELLS HISTORY'S COMING DISASTER

The effects of these overlapping pressures can be seen in the example of Sacramento, California, which infamously in 1862 endured a 45-day rain that turned the Sacramento and San Joaquin Valleys into a vast inland sea—as indeed it once was in the wake of six earlier, powerful storms that swamped the geologic region several million years ago.

The scale of the 1862 storm, that forced the removal of the California capital from Sacramento to San Francisco, is not only again plausible, but is perhaps an inevitable harbinger of future mega-floods. This tragedy could be ignited by an earthquake trigger from any of the five separate fault lines in Northern California which place the San Francisco Bay Delta and its human-built levees at the greatest risk. These levees are in jeopardy inside the Delta because the region's clay, sandy soil and peat shake up to 10 times more violently than areas outside it, increasing the chance of levee failure. If a 6.7 magnitude, or greater, quake occurs in the region's next 30 years, seismologists predict there is a 40 percent chance that it will destroy the levees.

Add to earthquakes a direct so-called ARK storm hit, and California may suffer a tremendous tragedy from destruction of old, weak Delta levees and a deluge in Sacramento, the nation's most flood-prone city. The nearby Delta is the configured link between two-thirds of California's fresh water supply—which originates in the Sierra Nevada Mountains and the Northern river network—and two-thirds of the state's population which resides in the South. In fact, this geographic design of terrain and occupation is well-known. (Opportunity had first appeared in the 1870s, as farmers began building 1,000 miles of levees around the Delta, both to control floodwaters and build farmland out of tule marshes.) Today the old levees are so decrepit and leaking that University of California, Davis, Professor Jeffrey Mount, predicts there is a 64 percent chance of a catastrophic Delta levee failure in the next 50 years.

Many forces alone or together could destroy the levees, setting off a mega-flood. With surface layers of the deepening Pacific, caused by the melt of glaciers and sea ice, the new water might be swept by wind and waves across the sea surface from Hawaii, arriving as a Great Wave at 450 miles per hour. Or as atmosphere and sea surface temperatures increase, they might expand the surface level of the ocean, with a rise up to 3.5 inches (9 cm.) by the year 2100, to be further pushed by a powerful storm across the Delta.[15]

So too might an earthquake, a massive waterspout, or a violent Pacific super-storm sweep water off the ocean face around Hawaii, powerfully dumping it on California's West Coast, while battering the seashore with high velocity winds and waves, heaving ocean boulders and submerging hundreds of square miles.

Picture the devastation in store: As salt water is sucked from the Bay and driven into the Delta, it destroys some of the nation's most productive farmland, washes away buildings, highways, gas lines and railroads—causing landslides and contaminating water supplies for 25 million people. In its wake, the California economy, sixth-to-eighth largest in world size, is crippled, global trade disrupted, and the downtown capitol of Sacramento becomes a diving bell under 20 feet of water.[16]

ANTHROPOCENE UPON US—BEYOND HUMAN CONTROL

Britain's Statigraphy Commission speaks of the Anthropocentic Age that is not quite upon us. Humankind has reached the moment where it no longer controls the damage that its own technology causes to natural cycles. In our time, by technological overreach—from fossil fuels, chemical addiction, nuclear power and glacial-melt viruses —we have created toxic climates and caused entropy and ecocide in one domain after another, igniting an endgame of global disclimax. There are several ways to explain the human descent into this abyss, which is without a path to recover species diversity and rehabilitate territorial destruction:

- Natural selection itself drives species to exploit resources to the moment of ecosystem collapse, ultimately destroying humans themselves (the Medea hypothesis).
- Then selfishness and competition undercut evolutionary cooperation that once stabilized the symbiosis between humans, other creatures and natural environments (the neo-Gaia hypothesis).
- Developing technology makes it harder for elites to conceal their despoliation of both the global commons and its species (the Flannery hypothesis).[17]
- Worldwide trophic cascades.

As for me, I believe, we can no longer rest on our past laurels of good will towards and safety for plants and creatures. For only Nature's symbiosis of the forces of the aereal oceans, the seas and terrains can cradle species diversity and save Planet Earth . . . likely one day without our destructive presence.

Other possibilities to secure Earth's species diversity remain speculative at best. Evolution moves forward, not backward; so the regeneration and cloning of hominin life that existed before the dinosaurs or giant mammoths remains speculative. Reviving dormant plants and micro-organisms is possible, though. Seeds as old as 2,000 years have been cultured back to life. Some organisms like lake-fleas and elderberry naturally revive after dozens, hundreds or thousands of years, possibly playing an important role in sustaining ecosystems, as in the South American *cordilleras*, or under Antarctica's ice sheets, or upon the Russian steppes.

Current research might also lead to reviving ancient microscopic and pathogenic species. Already genetic engineering has made possible three-party human reproduction. And the cloning of species of different phyla seemingly has no limits.

RECALLING THE PAST; JUDGING THE FUTURE

In the distant past, some 23-to-25 million years ago, South American geologic convulsions initially pushed the mid-continent in a westerly direction, uplifted the Andes, then reversed river flow northeasterly to the Atlantic Ocean. As high as 20,000 feet, far above the tree line, covering parts of the interior of Northern South America in the Andean *cordilleras,* thousands of vascular plants evolved, as Alexander von Humboldt discovered.

Elsewhere in the Americas and Antarctica, at the close of each Ice Age, melting and retreating glaciers left deposits of eskers composed of rocks, clays, soils, chemicals and trace minerals that lay the groundwork for emergent, encapsulated seeds, mosses, lichen and billions of molds and microbes, developing into previously unknown ecosystems.[18]

Discoveries continued. Nearly 200 years after von Humboldt's expeditions, ecologist Peter Convey and his colleagues dug up a core of permafrost from treeless Signy Island (about 26 miles from Antarctica), put it under a lamp in a lab in Britain and misted it from time to time with water. And *voila*, after a few weeks, the moss was sending up new green growth.

The deepest layer from which this resuscitated moss grew was three and a half feet deep lying at the gravel level, lying dormant without sunlight for perhaps thousands of years.

Vascular plants normally decay into organic matter as they freeze into permafrost. Looking at the ancient moss, however, Dr. Convey and his colleagues wondered if, after centuries of frozen darkness, it could grow again. For in the past scientists had not managed to revive moss that had been frozen for more than 20 years.

But this moss came to life. Based on radiocarbon tests, as Convoy reported in the journal *Current Biology*, the revived moss turned out to be more than 1,500 years old. It had been kept alive, in a state of suspension, since the age of King Arthur's court.

Now Peter Convoy wonders if moss, and perhaps other species, can survive under the ice for thousands of years and revive when glaciers melt. "[T]hat gives you a very different way of understanding the biodiversity of a region," he says.[19]

"You could use whatever is stored in ice or sediment as a sort of backup for biodiversity," says Luisa Orsini, lecturer in biosystems and climate change at the University of Birmingham in England. But, she warns, "one has to be really, really careful introducing something from the past."

Danger signs are not unwarranted. There are micro-organisms that, once revived, could strike down other species in the war of genomes. Could these be echoes of H. G. Wells' sci-fi vision of the *World State* being superceded by conflicts among lethal species in his *World of the Brain*? Could human-creation of a powerful electronic, brain-driven android snuff out other species and you and me?

Wells' prophetic powers stirred imagination and terror—as demonstrated when Orson Wells adapted H. G. Wells' *The War of the Worlds* (1898) for radio in 1938, causing a national panic that continued until the public learned that the fictitious celestial invaders had been struck down by the world's own micro-organisms, from which they lacked immunity.

RENEWAL OF DORMANT, LETHAL VIRUSES

Faced with melting icecaps and thermofrost, deadly dormant viruses have already been released with unknown consequences. Their size, judged by the number of genome units of DNA or RNA, are known as nucleotides, composed of the four letters of the double helix.

These megaviruses were discovered in 2010 with over one million parts in genome size. In 2013 pandoraviruses were also found, which are amoeba viruses. These had genomes up to 2.5 million bases, reached that of parasitic Eukaryotes—one of the three basic sources of life on Earth.

Should we fear their release into the environment? So far they have been found in shallow water sediments in Chile and Australia only in unicellular organisms called amoebas, not in other animals.

The most ancient Eukaryote infecting a revived giant DNA virus was also found among a Siberian permafrost layer that corrresponded to the late Pleistocene sediments, older than 30,000 years ago. Researchers used the virus to innoculate a particular culture of amoebas called *Acanthambeba castellanii.* By using electronic microscopy, they observed particles of this giant prehistoric virus multiplying in the amoeba culture! Then they raised the alarm, as these dangerous pathoviruses revealed more similarities to present-day viruses that normally infect humans and animals.

Could there be more viral pathogens that might be released, as the average temperature of the surface of Arctic permafrost has already increased by 3 degrees centigrade during the last 100 years, faster than the global temperature?[20]

HUMANS HAVE CREATED THE FOUNDATIONS FOR THEIR OWN DEMISE

There is need to awaken and act quickly, as worldwide trophic cascades and ecocide are upon us. Species' wars for survival are both imploding and igniting in clear sight before us.

The planet's balance is being denuded of trees and grasses that offer insects, birds and other creatures home and sustenance. The mountains are being cracked open and strip-mined, leaving toxic reserves of detritus, run-off with snows, winds and rains. As streams and rivers run down to the sea, these residues and their lethal taconites are undercutting Gaia's natural breakdown and absorption of volcanic rock and carbon dioxide through rock-weathering.

The oceans meanwhile become the repository of industrial waste and lethal chemicals destroying phytoplankton, the foundation for the entire sun-powered feeding chain. And the atmosphere is being loaded with greenhouse gases absorbing the Sun's heat re-radiated from Earth—the warming skies melting glaciers and sea ice, raising ocean levels washing over coastlines worldwide, displacing species wild, domesticated and human.

"I often wonder what Man will do with the mountains?" John Muir wistfully penned of his favorite Yosemite resting place:

> Will he cut down all the trees to make ships and houses? If so, what will be the final upshot? Will human destructions, like those of Nature—fire, flood and avalanche—work out a higher good, a greater beauty? Will a better civilization come . . . ? What is the human part of the mountain's destiny"[21]

More than a century later, paleontologist Tim Flannery also reflects on the human dimension: "We have trod the face of the Moon, touched the nethermost pit of the sea, and can link minds instantaneously across vast distances. But for all that, it's not so much our technology, but what we believe, that will determine our fate."

Belief in the technological way is the diametric opposite of belief in the Gaian way. The first is a paradigm of technical expansion; the second an understanding that the natural flows of energy and species evolution continue as the intertwined forces creating the conditions for existence and life itself.

Understanding this difference in the context of the realities of our time illuminates the present as history here on Earth, an unmistakable pattern in an impending close brush with ecocide now in process.[22] We must stop from turning Earth into a torrid zone exceeding the maximum thermal limits of species life, melting glaciers, and deepening seas overrunning continents and islands.

Without taking this step, it is possible that Anthropocentric heat will devastate natural cycles, loosening a last, Great Wave that will one day engulf all landmass on Planet Earth.

Until every fossil fuel resource is left in the ground; and every global cement factory is forever shutdown; until every moving vehicle is run by hydrogen fuel and every electrical utility operated by wind, geo, solar and sea power; and until every emission from greenhouse gas and waste ceases—all the hopes and voluntary promises to lower post-industrial temperatures by 2 degrees Centigrade, made by nearly 200 nations at the climate agreement in Paris in December 2015, will be for naught.

Planetary fate now rests in the everyday hands and the foresight, not only of national governments, but also of billions of ordinary people, albeit with different and varied cultures and visions. In the face of the unrelenting political, economic and ecological burdens placed on all of us by a handful of fossil fuel companies and their owners—accumulating vast sums of wealth and increasingly threatening our very lives—we must take charge of the means of fossil fuel destruction.

Otherwise, we face the annihilation of the world's biological community of plants, animals and fungi living in a sustainable state of being. That means a disclimax will consume us, too.

NOTES

1. Prud'homme, "California's Next Nightmare," *The New York Times Magazine*, July 3, 2011, 35.

2. Compare Dan Joling/Associated Press, "Alaska Battered by Storm," *The Detroit News*, November 10, 2011 http://detnews.com/article/20111110 /NATION/ 111100398/Alaska-battered-by-storm#ixzz1dQnVYJZk; with Brian Kahn March 18, 2014, editor in chief, *Climate Central*. For Kahn's following analysis, see http://www.climatecentral.org/gallery/graphics/keeling_curve-concerns-17187; http www.climatecentral.org; http:/www.climatecentral.org/news/floods-may-cost-costal-cities-60-billion-annually-by-2010-16356; http:/www.climatecentral.org/news/sea-level-rise-threatens-worlds-cultural-treasuries-17144; http://www.huffingtonpost.com/2014/03/18/400-ppm-co2-levels_n_4985580. html?ir=Green&utm_campaign=031814&utm_medium=email&utm_source=Alert-green&utm_content=Title. http://www.huffingtonpost.com/2014/03/18/400-ppm-co2-levels._n_4985580. html?ir=Green&utm_campaign=031814&utm_medium=email&utm_source=Alert-green&utm_content=Title.

3. Above six paragraphs are based on Jeff Masters, "The Jet Steam is Getting Weird," *Scientific American.com*, Vol. 311 No.6 (December 2014): 69-75. See also Rutgers Marine and Coastal Sciences, Jennifer Francis and the University of Wisconsin-Madison's Stephen Vavrus, "Evidence linking Arctic amplification to extreme weather in mid-latitudes." See *GeophysicaL Research Letters*, Vol. 39, L06801, doi:10.1029/2012GL051000, 2012, 17 March 2012. And see Carl Nolte, "S.F. Bay Area forecast: long-overdue rain, snow," *SF Chronicle*, January 11, 2012

4. Krooth, *Gaia, passim*.

5. Krooth, *Gaia, passim*. See also Flannery, *Weathermakers, passim*.

6. Krooth, *Gaia, passim*. See also Flannery, *Weathermakers, passim*

7. See Chapters 13 and 14. Also see Richard Krooth, Morris Edelson, and Hiroshi Fukurai, *NuclearTsunami: The Japanese Government and America's Role in the Fukushima* Disaster (Lanham, Maryland: Lexington Books, 2014).

8. Rachel Carson, *Silent Spring, 1962, Chapter1*, reprinted as Rachel Carson, "Fable for Tomorrow," Ann Krooth, ed., *A Unique World of Ideas* (Boston/New York: Bedford/ St.Martin's Custom Publishing, 2013), 127-128. See also Peter Tompkins and Christopher Bird, *The Secret Life of Plants, passim*. And see Richard Krooth, *Gaia, passim*.

9. U.S. Geological Survey, hypothetical conditions for a Superstorm based on sediment studies and metiorlogical modeling, 2011. For analogies, see also Edith Hamilton, *Mythology: Timeless Tales of Gods and Heroes* (Boston, Mass.: Mentor Book/Little, Brown & Co., 1940, 1942), 24-25, *et passim*.

10. See Krooth, *Gaia*, chapter 3. See also Richard Krooth, "Into the Florida Swamps," *Notebooks*, 2001-04. And see John Farrand, Jr., *How to Identify Birds* (New York: McGraw-Hill Book Company 1988), "Fresh Water," 36-37.

11. Richard Krooth, interviews with Lester A. Radke, adventurer parvenue, "Travels in Russia, the U.S. and Pacific Isles, 1971-2012. See also Richard Krooth, "World Travels, 1962-2014," *Note Books*, following Oxford Cartographers, Ltd., Oxford, UK, "New World Map"; American Map Company, Inc, "Europe," Map No. 9564; Rand McNally, "World Map," 2014.

12. Pernetta, *Guide to the Oceans*, 24. See also summary by Brian Kahn, editor in chief, *Climate Central*, March 18, 2014. For Kahn's analysis, see also http://www.huffingtonpost.com/2014/03/18/400-ppm-co2-levels_n_4985580.html?ir=Green& utm_campaign=031814&utm_ medium=email&utm_source=Alert-green&utm_content=Title. And see McGrath, BBC, "CO_2 levels," October 24, 2016.

13. See Pernatta, 56-57.

14. Richard Krooth, Hiroshi Fukurai, "BP: Case Study," Unpublished Manuscript, 2012. See also the report "On Shaky Ground: Fracking, Acidizing, and Increased Earthquake Risk in California," http://www.shakyground.org/ —concluded that if the state rapidly expands fracking and acidizing in order to extract oil and natural gas from the massive Monterey shale deposit in Central and Southern California, the results could be catastrophic. California, after all, is already one of the most seismically active regions in the world. "And the San Andreas Fault runs right down the middle of the Monterey Shale," said Shaye Wolf, who co-authored the study and is the climate science director for the Center for Biological Diversity, which produced the report, along with the environmental groups Earthworks and Clean Water Action.

15. Pernetta, *Guide to Oceans*, 25.

16. Prud'homme, "California's Next," 34-35.

17. See Tim Flannery, *Here on Earth: A Natural History of the Plant* (New York: Atlantic Monthly Press, 2010), *passim*.

18. See Chapter 1.

19. Esme Roads, Royce E. Longtton and Peter Convey "Millennial timescale regeneration in a moss from Antarctica," *Current Bio*logy, Volume 24, Issue 6, R222-R223, 17 March 2014. Of King Arthur's reign, see *Bulfinch's Mythology; Legends of Charlemange* (New York: The Modern Library, n.d.), 773, *et passim*.

20. Elana E Giorgie. "Icecaps Melt. Prehistioic Virus Escapes." Based on N. Phliippe, M. Legendre, M. Doutre, et al. "Pandoraviruses: Amoeba Viruses with Genomes Up to 2.5 Mb Reaching that of Parasitic Eukaryotes," *Scien*ce, 341 (6143), 281-286 DOI: 10.112/science. 1239181.

21. Donald Worster, ed., John Muir, *A Passion for Nature* (Edinburgh, Scotland, 2011), named a Best Book of 2008, *Washington Post Book World.*

22. See Caldeira, *Scientific American*, 78-83. These concepts are elaborated by Wendell Berry and Bill McKibben.

Chapter Thirteen

Remains of the Natural World

Vanishing Landscapes, Disappearing Species

> The economic system centered on the god of money needs to plunder nature to sustain the frenetic rhythms of consumption that is inherent in it.
>
> The system continues unchanged, since what dominates are the dynamics of an economy and a finance that are lacking in ethics. It is no longer man who commands, but money. Cash commands.—Pope Francis, October 2014[1]

> The eight-mile island I call my home may soon be underwater. My current home may not exist 10 years from now.—Millie Hawley, Tribal leader. Kivalina, Alaska, September 2, 2015[2]

This chapter details the underlying scope of international financial and economic organizations, which are often blind to the ecological consequences of their directors' goal to accumulate wealth above all else. They seem blind to the need to secure Earth's landscapes and sustain its species, on which rests the destiny of global life.

Searching for the causal forces for disappearing islands and continental fringes with diverse and vibrant species, we focus on the economic and material spheres that regulate the modern cycles of the mobilization of industrial technologies, fossil fuels and land use that directly impact the biosphere—interrupting Darwinian evolution and modification of species and possibly bringing on a future last wave covering large regions of the planet. How this is taking place and continues is shown through a factual account sifting through the forces shaping the remains of the natural world.

To investigate the causes, we return to earlier historical periods of aggrandizing wealth, leading to the emergence of financial organizations that

Figure 13.1. Waiting for Lunch.

today dominate economic affairs, create powerful political alliances, and foster environmental peril.

TRANSITION TO THE ANTHROPOCENE

During the transition to the Anthropocene, the centralized accumulation of vast sums of wealth were drawn from exploiting Nature and working populations, impoverishing them, destroying species diversity, tearing the skin of Earth for resources, poisoning land, atmosphere and seas.

Today Earth is being transformed by extracting and burning fossil fuels, creating torrid gases that are aerated and spread to the coolest regions of the planet, melting glaciers, sea ice and permafrost; raising sea levels overrunning islands and continental coastlines. The drive to accumulate wealth by a few comes at the expense of species diversity and impoverishment of billions of the planet's population.

As we will see, ecocide is upon us. Either we escape its grasp, or the five horsemen of the Apocalypse will plunder Earth, landscapes will continue to disappear, and species will diminish before our very eyes.

1. EMPIRE, BIOSPHERE AND SPECIES DIVERSITY

Species diversity preceded the emergence of humankind. As species gradually unfolded into a vast variety of living forms, the tree of life filled out its branches. Yet human evolution ultimately made Earth's biosphere, atmosphere, sub-terrains and seas vulnerable. Over the course of 23 centuries, since Aristotle invented the methods of science, our species has transformed the biosphere by what seemingly were benign forces: the emergence of technologies that transformed human labor-exertion into useful materials; the processing of natural resources and use of other species for human goals and consumption; and the pursuit of the accumulation and distribution of wealth among ourselves in increasingly skewed and unequal ways.

In the path of our multiplying technologies, *other species* have proven to be comparatively fragile, evolving by seed and egg; placental reproduction; networks of fungi and mycelia; as well as reproducing by sunlight and by chemosynthesis. As large parts of the natural world have been destroyed over time by environmental calamities and human travail, the remains of its biodiversity largely have been handed down to the present as truncated forms that often have lost their essential natural conditions for survival as well as the ability to reproduce as evolving, symbiotic communities.[3]

Climate and Species are Forever Interlocked

In today's relationship between human populations and their impact on climate, the central forces are pollution, inestimable mountains of waste, and the accumulation and discharge of heat from burning fossil fuel spreading from Earth's warmest and hotness regions to the lowest, as the second law of thermodynamics portends. The underlying question remains—are species exposed to these forces coming to an end in number and diversity?

Empires Past, Blinded by Wealth Overwhelmed Nature

The present moment has been built on and followed centuries of natural changes and human plunder, most often pursuing the conquest of territories and wealth.

So powerfully has humankind exploited and enslaved one another, as well as undermined natural conditions, that this wreckage led to massive species destruction.

The ancient empires were many: Mesopotamia centered in Uruk; Turkey, Syria, and Indus; Aztec, Olmec, and Maya; Grecian, Roman, and Egyptian. All ruined terrain, captured millions, and destroyed and domesticated wild species.

The drive to control and to accumulate wealth by conquest or through various institutions encompassed slave and feudal, territorial regencies and state helotries, oppressing populations though centralized dictatorships—all abusing the biosphere and extinguishing life's flame.

Conquests and Reversals

The early Middle Ages brought down earlier empires of Alexander the Great, Visigoths, and Vandals. After eleven centuries of Roman rule, dominating forces were led by Angles, Franco-Barons and, in 1066, William the Conqueror.

Each of their dictatorships bled and robbed the common people. According to Odericius Vitalis, circa AD 1130, William on his deathbed confessed that he took revenge on the entire English population, reduced them to famine, murdered thousands young and old, and commanded the Northern shires, taking their houses, corn, tools, and chattels, including great herds of cattle and other beasts.

For the next three centuries. Europe was then overwhelmed by mass dying from outbreaks of the Black Plague, as well as penury and starvation. At least one-third of the population died, and another quarter weakened, unable to farm. In their place, forests grew and wild life multiplied.

Then Europe was reborn in health, spirit, and material welfare.

Western Europe's reduced population had momentarily led to increasing urban wages and competition for limited supplies of food and goods, impacting and raising the prices of necessities and workshop wares, in turn limiting domestic and regional markets.

From the late 15th century forward, there was a developing link between Europe's working population, food production, the rise of the modern state, and overseas adventures focused on the quest for precious metals, spices, and food-producing plantations using slave labor.

These ties were shaped by the environment, especially during the Little Ice Age that off and on lasted from the 12th to 15th centuries. As sharply reduced agricultural output caused a corresponding decline in the decay of plant matter; and as carbon dioxide and methane emissions fell to a level they had not reached during the Ice Age approximately 190,000 years earlier, atmospheric temperatures plummeted, and polar glaciers stabilized or enlarged.[4]

The colder the climate, the fewer traditional European grains, vegetables and fruits could be successfully grown or cattle properly fed or husbanded. So, too, as fewer were successful harvests, the more people went hungry, became malnourished, susceptible to diseases, and lived shorter lives. The faster populations fell, the quicker did economic and social life cycle downwards.

Then came reversal. Europe's economically devastated areas, along with the expansion of trade and agriculture to colonies, seemed obvious ways to sidestep the overlapping crises of falling European food supplies and costly workshop production.

To accomplish this, though, required state centralization and its strengthening.

Five Centuries of European Domination and Plunder

During the ensuing 15th to 19th centuries, English and Continental feudalism fused with government power. European material advance, political influence and military power turned to the seven seas, capturing ever-wider territories, diverse populations, and their resources and species.

The center of Continental economic, political and military dominance meanwhile shifted from Venice in the 15th century to Spain in the 16th, Holland in the 17th, France in the 18th, and Britain in the 19th. Beginning with the merchant financiers of Venice in 1450, wealth was accumulated by global trade, buying goods at a low price in one geography to sell at a higher price in another—then reversing the process, maximizing gain. A century later the Iberian Portuguese and Spanish dominated the bounteous seas, setting sights on gold and silver, accumulating wealth for their crowned heads, leaving a trail of human destruction in the Caribbean and Americas. In fact, overall during the period between the discovery of America and the acceleration of England's industrial revolution about 1760, some $1,859 million in gold and $3,594 million in silver were produced in the world.

Within these 270 years, the periodic extraction of bullion rose steadily by 8 1//2 times, with the major sea powers competing to accumulate wealth to pay past debts to the Italian Lombards and Fuggers for imported foreign wares and to lay the base for the new capitalist industries.

Though England sought such overseas metallic wealth, most of this silver and gold was taken from the Americas by Spain and Portugal—not by England.

Fighting for territories by sea routes, by 1565 Holland displaced Iberian trade in spices and plant species, as well as plundered remote islands and their resources. And deploying her seadogs, the English Main sought to destroy Holland's fleet, as well as those of the other sea powers.

Pax Britannica, control by land and sea, proved not a peaceful affair. Transporting gunsmiths and cannon, cavalries and horses, British warships tacking into the wind brought vainglorious victories and sometimes devastating defeats. From 1558 to 1815, Britain endured 103 years of war to vanquish Spain and send the Spanish Main to the bottom of the English Channel; 33 years of war against Holland to plunder its colonies, destroy and take as prizes its 34,850 commercial fleet; and 83 years of war against France to establish hegemony in Europe.[5]

The Diadem Passes

In the late 18th century, Napoleon Bonaparte's citizen armies tried to conquer by arms and to dominate by trade a continental European sphere, as well as to enlarge its colonial holdings, infringing on northern Africa. And yet at Waterloo in 1815, France was soundly defeated by Britain's standing army and superior weaponry.

As all the Great Powers of the era had partitioned the globe, securing material wealth for themselves by exploiting the resources and labors of colonial populations, their struggle for territories destroyed inestimable species.

In Europe and beyond during the next century, Britain employed a twofold policy: to keep the dominating Continental nation too weak to unify the European land mass, but not so weak as to succumb to a new dominating nation.

Pax Britannica thereby kept the continental powers balanced and quarreling so none could unite the Continent and put itself in a position to try for naval, technological, commercial or colonial supremacy. And while Britain controlled affairs on the European Continent, she also created a vast colonial empire across the frontiers of Africa, the New World and Asia.

By 1939, Britain's empire itself covered 103 colonies of 42,818,000 square kilometers with 590,793,000 subjects; and a small home population of 46.487,000 living on an island covering 230,000 square kilometers—meaning about 8 percent of the English people theoretically ruled over the other 92 percent of the people in its empire; and the frail 5 percent of Great Britain's land mass dominated the other 95 percent in its total home and colonial sphere.[6]

Size was not everything, though. For all these aspiring empires seemed carelessly and senselessly blind to the destruction they rent: denuded forests; extirpated great, small and microscopic plants; destruction of animals and enzootic species; displacement of the wilds with intruding plantations, mines, and rail beds; ports operated by captives and subservient workers; export loading docks intruding on once-pristine beach-beads; great ocean fleets disrupting harbor sea life, waiting to ship foods and raw materials to Mother Countries, themselves already struck bare of Nature's wonders.

Ecocidal Myopia

Though the causal forces behind imperial expansion were many, all the Great Powers suffered in common from ecocidal myopia: tearing down forests, destroying mycelia networks of roots and microscopic animals; abusing planted soil; failing to rotate crops; neglecting to idle fields every few seasons; also neglecting to use of animal fertilizers; destroying wild creatures large and small; digging and mining for precious resources; leaving tailings and effluents to run off with the elements into steams and groundwater, or to effuse into the air.

At home, the Great Powers had neglected their domestic agriculture, leading to shortfalls in food supplies; rising internal conflict over distribution of food and wealth; and seething crises of too few resources to support local populations. Meanwhile they maintained the power position of elites among the discontented.

Vexing implosions and violent revolts recurred across Europe. Failing legitimacy of elite classes led to an escalation of their state spending for arms and armies, and imposition of lawless order on the discontented at home and abroad.

To replace diminishing domestic resources, crowned heads, grave court counselors and predisposed parliamentarians advocated foreign conquests, hoping to pacify home populations with bread, parsimonious amenities, and appeals to religious and nationalist patriotism.

Yet intensified exploitation of domestic and foreign populations foretold tragedies: more terrain stripped of forests; soil fertility eviscerated and declining; unadulterated food production putrefied and plummeting; resources and raw materials in short supply; skills and crafts undermined; and attempts at foreign conquest meeting resistance and sometimes defeat.

Imperial ruin then set up new equations of power between social classes, producing stresses that might only be lessened by conscious measures to secure populations and the remains of the natural order. Yet most empires rose and fell without such resolution, leaving behind weakened, dispersed peoples and environmental degradation.

2. CRISES OF THE MODERN ERA: TRANSNATIONAL FINANCE, UNLIMITED INFORMATION, AND CONQUEST

Past empires had sought wealth for the most powerful dominating nations. This contrasts with today's nationalist quest deploying the strongest armed forces and transnational finance capital to access unrestricted information about and to conquer Earth's resources and labor supplies.

Thus far, an emerging empire of high finance accumulating vast sums of wealth has been aligned against global populations and especially against the impoverished.

The use and misuse of this wealth has violated the principles and ethical and moral beliefs of many hundreds of organizations, including those of existing international agencies of the United Nations, and Pope Francis' current efforts to create a church for the desperate. This financial power has also spread and built a colossal production sphere—mobilizing and burning fossil fuels, leaving the remains of the biosphere and species diversity in tatters. Indeed, this financial sphere has evolved to dominate technologies that today are overwhelming the ecosphere.

To investigate the causes, we ask the reader's close attention as we go through earlier historical periods of aggrandizing wealth, leading to the emergence of financial organizations that today regulate most economic affairs, create political alliances and foster environmental peril.

High Finance and the Technological Sphere

The modern era has proven to be different than the past by degree and speed of transnational communications, strategic information, global financial organizations, and spreading crises.

All these were hyphenated by wars between states for territories and resources; by the maddened, revanchist quest of defeated nations and peoples to renew battles without end; and by massive population movements, everywhere creating hunger and sickness, lack of employment, and underpaid, common pools of workers.

It seemed every scrap of land as well as continental limits of the roiling sea were in contest between nations and emergent states, having been identified by the latest technologies of the day, along with the known location of thousands of resources and species once undetected—soon to be sequestered, trampled upon and abused.[7]

Since the parameters of the global economy and communications had been digitalized from the 1980's, the world's storage of knowledge was instantly available to almost anyone, anytime, anywhere.[8] And this treasure trove of information almost alone helped build the nationalist and corporate appetite for further discoveries and the extraction of the fruits of wealth—fossil fuels, gold, and silver; diamonds, titanium, other singular metals, and rare earths.

Under western laws, the integrity of protected private property and standards of territorial sea limits also were weakened, allowing the invasion of terrain and ocean boundaries to explore and take animals, plants and ocean creatures; seabed resources and, above all, petroleum.

Even ancient commons steadily were invaded by gun-toting marauders—places where timber, heavy metals, fossil fuels, and varieties of enzootic species were identified and taken at heavy risks.

Expanding Dominions

Almost no part of the planet was inviolable. Within nations, once protected parks, wildlife areas, and their drying and dying fringes came under attack. Often small, and too geographically isolated to support cloistered species diversities, they faced human occupation of the regions around and between them.

Plants and animal species in such isolated land reserves could survive only if their populations were able to expand into adjacent areas where temperatures, rainfall and food supplies once either existed or were changing favorably to their needs. But they were blocked by expanding urban centers, farmland and cattle ranges—as well as by forests under threat, oil wells drilled, and resources mined. These intrusions, that might otherwise have served as stepping-stones to larger protected areas, had completely replaced habitats between now isolated species.

The seeming solution to this dilemma was connecting the reserves to one another, creating wildlife corridors.

In Sumatra, for example, that would involve creating corridors for the orangutan population, now separated, with places to live without constant conflict with palm plantation owners who slaughter them. Corridors might also have handled storm water; prevented erosion; and literally saved a species close to us on the Tree of Life.[9]

Though such corridor plans in the present age of gathering wealth might seem unlikely, still this has emerged as a last hope of many biologists and conservationists, who are mapping where such connective passages most are needed and might be possible.

Unlimited Information and Exploitation of Nature's Resources

Given present technological GPS location devices, almost nowhere on Earth are wild species, raw materials and terrain completely isolated, inviolate, and safe from invasion and destruction by well-armed raiders and powerful nations.

The U.S. CIA has identified almost all resources on the planet. The National Security Agency (NSA), officially a public agency, has surveilled millions of people, their messages, and the world's resources and species, aided by partnerships with private sector corporations, with many NSA core functions outsourced. No one and no world resources are immune from oversight.

NSA itself roughly employs 30,000 people, but the agency also has contracts mobilizing some 60,000 employees of private corporations, who also provide essential services. Some 70% of the U.S. national intelligence budget is spent on the private sector alone.

At its headquarters, the NSA-contracted corporate partnerships carry out their surveillance and intelligence work for the agency. Current NSA targets include in its planning document—entitled "Identifying Challenges: Geopolitical Trends for 2014-2019"—lists the political leaders of Brazil and Mexico, countries in which oil resources are central assets for U.S. drilling and extraction. The NSA also catalogues citizen activities and resources in Egypt, India, Iran, Saudi Arabia, Somalia, Turkey and Yemen—now beset by genocidal water wars.

In fact, in 2006 NSA embarked on a program for "Large Scale Expansion of NSA Metadata Sharing," predicting that its metadata collection would grow by six hundred billion records every year, growth that would include one-to-two billion telephone call events every single day. Adding internet-based communications by 2014 the total number of events already stored was close to 1 trillion shared by NSA with other agencies. And March 7, 2017, Wikileaks documents also corroborated that the CIA had become the preeminent hacking operation using telephones and televisions to spy on people worldwide; and moreover nearly all CIA cyberweapons had been stolen and potentially in the hands of criminals and foreign spies.[10]

These collections can be used to expropriate private property, as well as violate personal rights and the right to privacy. Though today the 'cloud' supposedly stores so-called private intelligence, government surveillance agencies have undertaken every imaginable effort to crack the cloud and all types of encrypted internet communications—many forms of encryption still wired to avoid U.S. NSA detection.[11]

In their quest for privacy, millions of people, companies and agencies simply have not conceded that NSA should be in charge of controlling their communications and the safety of their private matters. Governments seem to have forgotten—for every attempted lock-down on the public, there is a key to escape surveillance.

Public records are different than personal records. Plans of the Federal Bureau of Land Management (BLM) are detailed; and digitalized NASA maps at the Library of Congress, can be viewed on your own computer. They also make up a codex for expropriation of what were once common territories, today empowering the U.S. Congress to identify resources such as coal in Colorado, and to potentially pass legislation auctioning-off and privatizing portions of the national, public domain.

The BLM auctions, some illegal, have been in process for decades. And the Congress in December 2014, literally gave away to a private corporation 70,000 acres of the Tongass National Forest. It also stripped away wilderness study-area-protection from thousands of acres of land in Montana; and handed over sacred Apache homelands in Arizona to a foreign mining corporation. As well, Congress exempted livestock grazing from complying with environmental laws on over 250 million acres of public land across the West; and allowed unlimited helicopter use in two wilderness areas in Nevada.

Anticipate that these invasions of Nature's frontiers bring added future oil drilling, fracking, coalmining and stock feeding on the open range. Imagine these are encompassed by the spreading system of world commercial relationships dependent on fossil fuels—entangled in ideological chains of denials of global warming and the decline of biodiversity.

Like others before them, these invasions of Nature's domain threaten to extend ecocide and dangers to workers in the mines—the old canaries once warning of deadly buildup of fire-damp methane replaced by guards at property gates alerting managers, calling foremen underground with code words like "it's a cloudy day," to warn that government safety mine inspectors had appeared unexpectedly, wrote reporter David Segal of *The New York Times*. Or inspectors simply lie.

There seemingly is no place to hide from the state's intrusive alliances with investors and fossil fuel operators set on turning natural environments into warrens for capitalizing assets through capture, privatization, and mobilization of labor, land, resources, seas and species.

Commercial and Military Competition in the Arctic Environment, Prophesying Nature's Destruction

Historically, U S. explorations have sought new lands and resources in the Caribbean, Latin America, throughout the Pacific, the Philippines, Hawaii, and as far north as the Arctic. Here in the early 20th century, U.S. Commander Parry described the potential resources which the United States might take over: "It would not at all be strange if in a region of the most striking contrasts . . . some materials were found which would make possible some yet unknown necessity of our future civilization"—a prophesy come true in the 1970's search for black gold.[12]

And impacting the top of the planet, the White House elaborated in 2015:

> Arctic ecosystems are among the most pristine and understudied in the world, meaning increased commercial activity comes with significant risk to the environment.
>
> The growth of human activity in the region will require stewardship to maintain the open seas necessary for land and resources use, . . . for global commerce and scientific research, for search-and-rescue activities, and provide for regional peace and stability.

But such Alaskan global commons do not exist. Nations claiming them as their own territory have moved to take charge, with Russia economically and militarily equipped with 41 ice breakers and 11 more planned, 10 bases and building four more; the U.S. with barely two fully-functional ice breakers, one medium and one heavy, nearly 40 years old; China with two equipped on the way; and a half dozen other nations with claim stakes over Alaskan waters and at least 40 percent of Earth's terrain, gas and other natural resources.[13]

Alaska as a location for intense battle over land, waters, species, and resources is new, but the struggle is hardly so.

And President Donald Trump proposed 2017 budget asking for $1.3 billion in funding cuts to the U S. Coast Guard would reduce monies for interdicting drugs, human trafficking and keeping a close eye on what Russia is doing in the Arctic. Ideally, says the head of the U S. Coast Guard Admiral Paul Zukunft, the marine service would need a fleet of 3 heavy and 3 medium ice breakers to remain competitive.

Underlying Present Dangers

Elsewhere the dangers from human missteps and ill-formed plans have jeopardized life as well.

Human toxic environments undermine conditions for the transmission of life sources by killing seeds, eggs, and fungal and mycelia networks that connect almost all plant and even small and microscopic animals on Earth.

Plants can harness underground fungal networks that cover great distances. And insects and worms searching for roots to consume can pick up network signals, like compounds produced by some plants to attract friendly bacteria and fungi to their root systems.

"These fungal networks make communications between plants, including those of different species, faster and more effective," says chemical ecologist Kathryn Morris of Xavier University in Cincinnati, Ohio. "We don't think about it because we can usually only see what is above ground. But most of the plants you and I see are connected below ground, not directly through their roots but via their mycelial connections."[14] Today these connective strands of land life are being severed by advances in technologies that are piercing territorial cover and burning fossil fuels.[15]

Bringing on Climate Crises

Earth encapsulates terrain, seas, species and their surrounding atmosphere that together guide species evolution and methods of survival; as well as human social structures; their nuances of community and state relationships; their means of exchange, of culture, of hatreds and compassions, as well as all artifacts and other evidence left for posterity to ponder.

Whenever humans attempt to reverse the process—to encapsulate Nature by dominating its species and means of survival, the common lands and seas of life—the framework for a disclimax sets in and leads to ecocide. Nowhere is this clearer than in current, hyphenated onset of the use of money, finance, and finance capital infringing on the biosphere. Here we explain how the different phases of capital evolved and impacted populations and the environment.

Thumbnail Sketch: Finance Capital Undermining Popular Rights and the Natural Domain

Today the overwhelming power of finance capital has undermined the natural domain and swept away common lands once held firmly by people throughout the world. On a global economic and political scale it has taken the form of neo-colonialism; and on a national, social scale unfolds as dominating lawless order.

Over at least the last several centuries, four stages of powerful developments have emerged as means of exchange—cumulatively impacting the world economy, as well as the distribution of social wealth among populations and denigration of the environment.

First count our species' accumulation of material values and means of exchange as wealth; then its centralization in the hands of social elites and institutions; next its deployment as financial means both within nation-states and as export as financial capital by banks and transnational corporations.

As the fourth stage, most recently, this financial capital has been linked to international agreements between the privileged from various nation states, functioning with their own rule-setting bodies, designed as an oligarchy to accumulate wealth in various forms from working and consuming populations worldwide.

This may sound extreme; so please allow a moment to understand why so many oppose the modern deployment of financial capital, including many of those in government, in the U.S. Congress, in other important institutions; in labor and consumer unions; as well as by the unorganized workforce, environmental organizations, Pope Francis, and the great majority of people worldwide.

Links

What is the precise link between the operation of finance capital and the environment? In itself finance capital is not simply the hyphenated power of money, but embodies a social relationship between ownership groups over the means of production and finance on one side and, on the other, common people, the disinherited and working populations—with the owners accumulating the value of unremunerated human labor-energy.

When accumulation of this surplus is centralized by a financial group or a corporate institution, its legally-endowed position is often enhanced and multiplied as an investor in, or lender to, other institutions.

A bank, for example, generally has a small equity ownership in its own organization—often a bare two-to-five percent of the entire capital it controls—drawing together the funds of thousands of small depositors and customers. This centralized capital is then often loaned or invested for profit in commercial, services or industrial enterprises. And these enterprises, in turn, employ a labor force paid varying wages or salaries equal to their socially-defined value—that is, the sum that classical economists designated as subsistence at various local, regional or national standards, or enough for survival and comportment, without social disruption of existing law and order.

A similar situation prevailed during the past, as in our time in many nations in which finance capital plays a central role.

Just as bank finance capital embeds itself in industrial or commercial enterprises, so too industrial companies deploying accumulated internal financial resources are able to take over the majority or controlling stock interests in banks. Both banks and industrial corporations then absorb still other cash-rich firms, adding to the power of their combined finances to take charge of say storage and delivery systems for raw materials or varieties of manufactured products—building marketplace control as monopolies or one among oligopolies.

Impact on Labor and Global Environments

Modern-day economists have ignored classical accepted wisdom: The nations able to set the fixed ratios between the currencies of different nation-states position themselves to determine the value of the exchange of their own manufactures—made by workers using superior technologies, embodying less labor-exertion—for more labor-exertion in subservient nations' goods made with inferior tools—becoming the foundation for dis-accumulation and impoverishment of subservient nations and their workers, and the accumulation and enhancement of wealth by the dominant nations and their various classes.

These relations of power are established by financial institutions that then impact global environments. Collectively they make up an apparatus for wealth accumulation from, and denigration of, subservient nations' populations and environments. And this balance of financial power plays out in the relationship between nations and Nature.

Today the two major economies and environmental polluters, the U.S. and China, have crossed swords over becoming the leading force in channeling finance capital worldwide, ensuring their abilities and power to profit without regard for the global ecosphere.

Still on top regulating the global economy is the U.S. Federal Reserve bank. According to David Rubenstein, founder and managing director of the Carlyle Group, one of the world's largest private equity holding firms, controlling $193 billion in assets:

> The Federal Reserve is really the central bank of the world. If the Fed raises interest rates a little bit, it will have an impact all over the world, particularly

> in emerging markets. [T]he truth is when interest rates go up, [production costs rise and competition pushes] prices . . . down in public markets. Therefore prices are less expensive and it is easier to buy things.[16]

AT the core of their collective schemes for stock and other prices are international agreements with other countries creating exclusive common market zones, awarding one another most-favored-nation status for trade and investments in channeling finance, capital machinery, service-labor, and manufactured goods between their partners—excluding or subserviating all other nations that refuse to comply with their pre-agreed terms and regimen at designated, fixed-currency-ratios for accumulating wealth.

In fact, under the control of capital exporting nations-state, the worldwide division of productive labor has been vertically partitioned into raw material producers, semi-manufacturers of raw materials, and advanced manufacturers of segments of finished products that are finally assembled in still other countries offering low tariffs, minimum taxes on output, secure export zones, and armies of subsistence workers often deprived of rights to organize and protest. It's a scheme driving wages towards subsistence—the so-called iron law of wages.

Geographic, horizontal trade zones like the European Union, NAFTA, the Atlantic Alliance and the prospective Pacific Alliance have been used or are being designed to promote the free, untaxed movement of finance capital, commodities and some classes of skilled labor between so-called First World nations—purposefully excluding economically weaker countries in order to deploy and integrate their subsistence working populations into the vertical production stages of extracting maximum labor toil.

The wage scales paid immediately affect the workers' potential condition—their purchasing power, their family condition and, with their relatively limited resources, the ways they treat their surrounding environments and other species.

Zones of Demarcation; Lead Forces in Exploitation; Interpenetration of Financial Capital

Until 2015, the United States was the dominant power broker in setting up finance capital zones. But as the dollar strengthened, its fatal financial weakness appeared: other nations could not acquire expensive U.S. goods; and there was no safe haven for profitable American investments, especially in the 10-year U.S. Treasury Bond market.[17] Now there was a loss of remaining security for capital invested in almost all U.S. Treasury bonds, or even state bonds, or traditional blue-chip stocks; and with no exchangeability of U.S. dollars for gold, the western standard for securing wealth was failing—escaping to Asia.

Thereafter China tried to take the lead, attempting to replace the U.S.-controlled World Bank and International Monetary Fund by setting up the so-called Infrastructure Investment Bank—an allegedly secure multilateral development lender with over 40 founding members. China was its lead partner; others like Australia, Egypt, and Ukraine were aspiring industrial nations. Members included the old standard bearers of autarchies: India and South Korea.

The interpenetration of financial capital in this new zone of demarcation could not be stopped. For the underwriters of this new global economic system were determined to replace the structure of the U.S.-led Bretton Woods combination established after World War II, as well as the World Bank and the IMF as the momentary unaltered, world lenders, who were still squeezing desperate borrowing nations, forcing them to pay their debts with interest charges by cutting their forests, mining ores, drilling petroleum, and extracting other resources. The old invasions of the biosphere might end, only to be replaced by the new.

New Invasions of the Natural Domain

In early 2017 U.S. President Trump ensured a new invasion of the natural domain was on his legislative agenda. He threatened to bring American investment capital home; wall up its geographic borders; stop immigration of both unskilled and skilled workers; use America's own domestic resources of oil, coal, lead, copper and mercury for production; erect tariffs to block out foreign-produced materials and manufactured goods; and reorient its trade to bilateral relations with other nations. It was a move towards U.S. autocracy. "A company that wants to fire all of its people in the United States, and build some factory someplace else, and then thinks that [its] product is going to just flow across the border into the United States—that's not going to happen," he warned. This was the sequel to Congressional Republican lawmakers laying the foundation to give away a large portion of America's 640 million acres of national territory to private corporations to make America self-sufficient in production resources.

China as Master Manipulator; U.S. on the Defensive

China meanwhile followed a different plan, extending its global footprint that would put its currency in reserve status comparable with the dollar, the euro, the pound and the yen—meaning borrowers, debtors, and importers of Chinese goods and machines would be forced to remit in Chinese currency sums due on their negative balance of payments.

The U.S.-controlled International Monetary Fund (IMF) also had planned to defend its sphere by creating a new artificial currency zone in October

2015 to control currency markets for $28 billion invested via the New York stock exchange, $191 billion in world equities markets, and $5.3 trillion in fixed asset markets.

But, alas, a decisive end had come to the prevailing system established by the 1971 Nixon Mandate ending the convertibility of the U.S. fiat dollar into Fort Knox gold for countries that sought to exchange their American dollar holdings.[18] And China was buying vast quantities of gold to back up its paper currency for global investments and lending. Here lay the new beachhead of nationalist and transnational finance capitals at war.

Strategic Paths for Carving Up the World

Competitive nations had laid out different strategic paths for carving up the world's territories, resources and the human expenditure of labor-energy producing surplus value. The main contestants during the years 2015-17 and beyond were the United States and China. But after a disastrous attempt to lead the Transnational Pacific trade group, the U.S. chose "America First" as a form of isolationism under President Donald Trump's Executive Order that withdrew the U.S. and left the field to China.

Autarchic self-sufficiency—politically forcing U.S. manufacturing and technological behemoths to produce at home, rather than export their operations abroad seeking cheaper labor, resources and nearness to markets—might enlarge American domestic employment, but in the long run would face workers being displaced by automated robots.

As well, U.S. bilateral trade agreements with vulnerable nations well might create tight-knit neo-colonial markets; yet the U.S. faced China, South Korea and Vietnam creating their own Asian trade groups with ever-cheaper labor and resources using copy-cat U.S. manufacturing techniques and robots.

Contradictions of Robotic Power Lead to Domestic Crisis

All the major global competitors faced the contradiction of introducing robots to enlarge production, yet momentarily displacing human laborers, reducing their income so they could not purchase their added output of commodities and services. Competitor nations introducing robots eventually would have to export to avoid an internal crisis of overproduction and under consumption.

To understand their predicament they had to know that equating mind-likeness of robots to human intelligence, or to the proprioception of other living forms, are inaccurate ways to explain robotic power as material entities. This is because human-intelligence is biologically inherited; while "skill" once built as steps into a machine is not the equivalent of human thinking.

Using this framework, the great British inventor Alan Turing (who broke the World War II German-encoded Enigma Machine and saved the nation from German air attacks) said the interaction between human agents and their environments is conducted through language, back and forth about anything; or through a computer program and its hardware conversing with another human, guided by human speech or written meaning.

Adapting this logic, Elon Musk, the CEO of Telsa and SpaceX, foresees a closer merger of biological intelligence and digital intelligence based on the bandwidth speed connecting the human brain and the digital version of the human brain concerning output and all else. Human intelligence is limited to communicating by the dimension of one bit of information; or by typing with fingers or by voice vibrations via a mobile device limited to about ten bits per second; or using computers capable of a trillion bits per second. Thereby a mega-band machine interfaced to the human brain will one day create a symbiosis between human and an artificial machine intelligence, creating a new layer of a brain power able to access information quickly and tap into artificial general intelligence, making this combined entity instantly smarter than the smartest human on Earth.

The future thereby lays open to aggressive competitive methods by mobilizing general artificial intelligence machines shaping production and delivery systems, displacing human labor and controlling the environment . . . for better or worse.

China today is meanwhile centralizing wealth to build robots for use in production and delivery of goods to its own domestic markets that are still limited. The robots are able to multiply output displacing Chinese workers in the production process. But as well it limits the population's domestic buying power; so the reimbursement or realization of the costs embodied in what is produced now largely comes from exporting to overseas markets at the expense of its own population. Displaced workers without money to buy face an increase of commodities that must be exported to consumers elsewhere. And when and if overseas markets are limited, the crisis will come home—forcing China to raise capital to cover domestic human deprivations and their personal rising maintenance costs. At this juncture, the consequences remain unscripted.

China's Dilemma Temporarily Resolved by Further Impoverishing Its People

In 2017, the Chinese state continued to add to its adaptations of robotic production and marketing, creating the largest stable of artificial intelligence machines in the world, followed by Japan and the distant lagging United States.

This was accomplished in part by the State's centralization of its capital resources—backing business corporate managers' expropriation from unschooled, naïve China's investors borrowing money to invest in high-priced stocks. The state itself sold part of its own financial stake in overvalued government-owned enterprises that had incurred huge debts.[19] As the fiasco unfolded in mid-2015, the central government urged people to buy stocks as a way to divert money from real estate in order to deflate a dangerous property-price bubble.

Then the scheme backfired. People initially had obeyed, sending the June 2015 Shanghai stock market soaring 150% since the previous summer. Many investors tried to sell to lock in profits, thinking stocks were reaching a peak. But the government could not stop the resulting plunge, the Shanghai market falling nearly 40%. As a cover-up, the government rounded up nearly 200 people, charging them with sedition for suggesting Chinese stocks might go down, as they did.[20] Some 250 billions were lost by investors who hung on to worthless stock certificates. Only the government and the very wealthy gained by buying their liquidated stocks—cheap.

For over a decade the State directed by the Communist Party had drawn the support of the wealthiest class to bolster its authority. In 2017, 209 entrepreneurs and business tycoons holding $500 billion were also lawmakers, representing about 20 percent of the nearly 3,000 delegates to Parliament. According to the Shanghai *Hurun Report* tracking the wealth of these delegates to China's March 2017 Parliament, the 100 richest had seen their net worth rise 64 percent in the four years 2012-2016 since XI Jinping became China's President.

The great public disillusionment and anger over the few getting wealthy then led to a new 2017 government strategy to draw investment capital from private Chinese conglomerates for building major infrastructural projects, creating the backbone for roads, airports, ports, railways and bridges linked to an export economy.

Shanghai already was China's entrepot to world trade and investment. U.S. Bloomberg investigators calculated that China spent more than $10.8 trillion on such infrastructure from 2006 to 2015, rising 17.4 percent in 2016. And the future outlays relied on still-unstable public-private partnerships.

One example reveals the risk. With a global portfolio in real estate, insurance and pharmaceuticals, as well as in Club Med and Cirque du Soleil, the Shanghai's Fusun Group became one of China's most active dealmakers, planning to consolidate the foreign and the domestic economy by taking a controlling stake in a proposed 270-kilomater $6.7 billon high-speed rail line linking the eastern cities of Hangzhou and Taizhou. Though China's 6.7 expansion of gross national product thereby would be exceeded by its 2016 rise of 17.4 percent in outlays for roads, airports, seaports and railways, China was in the process of consolidating its engine to deliver what its economy produced with delivery to foreign markets in a barrier-free world.

New Expropriation Zones Emerge

China sought to lead the world by trade and foreign investments, though over the long-term such tools to accumulate capital could not successfully be used by the communist state.

Just as socialist relations in communes or otherwise have collapsed when they turn to state capitalist methods, communist or socialist governments have been forced to reinstitute centralized state dictatorships over their economies to extract resources; to manufacture with a cheap labor force; to often sell inferior commodities cheaply on world markets; and to create fictitious stock exchanges to provide illusions of grandeur that are pre-designed for the expropriation of domestic investors' wealth.

By comparison, the struggle for employment and markets under the U. S. newly-elected Trump regime in 2017, included efforts to increase domestic production with American workers and moves to block many foreign, especially Chinese, exports into the U.S. Trump claimed the U.S. had a negative trade balance with every nation it dealt with. In fact, such balances were covered by U S. overseas earning on capital invested abroad, shipping, storage and insurance costs, as well as clandestine and laundered money operations.

Such U.S. operations also took place in a global framework in which European and other industrial nations followed a different economic script. Their competitive, nationalist financial institutions were positioned for world expansion, to both moderate employment at home and stimulate investments in debtor and subservient nations, creating momentary employment for low-paid workers abroad, as well as an extended overseas body of consumers for their nations' own products.

The future outcome seemed simple enough: industrial nations' imports from their own foreign investments in factories and services would become more valuable than their export of commodities. Like the United States, they would create their own negative trade balances, which would eventually be covered by repatriating profits and the costs of shipping, insurance and storage from operations in subservient nations.

Nothing new, this balance of finances was well known among Britain, France and the U.S. in the 1930s, and in recent times by U.S. expropriations from Mexico under the North America Free Trade Association.

This process of accumulation and dis-accumulation of wealth also has been enhanced during the modern era of finance capital, operating both within nations and through exclusive overseas trade and investment groups. In effect, dominating nations' institutions have been able to extract surplus-value from subsistence labor everywhere, as well as by exorbitant interest charges, knowing that if cheap labor resists, or interest can not be paid on borrowing nations' debts, they can withdraw their investments and recall their loans—often spreading unemployment, hunger, desperation, and fragmentation of the environments they abandon.

Global Reach; Degrading the World Environment

The disparity of wealth and poverty has been made more extreme in the international arena than domestically. Unpaid creditors like the IMF are powerfully positioned to cut employment in debtor nations, force them into bankruptcy, and make them dependent to repay their debts by mining their minerals and cutting down their forests to sell in world markets.

As well, creditor institutions can compel debtor nations to depreciate their currencies; force World Bank and IMF governmental borrowers to cut their social services; and spread extreme poverty and anxiety among subservient populations.

The largest foreign investors and lenders, operating within each investment zone as a demarcated, closed-end group of nations, also are positioned to drain debtor nations' gold supplies, as well as to place liens on their natural resources, including fossil fuels, forest woodlands, and deep jungle products.

There seem no limits on creditors able to expropriate debtor nations' farmland, invade the protected limits of their seas, and devastate wild and sea species. This can as seen among the dozens of animals that have become extinct in the last 100 years of colonialism, modern neocolonialism and poaching under the control of the major powers, the IMF, World Bank, Soviets, and now Vietnamese operating in Laos.[21]

Yet, the whole system of expropriation might go smash, if and when a future pressure point is reached, as debtor nations and their populations try to throw off the yoke of dominant nations' impositions and share their own economic resources among themselves.

The Unholy Alliance: Financial Investments, Lower Labor Costs and Natural Environments

Financial investors in industrially-developed nation-states, operating through tight-knit transnational organizations, play out their forces in labor markets and funding the movement of capital invested in raw material extraction, manufacturing, and trade in developing nations. Leading the process after the 1930's worldwide economic depression, the Great Powers established tight-knit colonial spheres of influence. Victors in World War II then undermined these spheres, with the United States dollar zone dominating world commerce and imposing so-called free trade in a widening area outside the communist world.[22]

Over the next 45 years, subservient states tried to break out of the U.S. control of their debtor status and currency rigidities. The U.S. had depressed their currency values *vis-à-vis* the dollar, making their exports cheaper and U.S. exports more expensive. Their balance of payments continuously bled red ink.

But during 2015, the conservative wing of the American Senate then sought to depress the dollar *vis-à-vis* other currencies to stimulate U.S. exports and otherwise curtail imports in order to strengthen the U.S. payments balance. To enforce a proposed new *régime* called the Trans-Pacific Partnership (TPP) under the U.S. president's so-called "trade promotion authority," they sought to fix currency ratios with eleven potential-member nations, including the most vulnerable, namely Malaysia, Mexico, Canada, Chile, Australia, and Malaysia. Malaysia was a human-trafficking, forced-labor, slave-permitting state; and to their dishonor, the U.S. Senate and the American President Obama closed a blind eye to the potential use of driven-labor to produce commodities for TPP commerce, allegedly to offset China's geopolitical advantage importing inexpensive oil through the narrow, rock-strewn Strait of Malacca, lying between Indonesia's Sumatra and the Malay Peninsula.[23]

The TPP agreement also focused on the components of modern day neo-colonialism and environmental destruction.

The dominant nations' negative trade balances and foreign earnings on investments were to be paid in the dominant countries' currencies. As under NAFTA, this meant all payment settlements were to be made in U.S. dollars or other hard currencies. Under the proposed TPP currency zone agreement, moreover, dominant nation currencies were to be held in bank "trust" funds by each subservient partner nation, payable on demand to dominant nations, like Britain, Japan and Germany.

These funds were generated from the difference between subsistence wages paid to workers in dependent nations and, above subsistence, the surplus-value of goods going to dominant nations, expressed in fixed-currency ratios, as well as in value above service wages that dependent nations' workers provided to the dominant nations.

The resulting balance of trade, including excess investment-earnings-payment made by dependent nations, thereby becomes the source of dominant-nations' financial companies' accumulation of wealth. This, at least, is the logic.

The Accumulation of Wealth: Emphasis on Subsistence Wages

Financial organizations operating within the prospective TPP and similar groups have surveyed employee pay scales in different member counties.

As First World industrial nations able to set the fixed currencies ratios between the currencies of different member nation-states, they are positioned to determine the exchange of more labor-exertion for less exertion embodied in commodities traded, as well as in investment values among their members—becoming the means for and enhancement of wealth by dominant nations, and the dis-accumulation and impoverishment of subservient nations.

Wages paid by subservient nations' employers or governments can then be set at a minimum based on the need of workers to live at given standards in different locations and societies, shaped by their differing levels of skill, as well as their class-won place in ordered hierarchies, and their complacent, compliant, or militant attitudes and actions toward governments, employers, politicians and others.

Labor's position in such relationships with capital and governments can and does change quickly, too, especially when the owners of the means of finance, production and marketing can not sell all that their own workers make—and the owners then may attempt to lower unit costs to secure their profits by speeding up production, stretching out the labor-day, cutting labor's hours, or discharging some or all of their workers.

Employers bidding their time are thereby positioned to withhold employment until the market recovers, then take their pick of the cheapest laborers. And the newly employed then might be paid still smaller wages for the previous jobs that have disappeared, further impoverishing the new labor force at less than subsistence.

As Pope Francis emphasized, the polarization of wealth between elite nations and their upper classes on one side, and on the other the underpaid and desperate who are employed in subservient states, is the source of unrelenting impoverishment of the poor.

Global Impact: Exporting Finance, Environmental Degradation

But what has all this to do with environmental degradation?

Would there be environmental degradation if leading nations did not export capital? The overall portrait of burning both domestic and imported fossil fuels has been drawn by Fred Pearce, British global population specialist, looking for the geographic disequilibria in forces polluting and heating the atmosphere.

Looking at carbon dioxide emissions, the biggest concern is consequent climate change. The world's richest half billion people—that's about 7 percent of the global population—are responsible for half of the world's carbon dioxide emissions. Meanwhile, the poorest 50 percent of the population are responsible for just 7 percent of emissions.[24]

The imbalances are worsened when there are declining direct investment opportunities in so-called First World nations, periodically leading those in charge of finance capital to export it for investment in petroleum exploration, drilling, shipment, refining and burning—enlarging environmental degradation in other countries, as well as at home. And in 2017, President Trump planned to back two new domestic oil pipe delivery systems that would worsen U.S. environmental pollution, yet not lessen it abroad where the major U.S. and nearly 100 other petroleum companies are heavily invested.

Investing Finance Capital Abroad; Threatening Species Life

The geographic backdrop is an old one. Finance capital has been directly invested abroad since at least the late 19th century.

Throughout Europe and the United States, national banks, trusts and cartels exported such capital for investment in foreign production and commercial facilities, as well as for financing and building colonies to find cheap raw materials for their home industries; inexpensive foodstuffs for their workers; and markets to sell their manufactures at a lower price than in the home market (known as dumping).

One of the main activities of European and U.S. overseas corporations and banks was to place investments in plantation output and mining raw materials—thereby expanding and concentrating commerce, industry and overseas banking as primary forces behind global colonial expansion, politics, conflicts and warfare throughout the periods before two world wars.

These forces often lashed the environmental domain, initiating a cascade of species disasters. Overseas plantations and mines destroyed forest woodland cover and unnumbered species, as well as cut Earth's frail skin, invading, overturning and drying terrain; exposed lethal mine tailings to the atmosphere, leaching toxins into water supplies that communities of indigenous people and wildlife relied on.

With the resulting shift from tribal and local food production to foreign-directed plantation crops that drew on un-reimbursed nutrients from weakened soils, indigenous peoples lost essentials they depended, and still depend, on for daily survival.

Finance Capital Past: Colonialism and the Export of Capital[25]

Pope Francis has apologized for the church's complicity in past colonization, that continues today as an association between the church and economic elites in many parts of Latin America.

Parts of Chilean socially conservative power elites, for example, remain loyal to *Opus Dei*, founded in Fascist Spain in 1928. They supported the reign of the Pinochet dictatorship that tortured and killed 60,000, both opponents and innocents.

The Pope views these organizations as a continuum of capitalism's systemic greed for money, an open or subtle dictatorship that condemns and enslaves men and women in poverty. "Working for a just distribution of the fruits of the earth and human labor is not mere philanthropy," he says. "It is a moral obligation. For Christians, the responsibility is even greater: it is a commandment." And he has called for a movement uprising against the new colonialism rooted in an inequitable economic order.[26]

In the past, empire had an enlarging modern counterpart. An expanded, more rigid colonial system and a vast effusion of foreign capital supplemented the mercantilism and free trade commerce of earlier times.

Finance capital export became the final dividing line between mercantilism and free-trade-imperialism. Indeed, large-scale industrial techniques, the integration of finance with industry, and the monopolization of industry often had pressured, shaped and literally privatized State policies—together pushing forward finance capital exports and the export of machines as capital goods.

Under heavily-lobbied state policies, for example, Britain, Germany and the United States together sent abroad ever-more finance in the form of capital goods, made up of 26 percent of all exports in 1800, but 39 percent in 1900, and 46 percent in 1913.

As the environmental drawbacks of empire appeared, the leading powers then made an effort to disguise the effects.

The prime example was Britain. For 100 years, her overall merchandise exports appeared as less valuable than her good imported—£12 million excess imports in 1830, £19 million in 1845, £45.5 million in1860, £90.5 million in 1875, £126.5 million in 1895, and £131.6 million in 1913.

It sounded like Britain was broke, tethered to colonies to survive.

But the net import bill was paid with imperial earnings. And so large were the excess returned revenues generated from earnings on capital exports—especially in the form of means of production—that these became the birth mark on the face of the new imperialism, which replaced the old limited colonial system; harnessed the labor of the peoples of one-fifth the world; as well as captured overseas natural resources, while destroying species making up the foundation of and for biodiversity.

An enlarged portion of the globe soon was covered with plantations, mines, railways, ports, and a vast merchant marine.

Only a studied actuarial plant geographer like Charles S. Elton, or a crack biologist like E. O. Wilson, even could guess the impact on species, their diversity, and reproductive fitness.

How Contemporary International Agreements Bypass National Sovereignty; Weaken Human Rights; and Destroy Efforts to Secure Nature

In the Preface to *The Palgrave Encyclopedia of Imperialism and Anti-Imperialism*, editors Immanuel Ness and Zac Cope count human death caused by abuse, war and human toil under various forms of empire at an estimated 448-to-548 millions from 1500 to our time.[27]

This caused environmental and massive species destruction during World War II and at least eighteen other neocolonial battles until 2015. The economic "value" of the loss as viewed by technocrats has been offset by finance capital's continued accumulation of wealth, albeit with different moral and ethical measures.

Today there are at least two basic paths leading to the accumulate of such wealth, both evaporating heat and funneling waste into the atmosphere.

On one hand, capital is investing in production and is reimbursed and realized as surplus-value from labor-exertion using machines and burning fossil fuels.

On the other hand, capital is invested in so-called stock and bond exchanges, investors preventing market prices from stretching too far ahead of corporate fundamentals that everyone claims are what truly matters for investors. While receiving dividends and trying to appropriate capital from one another, these funds often end up returning to production, burning fossil fuels to run factories and offices.

This double stranglehold of accumulating capital consummates the initial relationship, say, between an employer's capital *directly* paid in wages and salaries; and *indirectly* invested in stocks to secure dividends and interest on land-use, planting soils with seeds, securing water-rights, and burning fossil energy.

By directly investing in the technological realm that weighs on the biosphere, machines burning fossil fuels advance production and goods for markets. With employers paying labor at classical subsistence scales, the value of the workers' output of commodities or services is realized through the employer's market sale, effectively reimbursing employer's subsistence-outlays and any surplus extracted from labor.

Environmental Impact: Labor-Power, Resources, and Stock Exchange Capital

The separation of those exercising labor-power from possession and ownership of land, resources and capital also hinges on another platform: Finance capital invested directly in production is different than capital invested indirectly in the stock exchange.

Such indirect finance capital investment remains a powerful force for speculation, in which investors often produce no value, until they exit the stock exchange proper; and, if ever, again enter the technical production sphere—where they employ workers to produce in order to again extract surplus-value and produce commodities, with machines burning fossil fuels releasing heat and waste, that is then evaporated.

All relations in the stock exchange between an investor *vis-à-vis* other investors clearly create no worldly goods on one side; while by contrast on the other, capital invested directly in labor for production creates wealth as a stage of capital accumulation.

But the two spheres also converge, sustain one another, and eventually become the source of capital for production of commodities, heat and atmospheric warming; so that a portion of stock market expropriations and dividends are reinvested in the production of goods. Thereby, newly labor-created surplus-value re-enters the production sphere, its heated carbons and wastes again evaporating, negatively impacting the atmospheric living world.

India Past and Present; Environmental Rights

The process has classic roots in political economy, illustrated during India's subjugation by Britain, London's capital export enlarging from direct investors and the stock exchange, coming at great cost to both populations and the environment.

By conscious and calculated warfare and brutal subjugation, Britain made India the jewel in its colonial crown: undermining India's traditional crafts; displacing thousands of natural species; redirecting food and industrial production for profit; turning rural populations into low-paid laborers; and plunging into misery countless scores of rural and urban peoples.[28]

Past Oppression Continued into the Mid-20th Century

In the aftermath of World War II, Britain's empire was in sharp decline in Asia. And in 1950, to save its colonial shards, Britain established the Dominions of India, Pakistan and Ceylon, an area of 1,604,666 square miles, or a little over one-eighth of the area of the remaining empire, with a combined population of 449 million, comprising some 72 percent of the population of the imperial realm.

To call these expropriated "commons" might seem an exaggeration. But with its oversight, Britain handed over to Indian administration an alliance of Indian monopolists, princes and landlords; in Pakistan to large-scale landlords and big capitalists; and among Sinhalese to great plantation owners and traders associated under all-pervasive British control—a form of feudal-imperial exploitation at the lowest colonial income standards for both peasants and workers.[29]

India's physicist Vandana Shiva says eliminating the rural population's landholdings began much earlier with a 1865 law, that lifted protection for Indian forests as common lands, paved the way for commercial and industrial exploitation of land and forests, and impoverished millions who depended on these forests, sacred groves and grazing land.

This was followed by building dams, using ground-water for mining, and privatization of water resources; as well today seed-patenting by transnational corporations directly threatening biodiversity that thousands of communities rely upon for survival.

Shiva explains that the earlier commons served the whole community; that "for indigenous peoples, heritage is a bundle of relationships [a play on the western legal concept of a bundle of rights], rather than a [separable] bundle of economic rights."[30] Thereby historically protected common land rights were shattered.

Trans-nationals Evolve; Secrecy Prevails; Earth Suffers

All is not well in the house of transnational funding, either. And its operations are adding to Earth's suffering.

Financial transnationals operate under strict secrecy, designing the terms of modern international agreements to undercut remaining protective environmental and labor laws. These transnationals have created supranational organizations like the North American Free Trade Association (NAFTA), the South Korean Free Trade Agreement (CAFTA), the Normalized Trade Agreement with China (PNTR), and the potential blockbuster Trans-Pacific Trade Agreement (TPP), since early 2017 without the United States membership. It is still unknown if other nations will withdraw, and China will become the lead partner.

Such agreements are designed to regulate worldwide production processes, transportation systems, raw materials access, and deployment of labor as wage and salaried workers, even as slaves or in other forms of servitude.

Operating in selected nations seeking ever-lowering wages, they sell those commodities like pharmaceuticals and sophisticated technologies in high-priced markets. As well, they operate in lowly-paid, outsourced services areas such as call-centers, computer programming, engineering, accounting, and medical diagnostics.

These financial alliances have spread their global reach through supranational cartel organizations, creating geographic "free trade and investment" areas that override national anti-trust, commercial, labor and environmental laws and regulations.

They bind cartel member nations to resolve matters in their own "international tribunals" and through United Nation agencies—to make decisions that bypass national laws; to resolve conflicts over what can be produced, how produced, and sold at what price. They set environmental standards, if any, that must be followed. They determine which labor regulations can be ignored and which human rights laws bypassed. These cartels are literally able to set global "principles" and policies for their own member-state operations.

Within each cartel, member nations advance the interests of transnational financial organizations, funneling investment funds—maximizing their accumulation of revenues at the expense of working populations and the global environment.

Some liken these organizations to Trojan Horses opening national or regional gates to transnationals invading a sleeping world. The financial world they enter expands control over both the relations of production and social relationships. Production linkages to financial capital allow in only a handful of companies, including 600 transnationals that have registered their investments overseas. They are empowered to control labor conditions which they impose. They decide whether or not local suppliers and industry can be favored, and which banking services shall prevail. In cases of conflict of interests with local institutions, they also control the treaty or tribunal enforcement of expected future corporate profits from local interference.

Social linkages that can be controlled by multinational or bilateral treaties between member nations also cover environmental laws and health regulations, which act as corporatized controls undercutting any democratic participation.[31]

Trans-Pacific Partnership—A Neo-Mercantile Organization

The Trans-Pacific Partnership is a neo-mercantile organization, meaning it is a policy regime that encourages exports, controls imports and capital movements, as well as centralizes currency ratios in the hands of a central command group. Within its treaty structure, dominant nations are able to use their "reserve currencies" (dollars, yen, pounds sterling, francs and yuan) to invest in one another and control the commercial prices of commodity and services flows, as well as the relative condition of labor in each subservient member nation.

These "superior" nations also control subservient nations' scales of freedom, human restrictions, and slave status, the levels of subsistence approximating the "iron law of wages" and the revelant standards for survival. They regulate labor mobility or confinement to restricted areas, using pass-laws to oversee where laborers work and live. They mobilize police to impose lawless order, or use so-named, democratic organizations controlled and overseen by social elites, classes or castes.

All these are forms for controlling a population's living condition through labor regulations; degrees of freedom or un-freedom; and the wages assigned to workers for their labor-exertion—refined and monetized in relation to labor-embodied commodity values and relative prices for the neo-mercantile trade.

This relationship underlies Congressman Bernie Sanders' explanation of the potential impact of the Trans-Pacific Partnership, with Japan and ten other growing nations. "The Trans-Pacific Agreement is a disastrous trade agreement designed to protect the interests of the largest multi-national corporations at the expense of workers, consumers, and the environment, as well as to negatively impact some of the poorest people in the world."[32]

Rebirthing Spheres of Influence

Considering the PPT agreement that he promoted, President Obama said Japan would have to open the door to American cars and remove tariffs on U.S. agricultural products, including rice, wheat, pork, beef and dairy. Though that might not happen in the short-term, if ever, Obama sought to control the Trans-Pacific Partnership to bolster his flagging legacy, telling *The Wall Street Journal* that "if we don't write the rules, China will write out the rules in the Asian region."

This focus was a return to past strategies of 19th and 20th centuries' spheres of influence in the Pacific, with Beijing suspecting the U.S. of trying to encircle China—trying to blunt, if not reverse, China's rise.

As China has flexed its muscles in the East and China seas, and moved to exert ever more influence to the West through massive government-led investment plans to create a new "silk road" (providing an April 2015, $46 billion aid and investments for Pakistan), for example, this was just a beginning to 21st century economic competition with the U.S. that could dangerously turn on a future naval arms race.[33] And the future military impact on the ecological life of the seas is virtually unpredictable.

Ecological Debt Imbalances

"Finance overwhelms the real economy," says Pope Francis; "and the economy accepts every advance in technology with a view to profit, without concern for its potential negative impacts on human beings."[34]

He cites the chaos already loosened by the so-called "ecological debt" between the global North and South, connected to commercial imbalances impacting the environment and the disproportionate use of natural resources by Northern countries over long periods of time, emphasizing the export of non-industrialized nations' raw materials to satisfy markets of the industrial North causing harm to the South, including mercury pollution in mining gold and sulfur dioxide pollution in copper mining—two effluents U.S. President Trump also plans to use domestically for mining in the Grand Canyon and the Wisconsin-Canadian Headwaters Waters.

Environmental devastation also follows industrial nations' export of solid waste and toxic liquids to so-called developing countries.

"We note that often the businesses which operate this way are multinationals," Pope Francis cites the Bishops of the Patagonia-Conahue Region (Argentina) in their Christmas Message (December 2009), page 2: "They do here what they would never do in developed countries or the so-called first world. Generally, after ceasing their activities and withdrawing, they leave behind great human and environmental liabilities such as unemployment, abandoned towns, the depletion of natural reserves, deforestation, the impoverishment of agriculture and local stock breeding, open pits, riven hills, polluted rivers and a handful of social works which are no longer sustainable."[35]

Flash Points; Technological Inequities

The third stage of the export of global finance-capital sustaining technological projects is increasingly undermining the biosphere.

Pope Francis foresees fragmentation of the natural orientation and evolution of species and human communities, undermining the conditions necessary for both organic life and social production.

Not only are natural networks of species undermined by human fragmentation of knowledge and infrastructures, but so too are social institutions. Society viewed as part of nature makes both species and humans interdependent, so much so that social production and social structures turn on solidarity of primary social groups—from family to local, national and global communities. Human communities thereby are wedded to their traditional means of sustaining themselves in given natural environs.

But these linkages are broken when outside pressures force them to abandon their homelands. Foreign invasions, deforestation, colonial agriculture, and mining for resources not only unsettle and degrade ecosystems, but also displace indigenous peoples, destroying their sustenance and culture. Those forced out or remaining suffer physically as well as emotionally: loss of traditions and stability; degraded environmental conditions; loss of dignity under inequitable foreign laws or alien police states.[36]

Such changes have taken place amidst rapid technological advances that both enlarge output and disintegrate natural environments—weakening and undermining the foundations for communities' structural, economic, and social equities.

Culture Clash: Advancing Science Infringing on Ecosystems

This has become the penultimate clash of those controlling the global means of production with existing societies in the so-called Third World.

And the crises has worsened as unrelenting current wars in Syria, Afghanistan, Iraq and elsewhere in Africa already have sent outflows of refugees as the massive destruction of cities and rural environments continues to uproot many thousands, often leaving them starving amidst agonizing pandemics.

Inventions and Nature's Tenuous Balance

What and who are to blame for the current state of disasters? Consider that, in part, technological improvements are often linked to geographic expansion that has infringed on and often undermined traditional communities' environmental integrity and balance.

Looking to the past during the last two and a half centuries of industrial advance, humans have rapidly mobilized science and tool making, inventing mills run by wind, water, steam, Sun and electrical power that changed the environmental, social, religious and military order.

Steam-power drove railways and factories. Steel mills were powered by coal, gas and electrical currents. Telegraphs, telephones and cellular communications were operated along electrified copper wires, then fiberoptic lines. Automobiles and airplanes were run by motors using gasoline and oil, compressed gases, vacuums and solar panel arrays.

Chemical compounds were deployed to fuel advanced industries. Laboratory technicians synthesized medicines, building giant complexes supplying local drug distributors and hospitals worldwide. Biological and offensive weapons were developed. Nuclear-power was released by scientists and engineers for uses in medicine, producing electricity, war, and deadly arsenals of weapons used as threats of some nations against others.[37]

Garage inventors initiated the digital revolution and the worldwide web, distributing information and advancing industrial robotics, biotechnologies, nanotechnologies, and a raft of other sophisticated technologies and applications available to many millions across the globe.

These classic inventions merged integrated circuits with the connectivity of previously isolated electronic functions shaping future computer structures—extending their reach into the environment, sometimes disseminating information that has jeopardized the conditions essential for the survival of species.

Considering the so-called law of scientific advance, integrated circuits are expected to double their components every 2 years. Yet this doubling may come on more slowly as it demands the investment of more financial capital, either internally generated by technological firms or by the takeover by larger firms of others that have already developed complementary technologies. The enlarged capitalization—raising the technical composition of capital—has led towards the combination of ever-new companies outflanking smaller capitalized firms. And the new technological base continues to infringe on the environment as well as speed up pressure on the workforce.[38]

The Pope's Account

Pope Francis says:

> Many problems of today's world stem from the tendency, sometimes unconscious, to make the method and aims of science and technology an epistemological paradigm which shapes the lives of individuals and the working of society.
>
> The effects of imposing the model on reality as a whole, human and social, are seen in the deterioration of the environment. We have to accept that technological products are not neutral, for they create a framework which ends up conditioning lifestyles and shaping social possibilities along the lines dictated by the interests of powerful groups. Decisions which may seem purely instrumental are in reality decisions made about the kind of society we want to build.[39]

In effect, technological improvements are often linked to geographic expansion that infringes on social conditions, vulnerable resources and the environment's integrity. If we believe there is a way to stop the technological system from eating a way through the biosphere, piece-by-piece, we had better mobilize our still disunited forces.

The question remains: Can we?

3. IS THERE REASON FOR HOPE?

Leading scientists believe human existence is now just three minutes away from an apocalypse at a theoretical midnight point on the Doomsday Clock that the *Bulletin of Atomic Scientists* set up in 1947. Then the founders were concerned with the danger of atomic arms, while today they view the impending crisis as a composite of the spread of nuclear arms, increasingly sophisticated technology, strident nationalism worldwide, the growing disregard for scientific expertise, and Earth's increasingly torrid climate.

The deeper fundamental question remains—is there a path to save Earth, our only home?

The great transformation of global ecosystems requires human reorientation away from the competitive financial accumulation of capital for a few and towards production by, and distribution of goods and services to, the many.

Along with millions of others, Pope Francis is disappointed with national and global attempts to stop environmental degradation while global warming worsens. He seeks to stop the fragmentation of the planet's "commons" and seeks to prevent the economic and financial transnationals from controlling state-set political conditions that are overwhelming efforts to save Earth's ecosystem.[40]

The advance of the technological complex driven by burning fossil and nuclear fuels has led to ecocide on the entirety of Earth. This has been worsened by multi-national, invasive finance capital, in turn leading to the demeaning of and near-helplessness of labor in most farming, industrial and commercial activities. This invasion has led to a disastrous impoverishment of workers, consumers and the environment on a global scale.

Yet labor solidarity within nations and across national boundaries remains frail or nonexistent. It may take a nearly simultaneous global collapse of economic life and a disclimax of the biosphere to ignite collective worker and consumer determination to change the international social order protecting the planet. And such resistance may come too late.

During the remains of the Anthropocene, then, our survival focuses on the seemingly irreversible destruction of the ecosphere and species, weighing on human dignity, territorial integrity, and species diversity itself.

Technosphere; Biosphere

The overlapping technological and manufacturing spheres, together with the Sun's direct radiation and re-radiation from Earth, act to evaporate gasses, heat and effluents into the atmosphere from land and sea—compounded by their redistribution to cooler regions of the planet, as the Second Law of Thermodynamics foretells.

To secure the biosphere, however, a ray of hope has appeared in today's path of production—from refining and burning fossil fuels to increasing use of solar-, geo-, wind- and water-power.

In still minimal fashion, this change from dependence on fossil fuels in the operation of factories, mines. homes and other places will rely on solar hot-water systems and voltaic cells sustaining the energy needs of most places of work, recreation and residences, holding a cost advantage over fossil fuels for power generation.[41]

Transnational "Gods of Money" Plunder Nature

Fossil fuels are now the number one public enemy .Organizations controlling transnational financial capital investing in world production, extracting and burning fossil fuels—jeopardizing Nature—are contrary to the beliefs of millions of people as well as the Vatican's new efforts to protect the global ecosphere.

As Pope Francis, told a meeting of Latin American and Asian landless peasants and other social movement activists in October 2014, the "economic system centered on the god of money needs to plunder nature to sustain the frenetic rhythms of consumption that is inherent in it."

"The system continues unchanged," he insists, "since what dominates are the dynamics of an economy and a finance that are lacking in ethics. It is no longer man who commands, but money. Cash commands."[42]

Cash has a double edge as the means of both investment and exchange. Ironically, this also affects Vatican ethics under Pope Francis' efforts to clean up the Holy See's own questionable past financial operations.

As savvy investors in security trading—especially bonds—Vatican Bank traders secured a rise in profits from $3 million in 2013 to nearly a whopping billion in 2014.[43] On an estimated capital-value earning 3 plus percent, the Vatican Bank holds $3,589 billion.

The Pope' opponents and faceless wealthy have also mobilized economic and political power. The 20-years' accumulation of $3.6 trillion by 917 billionaires worldwide may vary somewhat, with their concern with inheritance, retirement, health, charity, and leaving a personal legacy.[44] Even Microsoft's politically liberal Bill Gates is expected to become the world's first trillionaire in 2017. But the scale of accumulated wealth clearly favors the billionaires against the Vatican Bank. So the Pope is up against the power of their money and their social-political biases impacting global warming.

What is the chance of the Pope's plan for fundamental reforms of capitalist production and marketing that essentially depends on burning fossil fuels causing global warming—to prevent the immense power of money from overwhelming Nature's essences, as well as the survival and reproductive success of most species on Earth?

Why Don't Populations Rise Up Against These Dangers?

Vain political denials of scientifically-proven evidence of escalating planetary heat have failed to convince Pope Francis among other luminaries. They foresee that ecocide is on the horizon, and that immediate measures are needed to slow or stop the oncoming future changes in terrain, atmosphere and seas.

They understand that populations now and in the past often have relied on politically-inspired government pronouncements and prescriptions of what they must do to protect and secure themselves. But what if government is misleading the very population that the political and social-economic system is placing in danger? What if those endangered one day literally awaken and revolt against the existing government, seeking their own self-determination?

An historical example, is the July 4, 1776, *Declaration of Independence of the United States of America*, when the colonial population sought to secure unalienable rights of Life, Liberty and the Pursuit of Happiness. The representatives in General Congress Assembled proclaimed they had derived their just powers from the consent of the governed, saying that whenever any form of government became destructive of these ends, it was the right of the people to alter or abolish it, and to institute new government, laying its foundation on such principles, and organizing its power in such form, as to the people seemed most likely to affect their safety and happiness. Yet, they noted, prudence would dictate that governments long established should not be changed for light and transient causes; that all experience had shown that humankind was more disposed to suffer, while evils are sufferable, than to right themselves by abolishing the government forms to which they were accustomed.

"But," they cautioned, "when a long Train of Abuses and Usurpations . . . evinces a Design to reduce them under absolute Despotism, it is their Right, it is their Duty, to throw off such Government, and to provide new Guards for their future Security."

Security of the populace has remained the central ingredient of self-determination throughout U. S. history.

For example, this was the key political matter that the famous linguist E. B. White questioned in 1935, and political investigator Naomi Klein queried eight decades later?

White had questioned the population's blind obedience to government air raid preparations in New York City during A-bomb drills preparing for Russian missiles that placed eight million well-behaved citizens huddled in hallways and tunnels as a hush fell over the metropolis. It was a ceremonial exercise of conformity to fear of mass annihilation, that a handful nuclear authorities and scientists together had formulated and imposed by position and an apparent bow to Medieval witchcraft —no 13th floors existing in high-rise Manhattan buildings as potential A-bomb survival chambers.[45]

In our time there is a new scale of blind obedience to government.

Klein says ecological amnesia turns on fear that letting in the full reality of the rise in global heat and sea levels would change existence itself, bringing forward unthinkable truths about the causes—an economic-driven profit system and consumption patterns, both requiring deep transformative changes—as Pope Francis reaffirms.[46]

Klein also says the presidential victory of Donald Trump was a corporate *coup d'etat*, as he appointed billionaires to his cabinet and Federal agency posts, potentially jeopardizing popular rights of the population to Life, Liberty and the Pursuit of Happiness.

Popular awareness of the fragility of our democracy might lead some individuals to pursue self-survival, alone, those already physically secure blunting their compassion, altruism and generosity towards the plight of others; all of them living in a world befouled by a two-or-more percent yearly rise in carbon emissions now streaming from the elevated extraction and burning of fossil fuel energy.[47]

Other dangers to the ecosphere and species are obvious, but in early 2014, President-to-be Trump twittered, "This very expensive Global Warming bullshit has got to stop." His belief that thousands of scientists are wrongly depicting Hothouse Earth as a menace to future species survival is an unforgivable blind spot. He has not considered the dangers, for example, that Indonesia as the most corrupt worldly government has huddled with palm-growing corporations to torch hundreds of thousands of acres of ancient forests—along with their elephants, human communities, tigers, birds, monkeys, orangutans, and insects, in order to plant palm oil trees for ingredients and lubricants essential for products like Lays's or Nabisco's line of foods—creating funnels of smoke, debris and heat, raising immediate area temperatures by over 2 degrees centigrade, spreading and elevating global heat.

Western demand for palm oil for foods, cosmetics and industries thereby has set loose the fires this time that also threaten the future *balance* of atmospheric oxygen, carbon dioxide and methane essential for existing species life. From his bully pulpit, Mr. Trump is missing the essential facts.

Understanding Source of Global Warming

"We need to care for the Earth so that it may continue, as God willed, to be a source of life for the entire human family," Pope Francis in 2015 reminded believers, especially Catholics, in a teological declaration—meaning, first cause, uncaused, "as God willed." The Pope hoped to initiate a global ground swell to urge world leaders to enact a sweeping United Nations climate-change-accord that would commit every nation to enact tough new laws to cut the emissions that cause global warming.[48] This demands full knowledge of the dangers to Earth and all of creation in order to begin the transition to renewable, sustainable environments and politically responsible economies.

To pull off a movement like this, great sums of money and Catholics bred to faith were needed—far more than parisoners saying their daily Seven Holy Mary's could afford. So the Pope had to go beyond the God-willed teological mandate—to wait for Christ's return or Saint Peter's authority—blocking powerful, cash-rich and well-heeled political opposition.

Opponents of climate policy in the United States were enraged, including powerful Catholics among the 2016 runners for the Republican presidency, as well as the multi-billionaire House of Koch, the brothers running the Charles G. Koch Foundation. Conservative opposition was being solidified. Former *Breitbart News* director Stephen K. Bannon, who President Trump had placed above Pentagon Generals on political-military matters, earlier had gone to the Vatican to cover the canonization of John Paul II and to bond his worldviews with those of arch-conservative American cardinal Raymond Burke who had openly clashed with Pope Francis. At the center of the American ideological battle, Joseph Blast, the president of the conservative Heartland Institute, ventured that: "The Holy Father is being misled by 'experts' at the United Nations who have proven unworthy of his trust. Though Pope Francis' heart is surely in the right place, he would do his flock in the world a disservice by putting his moral authority behind the United Nations' unscientific agenda on the climate."[49]

Swords Were Clearly Crossed

What will happen in the long run if there is public awareness that today's ecological dangers stem from accumulated and centralized financial wealth that looms large in the equation of national and global power? What if this consciousness turns on an understanding that the public is subservient to those in control of technology causing global heat; that agglomerated finance, political power, and State favoritism to corporate and transnational interests regulates international investment arrangements?

Consider the now-failing North American Free Trade Association linking Mexico, the U.S. and Canada, the weakened Atlantic Alliance with the U.S., the *Brexit* remains of the Common Market, and the U.S. exit from the pending Trans-Pacific Trade Agreement among nations of the Pacific Rim. What if others nations also exit?

These so-called free trade agreements have embodied provisions for stabilized currencies for regulated trade and investments that *allegedly equalize* the rights of industrially-advanced nations and those countries receiving services and commercial concessions for their raw material and mass-production of foods.

Illusions of equality are exposed, though, when considering the facts that the raw material and food producers often look to their countries' Gross National Product (GNP) as a benefit, which in fact involves such negative consequences as deforestation; destruction of soil-and micro-organisms; and indigenous workers held in slave-like conditions or driven as semi-free labor.

Middling states offering services also bolster their GNP; but this is accomplished only with their populations providing peonage or lowly-paid wage labor negotiating for air-and-ship transport, hotel accommodations, insurance and cargo ratings and metrics, diagnostic dental and medical analysis, and the like.

And for corporate-and transnationally-sustained states, GNP represents the surplus-values drawn from their own working classes, as well as those of the other states.

The Lions Take Their Share without Obligations to Secure Earth

With the lion's share of inflow of surplus largely going to the ownership groups in cartel-like organizations, they are allied with other financial groups and industrialized nations to nominally promise to regulate greenhouse gas emissions, yet drag their heels in setting penalties on one another if they fail to meet the quotas set by the U.N. sponsored Global Warming pacts that began as early as 1962.

While these environmental pacts have no teeth, or very weak ones, for enforcement, the international trade agreements between the leading nations and oligarchies have 'steel-jaws" to enforce their provisions. These allow a single market competitor to stop another member from gaining cost-advantages by using internal resources and provincial raw materials that best secure the environment. They also allow one member to stop another from deploying cheaper and more-highly skilled local workers advancing employment, as well as the deployment of advanced technologies with a low-carbon footprint.

These limits may prevent the turn from carbon-based production to using solar, wind and sea power lowering the output of greenhouse emissions, and they also mean that populations live in local environments under a pall of pollution and methane-spewing waste, at low wage-and-salary scales set by the regulating trade ageements.[50]

Transnational Operations vs. Environmental Security

What are the precise mechanisms that have allowed such measures to undermine environmental safety?

Transnational financial competition has speeded oil exploitation and drains government programs protecting the environment. Oil revenues expressed in dollars, called petrodollars, allow nations accumulating them to buy the energy sources they require to support their economies.

Such revenues also lead to periodic currency adjustments that push nation states to print money without gold or economic backing, in order to buy oil to support their industries and military—enlarging petroleum debts to oil providing cartels, while taking away funds meant to sustain the unemployed and impoverished under-classes, as well as to spend on environmental projects.

Financial transnationalism thus acts to control the extraction and price of burning fossil fuels. And when the overall competitive market price of fossil fuel is forced down, the transnationals are positioned to enlarge sales, consumption and a heated global atmosphere.

Earth's Temperature Rise Unsettles Financial Accounts and the Balance of Species

Interlocked species varieties supported by traditional natural conditions cannot quickly adapt as financial market gyrations impact environments.

We observed this during 2014-15 when trillions of U.S. dollars were transferred, actually laundered, from Washington to Saudi Arabia, increasing the output of inexpensive oil to burn, evaporating massive additional carbon dioxide to the atmosphere.

The Federal Reserve literally purchased 12 billion of dollars' worth of Treasury Bonds from America's largest banks, in turn empowering these banks to expand the money supply five-times by printing dollars to bail out the Saudi princes—who were cutting the price of a barrel of oil below $50 by overproduction, so that total Saudi revenues would remain the same as in years past.[51]

Whatever finance capital's logic or lack of it, the fact is that as more oil is burned, it evaporates the added carbon dioxide and particle matter into the atmosphere—capturing solar radiation re-radiated from Earth, elevating global heat, melting glaciers and tundra, and raising sea levels inundating islands and continental fringes.

If others follow a similar future path, world temperatures will exceed another two degrees Fahrenheit by 2020.

If Russia holding the largest world oil supply reserves, retaliates, cutting off gas supplies to Western Europe in the winters—followed by Iran and Saudi Arabia increasing oil output to maximum capacity, and other OPEC countries following suit—it might lead to a cut-price oil-war that promises to load the market with surplus fuel to burn, again pushing up global heat.

During such a grand display of oil flows and disruptions, more or less revenues going to each competitor nation to maintain revenues in the form of petrodollars might also lead to manipulations of transnational finance capital, in turn ultimately starting a twenty-five-year economic crises and collapse, followed by volatile political and military engagements, says one global expert on such matters, James Rickards, in his 2015 study of *The Death of Money*.

The Pope Moves the Chess Board

Pope Francis has stepped into the crossfire attempting to block competitive, transnational finance capital's focus on increased exploration for and burning of fossil fuels.

And Bishop Marcelo Sánchez Sorondo, as chancellor of both the Pontifical Academy of Sciences and the Academy of Social Sciences led the November 7, 2014 attack, saying human action using fossil fuels was having a decisive negative impact heating the entire the planet. This, he said, was coupled with an economy driven by the banal logic of profit accumulation.

He foresaw that the grand display of finance capital would ultimately lead to crises, as "games finance plays in order to profit from money itself, without a clear orientation to the production of goods . . . lead[ing] to social exclusion [of the poor] and the new form of slavery such as forced labour, prostitution, organ trafficking, and the use of drugs as a method of corruption." Offsetting the unjust social structures and the injustice of the "globalization of indifference" towards others in need, Bishop Sorondo quoted from the venerated Pope Paul VI's goal—that peace and justice could be achieved by the "civilization of love": a healthier, safer, fairer, more prosperous and more sustainable world order, with humankind acting as the steward of Nature and of its environmental and human development, derived by values and faith more than data.[52]

"The monopolizing of lands, deforestation, the appropriation of water, inadequately [secured] agro-toxins are some of the evils that tear man from the land of his birth," Pope Francis added—"Climate change, the loss of biodiversity and deforestation are already sowing their devastating effects in the great cataclysms we witness."[53] Neither Saint Peter nor Francis of Assisi, nor Pope Francis could countenance these.

Closing Centuries of the Orbital Stage of Earth's Global Cycle

In chapter I we made the contrast between the Elliptical (egg-shaped) Stage of Earth's orbit around the Sun bringing on a frozen planet and the death of most species, and the oncoming closing centuries of the Circular Stage of Earth's global cycle, when the Sun is closest to Earth and the temperature rises. During this phase, the glaciers and ice fields melt; the seas rise; lethal amoebae of the past are revived; rock–age debris lie exposed at inundated continental fringes and nearly-submerged islands; land species are either destroyed by the spreading tropical heat or driven to polar or higher elevations; and sea species are bathed in acidic waters undermining their life-essential formation of protective, calcified shells.

Species Survival

During *this moment* of Earth's still lasting endurance, Charles Darwin wrote: "It is not the strongest of the species that survives, nor the most intelligent, but the one most responsive to change."

These alternative possibilities have been discovered today in many forms by paleo-anthropologists and oceanographers. Seascapes and varied species once reached hundreds of miles inland from their present settings in today's landlocked Mesopotamia; and even New York marshes reveal sediments of planetary prehistoric life and thousands of years of tidal-flow sediments silting up shorelines, as well as more recent anthropocentric discard of detritions, including lead and incinerated urban refuse—residues of rock evidence of the stages of human civilization during the final, circular move of Earth around the Sun. The prehistory of Colorado Plateau also reveal thousands of years of routine occurrences of supefloods. The paleo-flood record across a broad swath of California happen every 200 years or more frequently, drowning all in their path.[54] And if Atmospheric Rivers scoop up surface moisture from the Pacific face, powerful mountain flooding may occur every two-to-eight years during *El Niño.*

These evidentiary events mark the path from Darwin's evolutionary walk towards a possible future planetary disclimax—a last wave inundating both islands and continents, destroying most species on Earth.

Blinded to Great Risks

Yet, so long as Earth's orbit around the Sun remains circular, it still may be possible to prevent burning fossil fuels releasing greenhouse gases, extreme heat waves causing more evaporation and dry spells, more intense and threatening storms, and the loss of landscapes and diminishing species.

We are facing an oncoming battle royal against those blinded to the great risks of spreading global heat, rising seas, and a circle of submerging territory and species death. To stop the coming destruction will take a Herculean effort by billions of people, led by powerful, determined leaders.

Even the Pope has taken on the Curia to reform it, as well as to take charge of the Vatican Bank's trillions, foretelling the Curia resistance and a new phase of skullduggery unequalled in the rapacious past.

Is it possible that the Pope can or will mobilize a global religious crusade to prevent the onset of entropy on Earth.

Would the Vatican use its financial power against the strength of seemingly overwhelming transnational financial capital?

The opposing forces are formidable. The Pope holds uncertain centralized elite influence and control over the Curia's material and financial strength. He is pursuing diplomatic panoply within the universal Church, while using his cultural leverage, modulated by personal ambitions within the Curia to change the Church institutional bureaucracy that has been handed down for some 600 years.

Today these forces intersect with the competitive quest of capitalist, socialist, commercial and imperial domination over planetary territory and life. Political, ecological and religious organizations have drawn battle lines—igniting a fierce environmental crusade for awareness that likely will consume our remaining presence on Earth. The central matter remains this: Can

Pope Francis convince the Curia and, with thousands of others, act in time to help reverse climate change, to save life on Earth?

Will we support his effort and, if so, how?

Can he mobilize the billions held in the Vatican Bank (the *Instituto per le Opere di Religione*) to help shield and protect wildlife, species diversity and humankind from the avaricious quest for wealth by a well-organized 97 new billionaires with vested interests in 100 giant cement and fossil-fuel companies, poisoning Planet Earth and raising global temperatures—as the Second Law of Thermodynamics postulates?

What is certain is that a life-and-death struggle is already on the rise between worldwide populations and the rarefied few controlling financial wealth, with the means and power to release greenhouse gases to the heavens, searing Planet Earth and decimating species diversity.

NOTES

1. Pope Francis, "Meeting of Latin American and Asian landless peasants and other social movement activists," October 2014. See also Caroline Wyatt, 'Pope of the poor makes big impact in Africa," *The New York Times,* December 1, 2015.

2. Millie Hawley, Tribal leader, Kivalina, Alaska, *The New York Times*, September 2, 2015. For comparable loss of territory see Armand Marie Leroi, *The Lagoon: How Aristotle Invented Science, passim*. See also Diane Ackerman, *The Human Age: The World Shaped by Us* (New York/London: W.W. Norton & Company, 2014), *passim.*

3. Will the world of contentious politicians representing 200 nations actually use the near-congruous findings of supercomputer models covering present and future climate change established during their Paris COP 21 negotiations in December 2015? See Nell Greeenfield Boyce, "Big Data Predicts Centuries of Harm if Climate Warming Goes Unchecked," *NPR*, November 30, 2015.

4. See "Global Warming," *National Geographic*, Vol. 206 No. 3 (September 2004), centerfold. See also Henry Fountain, "Australia Heat Tied to Climate Shift," *The New York Times*, March 3, 2017. During November and December 2016 Southeastern Australia suffered through a series of brutal heat waves, temperatures reaching a scorching 113 degrees Fahrenheit in some parts of New South Wales, 10 times more likely than a century ago, before significant greenhouse gas emissions from human activity started warming the planet. These effects are occurring worldwide.

5. Peter Whitfield, *Sir Francis Drake* (New York: New York University Press, 2004), Chapters I, II, *et passim*. See also Krooth, *Arms and Empire*, Chapter I, 121. See J. Ella Barker, *Rise and Decline of the Netherlands* (London: Smith Elder, 1906), *passim*. See Krooth, *Gaia*, "Introduction," 419-23, *et passim*. See Krooth, *Empire: A Bicentennial Appraisal* (Santa Barbara: Harvest Publishers, 1976), 2.

6. Krooth, *Arms and Empire*, 121. See also Krooth, *Gaia*, 419-423, *et passim.*

7. Glenn Greenwald, *No Place to Hide: Edward Snowden, The NSA, and the U.S. Surveillance State* (New York: Metropolitan Books/Henry Holt and Company, 2014), 101 141,151 153, 172-173. See also Ackerman, *The Human Age*, 28, *et passim*.

8. Greenwald, *No Place to Hide, passim.* Ackerman, *The Human Age*, 28. For corridors, see also Edward O. Wilson, *The Meaning of Human Existence* (New York/London: Liveright Publishing Corporation, Division of W.W. Norton & Company, 2014), 128-131, 140, *et passim*. And see Leroi, *The Lagoon*, 9, 16-17.

9. Ackerman, *The Human Age*, 28. For corridors, also see Wilson, *The Meaning,* 128-131, 140. And see Leroi, *The Lagoon,* 16-17.

10. Glenn Greenwald, *No Place to Hide*, 101 141,151 153, 172-173. U.S. Supreme Court Justice Louis Brandeis and Samuel Warren once wrote of such rights to property and privacy in a *Harvard Law Review* article in 1890. See also Jose Pagliery, "Wikileaks claims to reveal how CIA hacks TVs and phones all over the world," *CNN*, March 7, 2017. See Wilson. *The Meaning,* 131.

11. Der Spiegel, "New Snowden Documents Reveal NSA's War on Internet Security and Tools to Protect Yourself," December 29, 2016. See also Shepard Fairey, "Edward Snowden Wired," *ACLU*, Summer 2015, 13-17.

12. H .G. Draymond, "The Fascination of Arctic Exploration," *The African Monthly*, February, 1907, 172. See also Richard D. James, "Boom State," *The Wall Street Journal*, July 25, 1974.

13. Julie Hirschfeld Davis, "Obama to call for more icebreakers in Arctic as U.S. seeks foothold," *The New York Times*, September 1, 2015. See also Andrew C. Revkin, "The U.S. Icebreaker Gap is about Arctic Needs, Not About Chasing Russia," *The New York Times*, September 1, 2015. For the following two paragraphs, see Paul Szoldza, "Trump's Cutting $1 Billion from Coast Guard at a Time When It's Doing More than Ever," *Business Insider, SF Gate*, March 3, 2017.

14. Nic Fleming, "Plants Have a Hidden Internet," http://www.bbc.com/earth/story/20141111-plants-have-a-hidden -internet.

15. See University of Wisconsin, Green Bay, Studies, "Biodiversity of Macrofungi in Northern Door County, Wisconsin," 2015.

16. Tom DiChristopher, "Why Fed Won't Raise Rates Now," *Yahoo Finance*, September17, 2015.

17. Sonali Basak and Doni Bloomfield, "'Once-in-Three-Billion-Year' Jump in Bonds Was a Warning Shot, Dimon Says," *BloombergBusiness*, April 10, 2015.

18. Murrray, "Summers Says U.S. IMF Inaction," *BloombergBusiness*, April 6, 2015. China's legitimacy to lend the yuan through the IMF followed IMF-approval of its inclusion among the basket of hard currencies—including the dollar, British pound. French Franc and Japanese yen. Approval was based on China's accumulation of vast amounts of gold backing its currency and its investment position in Latin America and Africa. China could now lend its currency from the so-called basket, and demand repayment in kind; if the debt could not be repaid, China could take an investment interest in the debtor nation's assets and resources.

19. Murrray, "Summers Says," *BloombergBusiness*. April 6, 2015. See also "Rates a Day After a Market Plunge," *The New York Times;* Rick Newman, "It's amateur hour in China," *The New York Times,* September 3. 2015. See Keith Bradsher, "China Cuts Interest Rates a Day After a Market Plunge," *The New York Times*, June 28, 2015. And see Michael Santoli, "Is this the 'stock picker's market you've been asking for?'" *Yahoo Finance,* April 6, 2015 and April 24, 2015.

20. Newman, "It's amateur hour in China," *The New York Times,* September 3. 2015. See Bradsher, '"China Cuts Interest Rates a Day After a Market Plunge," *The New York Times*, June 28, 2015. See Sui-Lee Wee, Chinese Lawmakers' Wealth, Soaring Under Xi, Reaches Half a Trillion Dollars," *The New York Times*, March 3, 2017.

21. Krooth, *Arms and Empire, passim*. See also Mitchell Friedman. "Here's Every Single Animal That became Extinct In the Last 100 Years (Photos)," September 16, 2014, utm_source=zergnet.com&um_medium=referral&utm_ campaign=zergnet_434454). See also William DeBUYs, *The Last Unicorn: A Search for One of Earth's Rarest Creatures* (New York: Little, Brown and Company, 2015), 233-234. For a literary reference, see Christian Kiefer, *The Animals: A Novel* (New York: Liveright Publishing Corporation/W. W. Norton & Co., 2015), *32.* For plants surviving in the Andean *alto plano*, see our chapters 4 and 5. Also see Andrea Wulf, *The Invention of Nature: Alexander Von Humboldt's New World* (New York; Alfred A. Knopf, 2015), chapters 17 and 18, *et passim.*

22. Krooth, *Arms and Empire, passim*. See also Krooth, *Empire: A Bicentenial Appraisal, passim.* See also Ness and Cope, *Imperialism and Anti-Imperialism* (London; Palgrave Macmillan, 2015), *passim.*

23. Huffington Post, "Why is the U.S. Desperate to OK Slavery in Malaysia?" May 26, 2015.

24. Fred Pearce, *The Coming Population Crash and Our Planet's Surprising Future* (New York: Beacon Press, 2010).

25. For providing for world population needs *vis-à-vis* production of raw materials, foods, etc., see Krooth, *Arms and Empire*, 85-86, tables. See also Charles S. Elton, *The Ecology of Invasion, passim*. And see Wilson, *The Diversity of Life, passim*

26. Jim Yardley and Binyamin Appelbaum, "In Fiery Speeches, Francis Excoriates Global Capitalism," *The New York Times*, July 12, 2015.

27. Immanuel Ness and Zac Cope, "Preface," *The Palgrave Encyclopedia of Imperialism and Ant-Imperialism*, vii.

28. Krooth, *Gaia*, Chapter 25. "The key of India is London," Benjamin Disraeli railed the House of Lords, 5 March 1881.

29. R. Palme Dutt, *British Empire*, 44-45, *et seq*.

30. See Vandana Shiva, *The Enclosure and Recovery of the Commons: Biodiversity, Indigenous Knowledge, and Intellectual Property Rights* (New Delhi: Research Foundation for Science, Technology and Ecology, 1997), *passim*.

31. See Democracy Now: 'Secretive Deal Isn't About Trade, But Corporate Control," May 27, 2015.

32. Bernie Sanders, "The Trans-Pacific (TPP) Agreement Must Be Defeated," *Reader Supported News*, January 3, 2015.

33. See Julie Hirshfeld Davis, "No Deal, but Progress on Trans-Pacific Trade, Obama and Shinzo Abe Say After Meeting," *The New York Times*, April 28, 2015. See also Bill Powell, "In Washington, A Strategic Shift on China—Towards Containment," *Newsweek*, April 29, 2015.

34. Pope Francis, *Encyclical Letter: On Care for Our Common Home, Laudato Si'*, page 53, paragraph 109.

35. Francis, *Encyclical Letter*, page 24, para. 51.

36. Francis, *Encyclical Letter*, 67-76. See also Ness and Cope, *Imperialism and Anti-Imperialism, passim*.

37. Francis, *Encyclical Letter*, page 22, para. 57; page 50, para. 104. See also Krooth, *Nuclear Tsunami*, Introduction, *et passim*

38. Tim Mullaney, special to CNBC.com "Apple car: 5 reasons it's intriguing; 5 reasons it's nuts," *CNBC*, February 26, 2015. Miniaturization of technology has its limits with Moore's Law so-called, holding that the number of transistors in an integrated circuit would roughly double every two years—as held true throughout decades of computer development—but now faces limitations. To overcome the barriers, finance capital comes into play to build a new generational manufacturing base among computer companies.

39. Francis. *Encyclical Letter*, excerpts. 80-98. See also "Pope Francis steps up campaign on climate change to conservatives alarm," *The New York Times*, April 28, 2015.

40. Francis, *Encyclical Letter*, 80-98.

41. SunEdison, one of the largest solar manufacturers in the world, momentarily raised its corporate stock value despite the fact that the price of oil dropped to $50 from the $80s between October 2014 to April 2015, while SunEdison stock rallied 74 percent during the same period. But the extended lower oil prices, then led to Sun Edison's stock price crash, as solar could not compete for delivering electricity. See Abigail Stevenson, *CNBC*, April 9, 2015] "Cramer Remix: This Stock may be ready to breakout," said a leading investment advisor. But it didn't happen, as the price of oil in late 2015 fell to $40; and SunEdison stock plunged to $3—as solar panel production of electricity could not compete with the ever-lower cost of burning fossil fuels. And yet Morocco's desert-reach is slated to be covered with solar collectors starting in 2016. See COP 16, 2015. And SunEdison would have been the perfect supplier.

42. Pope Francis, "Meeting of Latin American and Asian landless peasants and other social movement activists," October 2014. See also Coral Davenport and Laurie Goodstin, "Pope Francis Steps Up Campaign on Climate Change to Conservatives' Alarm," *The New York Times*, April 28, 2015.

43. Yahoo Finance, "Vatican investing," May 26, 2015. Also see Bill Chappell, "Vatican Charges 5 Over Leaked Documents, Alleging Organized Crime'," *NPR*, November 21, 2015. No one really knows the amounts or where the Vatican invests its billions. Its canny bond traders do not wear church vestments during telephone-transactions.

44. Jeffrey Voegell, "Growth in billionaires wealth to taper off," *Bloomberg,* May 27, 2015. See also *Yahoo Finance*, "Gilded Era Ending," May 27, 2015.

45. E. B. White, *The Second Tree From the Corner* (New York: Harper & Brothers Publishers, 1935),128-131.

46. Naomi Klein, *This Changes Everything: Capitalism vs. The Climate*, 1-63. See also "Pope Francis steps up campaign on climate," *New York Times*, April 28, 2015.

47. Garrett Hardin, "Lifeboat Ethics: The Case Against Helping the Poor,' *Psychology Today*, September 1974. Update 10 June 203. See also Klein, *This Changes, passim.*

48. "Pope Francis Steps Up Campaign,*"* *New York Times*, April 28, 2015.

49. Jason Horowitz, "Vatican Traditionalists See Hero in Trump Aide," *New York Times,* February 7, 2017. See also "Pope Francis Steps Up Campaign,*" New York Times*, April 28, 2015.

50. Klein, *This Changes*, 74-76. See also Sanders, "The Trans-Pacific (TPP) Agreement." And see Senator Russ Finegold, *Newsletter*, November 2015.

51. See *Barnhardt Bz* posted Jan. 10, 2015 (Cached) http://www.barnhardt.biz/2015/01/09/.

52. See Andrew C. Revkin, "Tracing the Roots of Pope Francis's Climate Plan for 2015," *The New York Times* December 31, 2014. See also Leroi, *The Lagoon*, 86-87*, et seq*. The Pope's plan was a brave, if unheard, defense of Nature, the Holy Bible saying only Saint Peter as head of the universal church was empowered, before Christ's return, to spread the Gospel, save souls, forgive sins, and make infallible pronouncements on Christian faith and moral matters.

Centuries ago, Saint Francis of Assisi had gone steps farther, taking up the mantle—to secure natural species. Today the Pope has anointed himself Francis, after the 1182-birth of Francis of Assisi, protector of the poor, lepers, animals and Nature's diversity—now proselytizing among the world's religious communities, determined to save the planet and its species.

The teleological view, that everything exists for an end, dictates that the end itself is the causal force of the present coming into being. This means that intelligent design of the world, its creatures and plants is prefixed; so no Church pronouncement on the existence of these conditions need be stated by either Peter or later Popes claiming Peter's authority.

Only Aristotle, writing 23 centuries earlier from today had upstaged the Church and challenged its later proclamations. Aristotle effectively questioned intelligent design of both living matter and machines, with the subtle implication that the biosphere itself turned on scientific evolution and that the technological sphere was created by human toilers of material things, like building houses with ax and hammer.

And now in our moment the Church backtracked. Its plans to save the planet were no longer simply ethereal matters of intelligent design.

53. Gerald Posner, *God's Bankers, A History of Money and Power of the Vatican* (New York: Simon & Schuster, 2015). See also Bill Chappell, "Vatican Charges 5 Over Leaked Documents, Alleging 'Organized Crime'," *NPR*, November 21, 2015. See various editions of Murray Bodo, O.F.M., *Tales of St. Francis* (Cincinnati, Ohio: St. Anthony Messenger Press, 1988). See also variously-dated, Felix Timmerman, *The Perfect Joy of St. Francis*; G.K. Chesterton, *St. Francis of Assisi*; Paul Sabatier, *Life of St. Francis.*

54. See Chapters 1, 12 and 14. See also Mark Santora, "Seeing City's Perilous Path in Ancient Patch of Marsh," *The New York Times*, January 20, 2017, A 26-27. And see Nkrosta Langlois, "West's Superfloods May Wash Over Aging Dams, Scientists Warn," *High Country News*, reproduced at *Yahoo*, March 7, 2017. Elsewhere life evolved. Ending the Permian Period about 252 million years ago, after the largest mass extinction of 90 percent of all species on Earth, there was a fairly rapid, 1.3 million years evolutionary recovery of 750 individual species at Paris Canyon Lake County in southeastern Idaho. See Nicolas St. Fleur, "After Earth's Worst Mass Extinction, Life Rebounded Rapidly, Fossils Suggest," *The New York Times*, Science, February 16. 2017.

VII

End of an Epoch

> A confused noise issues from every bush, from the decayed trunks of trees, from the clefts of the rocks, and from the ground undermined by the lizards, millipedes, and ceclias. There are so many voices proclaiming to us, that all nature breathes; and that, under a thousand different forms, life is diffused throughout the cracked and dusty soil, as well in the bottom of the waters, and in the air that circulates around us.—Alexander von Humboldt, *Personal Narrative*

Those who followed in the footsteps of Alexander von Humboldt have watched the Anthropocentic Era gradually slipping to a close. Advancing technology has had unimaginable consequences, lighting the fuse for unending implosions of the natural order and rising social conflicts.

Our species is not alone responsible, as unforeseen natural forces also periodicallyintervene and shape the cycles of the biosphere. But during the Anthropocene, humans are increasingly the central force in stripping Earth of forests, invading the seas with toxins, and burning fossil fuels loading the atmosphere with greenhouse gases heating the planet.

Human conflicts also lay at our doorstep. We may genetically inbreed compassion for others and recognize the need to care for Nature. But when the means for survival and resources are limited at the very moment our population expands on a geometric scale, the drive for self-preservation overwhelms inbred compassion, social customs, impartial institutions, and the ability to preserve the surrounding environment.

Competing and destroying one another for living space and sustenance, humankind is now scouring the planet for fossil fuels, their burning raising earthly temperatures beyond human and other species' thermal limits to survive. For the increasing billions now being born, the human capacity and limits for extreme privations and psychological stress have already overreached Nature's storehouse of resources and cures. Our species unrelenting push to further extract and process Earth's fossil fuels and other resources are leading towards deepening seas, disappearing landscapes, declining species, and a global disclimax.

Although these are impending catastrophes besetting our era, all life on Earth will not perish. Though a decisive sixth extinction on Earth will not take place, surviving species will likely initiate an evolutionary re-emergence of still unknown forms of life.

Chapter Fourteen

Darwin's Walk and the Last Wave

> To make a case [for the existence of the Anthropocene Epoch], scientists must show that human impacts will leave a clear mark, fossilized in strata, that could be readily recognized tens or hundreds of millions of years from now by some geologist in the far future. The focus on strata is important. To a geologist, geologic strata equal geologic time. The key is a "time-rock" interval—a layer of strata that can be hammered, sampled or dug in (as for dinosaur bones) and that defines a new course of history. For the Anthropocene Epoch to have such deep geologic meaning and to have any chance of being made [scientifically] formal, it must show its own time-rock unit.—Jan Zalasiewicz, "What Mark Will We Leave on the Planet?"[1]

In 1859, Charles Darwin took a random walk in a forest basin in Australia's New South Wales. Here he perceived something clearly part of the past cycling of matter and species evolution and development. "If we imagine a winding harbor, with its deep water surrounded by bold cliff-like shores, laid dry, and a forest sprung up on its sandy bottom," he wrote, "we should then have the appearance and structure" created in the distant past.

Your author has walked along Darwin's path. And seeking a deeper understanding of the forces then and now shaping planetary life, I am returned once more to Darwin's harbor waters surrounded by sandy bottoms.

LOOKING BACK

Looking back in the aftermath of environmental destruction, Nature appears very different from what it once was; and survivors are tested to understand what happened to bring about the present condition. How, for example, will future botanists comprehend Australia's New South Wales beaches, today enduring hurricane-force waves eroding shorelines and tearing at wood-and

Figure 14.1. Author on Darwin's Walk.

steel-structure homes? Will they sense the remaining wood fibers once were the storehouses of carbon in forests, or that the iron girders were once part of sulfide ores and oxide land ridges, time-rocks now being returned to the sea?

And what if a future Last Wave lays almost the entire Australian continent awash? Will a future intelligent being understand how any remaining contours of beaches appeared and evolving sea species emerged?

1. LAST WAVE: AUSTRALIA & BEYOND

Today in the wake of Australia's evolutionary drama, its magnificent natural domain and industrial chaos have overlapped and clashed. Its ever changing environmental surroundings, materialized over four billion years as the first, reconstituted continent on Earth, after the breakup of the single landmass of

Pangaea, passed through five great moments of natural changes and species extinctions, each highpoint punctuated by the following distinctive evolutionary living forms and changing conditions . . . and then declined.

- Lungfish with evolving legs once crawled onto the beaches of New South Wales.
- Great forests raised oxygen levels one-half greater than today, the oxygen rusting ferrous soils and leaving behind breathing species strange and beautiful.
- The last Permian forests were transformed into underground coal through millennial geologic pressure.
- Soon a new Ice Age gouged out half the continent, leaving a giant inland sea basin with evolving voracious predators which in time expired when the sea turned to desert.
- And thereafter volcanoes disgorged hydrogen sulfide that destroyed 95% of the continent's remaining land and surrounding sea species.[2]

From New Guinea then came an Aboriginal population some 50,000-to-100,000 years ago. And nearer our moment, British and other European immigrants arrived searching for land, materials to mine, and later, to build industries, trying to emulate Great Britain's successes. In one-and-a-half centuries these European invaders' violent culture practically destroyed the Aboriginal population.

Today the nation has become a vast industrial and commercial complex that has overlapped with Nature's unequaled beauty and terrifying natural cycles periodically enveloping the continent in intense heat, fierce storms, blazing fires, and clouds of nearly impenetrable dust and fury.

The Future

The future remains uncertain, as these forces one day might converge as an undulating typhoon with incalculable destructive power—bringing on a last wave that once more will gouge out nearly the entire continent, leading again to a giant inland sea swarming with unimaginable, evolving voracious predators.

Such future events might seem unbelievable, an unfair warning from a foreign visitor like myself. But as metaphor, I am speaking about every continent and island on Earth that will be deluged as a torrid planet melts polar ice caps and permafrost; raises future ocean levels as much perhaps 30-90 meters (90-270 feet); evaporates water vapor and thermohaline moisture; causes violent downpours; and loosens powerful winds, atmospheric rivers of water, and waves engulfing both islands and entire continental coastlines.

These predicted events would follow the Anthropocentric Epoch, our human-directed, dominant geologic force, barely self-aware, shaping the planet.

Conditions for Life on Earth[3]

The geological timescale of Earth, defined by the International Geological Congress, is divided into units, the longest of which are billion-year-scale Eons. These Eons run from the Hadean's blistering heat to the cooler Archean nurturing the emergence of life; then to the Proterozoic flooding the atmosphere with photosynthetic oxygen, chemically poisoning much of the atmosphere, yet leading to complex multi-cellular life. This set the foundation for the current Phanerozoic Eon. And a fifth Eon might turn on human thought, deeds, and ingenious creations of a Superman recently named Sapiezoic, lasting as long as humankind itself.[4]

Periods, then Epochs and Ages, are lesser time-rock dividing lines. Scientists measure them by studying such things as fossils and rock layers. Here we review some.

Proterozoic Eon and More

The Proterozoic Eon spanned almost half of Earth's entire existence from 2.5 billion to 541 million years ago, in which three lesser Periods appeared. The Proterozoic, itself lasting 500 million years or so, was punctuated by a trilogy of forces, including: photosynthetic algae exuding atmospheric oxygen; iron bands forming ridges on the sea floor; and biological eukaryote cells as precursors of all earthly plants and animals, their DNA sequestered in a nucleus.

The Proterozoic's next 1.85 billion years to 850 million years ago witnessed animal migrations and disappearances of landed territories across the globe. This involved the many-sided assembly of Pangaea, then separation of super-continents; the relative stability of sea chemistries and minerals; and small-scale evolution of ocean life.

The third period proved to be life's major starting point. Up to 850 million years ago, most continental territories were locked together around the Equator's dry, denuded, rust-coated surface, blocking species life. These territories were surrounded by the Mirovian Sea, lacking sufficient oxygen and releasing a flow of gases unable to build a protective layer to ultraviolet ozone (O_3) for surface life existence. Complex geologic and atmospheric forces shaping evolutionary life were about to begin.

Life Evolves on Earth[5]

Life had many evolutionary beginnings. And periodically Earth reached a disclimax—the end of a biological community of plants, animals and fungi living in a sustainable state of being—about 740 million yeas ago.

The planet had endured alternating sequences of glacial conditions and torridity, surviving species avoiding extinction by evolving into competitive intermediary species. In fact, the evolution of ecosystems and microbial growth had been interlocked—directly tied to the availability of phosphorous nutrients in neo-proterozoic seas. It is likely this phosphorous formed the backbone of the genetic molecules DNA and RNA—stabilizing living cell membranes, as well as storing and transferring cellular chemical energy.

According to Robert M. Hazen's *Story of Earth*, and Richard Smith, Australia's geologist *par excellence*, these new microbes colonized coastal landscapes and turned hard rock into soft clay. The organic molecules then led to a clay-mineral production cycle contributing to the rise of atmospheric oxygen as the foundation for evolution on Earth—the burial of carbon; the rise of oxygen; the accelerated chemical production of more clay minerals on land, and still more carbon burial, followed again by the production of oxygen. It's a viable theory.

Yet another theory turns on discovery of ancient lava flows from Iceland's Hengill volcano that combined carbon dioxide with the elements of basalt and became calcite, a solid crystalline mineral—turning carbon dioxide into stone.

Could our species find a way to mimic this ancient volcanic process? Today we likely will not have time to save the biosphere by trying to bury sufficient atmospheric hydrocarbons exposed to the Sun's heat, reradiated from Earth, in order to lower the planet's temperature. Volcanos did that. But this method probably would not now allow Earth to return to its past geologic wonders that once saved diverse species and brought us to our evolutionary moment in time.

Past and Present Evolution

Our burning of Earth's fossil fuels continues, likely one day to snuff out most oxygen-dependent species. For now, though, life's story is hardly over, sidestepping a sixth extinction. Reviewing the past, we see that continents were formed and torn apart; at least three times, ice and torrid ages came and went; species evolved and were destroyed; new species evolved from those remaining.

Coming closer to our moment, new environmental conditions emerged after the second of three great Ice Ages; natural mineral resources were in abundance, sustaining life's vibrant evolution. About two billion years ago, single cells began to develop symbiotic relationships with other cells, reaching the point where they merged, yet retained their individual character and special functions—larger cells taking smaller ones under their wing, an evolutionary form of endosymbiosis, where one organism lives inside the other.

Then about 630 million years ago the rise of atmospheric oxygen drove single cell evolution, as well as multi-cellular cooperative life among worms and jellyfish. The cycles of neo-proterozoic ice-to-hothouse Earth, leading to this rise of planetary oxygen, supported a vast diversity of animals, plants and fungi. Likely, all life came from the seas. Ocean life then invaded the land some 580-to-545 million years ago. The Edicara beach heads of southern Australia brought sea and river life ashore; and soft-bodied animals appeared in shallow seas throughout the world.

Yet the planetary temperature cycle from an Age of Ice to an Age of Thermal Heat, every 5,000-to-55,000 years or so, had little to do with the level of carbon dioxide in the atmosphere, as the Milankovich cycle reveals. In contrast, in our time of Earth's orbit around the Sun, the burning of fossil fuels creates the source of thermal hot-house conditions, creating a heat blanket surrounding the Earth.

2. DISCLIMAX AND SPECIES EXTINCTION

Which converging forces on Earth might speed entropy and a loss of resources and favorable conditions essential for species survival and reproduction? Between 200,000 and 160,000 years ago, after a devastating Ice Age, anatomically modern humans appeared in Africa. Our forebears, the hominins, evolved more than 65,000 years ago, even before the age of the dinosaurs. Even earlier, around 75,000 years ago, a so-called "volcanic winter" spewing a sulfuric cloud may have blocked plant life, also leading to the death of most emergent humans.

Anatomically modern humans out of Africa appeared earlier than 75,000 years ago, revealed by western Asian fossils and a so-called admixture between Africans and Neanderthals. As well, anatomically modern human Papuan New Guineans migrating from Africa have an even earlier travelogue, 2% of their genome different from those of later migrating Africans.

Some 200,000 years ago modern humans had evolved in Africa. Their exodus out of Africa and into Europe between 50-120 thousand years ago witnessed them crossing vegetative pluvial corridors from northeastern Africa into the Arabian Peninsula and the Levant, then expanding further into Eurasia, Australia and the Americas. Over massive periods of time-varying climate and sea level conditions during the past 125,000 years of the late Pleistocene, the exodus of African populations was thus distributed globally.

Though there were many dispersals out of Africa, modern humans trace their ancestry back to the single surge between 50,000 and 80,000 years ago. Those that departed well before 50,000 years ago possibly went extinct or were not part of the dominant strain of genomic inheritance of non-African Eurasians. No one knows where they came from or went.

Dispersal routes of modern humans from China started 30,000-to-70,000 years before those from the Levant and Europe, for example. But Neanderthals may have blocked their entrance into Europe before 45,000 years ago.

Some argue the Neanderthals merged their genome with invading Africans. Still others argue that from a single lock of museum-century-aged Aboriginal Australian hair, held in a museum, lies proof that its DNA variants are absent from Eurasians, but similar to, by some 98 percent, of the genome of peoples of Papua New Guinea—as paleoanthropologists over 100 years ago documented their exodus tracing their single-out-of-African origin. The remaining 2 percent of the DNA of Africans exiting in Africa, as long ago as 140,000 years past, perhaps merged with that of ancient people, or destroyed them.

There is no doubt, though, that between 100,000 to 200,000 years ago, African populations were searching for routes to cooler climates, where the plant life survived in plentiful rain, animals survived on plants, and some carnivorous animals survived on others.

Those people who already lived in regions like Southern Africa with adequate plants and wild animals, such as KhoiSan hunter-gatherers, perhaps split away from other living humans, leaving their own DNA variants among survivors.

But DNA samples are relatively consistent among other survivors from many hundreds of indigenous populations in Spain's Basques territories; among United States Indian tribes, South American Myans, North American Aztecs, North African Bedouins, India's Sherpas, and Chinese, Saudi Arabian and Israeli peoples.

Does this imply that those with the same DNA tracers in different geographies were satisfied with the resources they possessed or needed, and did not feel the need to seek those resources from others by aggression or war? Or is the need for adequate food, pure water, and possession of land to hunt and plant a DNA-driven human harbinger of local, regional and eventual world division and conflict? Did people enslave one another and willingly or forcefully exchange their seed to produce progeny who later also were enslaved or abused by hominins with 3.1 million old stone tools?

After the last Ice Age warming had begun, a new wave of African invaders possibly combined their genome with those of Neanderthals who, for 250,000 years, had occupied frigid continental Europe, Gibraltar, and England. Hunters and gatherers in Northern France and Britain had meanwhile survived by developing Mesolithic tools and languages promoting social cooperation, and the trade of wheat with Neolithic farmers living in the warmer climate of Southern Europe.[6]

As the last Ice Age gradually watered soil in a Sun-splashed Northern Europe about 8,000 years ago, farming took hold; human populations grew on a wider diet of carbohydrates. And yet during our moment of the Anthropocene, we have developed a technological sphere overwhelming the biosphere, bringing on regional ecocides and the disproportionate distribution of worldly land, resources, and material wealth.

Moving beyond the evidence of the rock-ages 3.1 million years past, and 200,000 to 160,000 years ago, today several major forces are overlapping, leading towards a possible destruction of conditions essential for life existence, and bringing on a global disclimax—again meaning the end of a biological community of plants, animals, fungi and humans living in a sustainable state of being.

The possibility of a new Lethal Heat Age or even another Ice Age could destroy most of Earth's remaining agriculture, leading to conditions so adverse that most existing species would be destroyed at a faster rate than new species could evolve.

Heat Rising

On June 2, 2016, the world already was about 1 degree Celcius (1.8 degrees Fahrenheit) warmer than in pre-industrial times. Stunningly, today's heat puts the world on the path to 3 degrees to 8 degrees Centigrade (5 degrees to 15 degrees Fahrenheit) more. It was the hottest year ever recorded in modern times.

Arctic permafrost is now fast melting. If global warming increases another 2 degrees Celsius, it will force some tropical plants, animals and humans to try to avoid even a small increase in heat by moving as far as 1,000 miles to cooler regions—avoiding rising isotherms, moving to higher isolines, longitudes and upward elevations in "an almost complete evacuation of the equatorial band."

A mid-2016 study (*Nature Climate Change*, May 18, 2016), led by Katarznia Tokarska (a climate scientist at the University of Victoria, British Columbia) says that by the year 2300 if we continue to burn our existing fossil fuel resources, the Earth will encounter heat of 6.5 to 9.5 degrees Celsius (8-12 degrees Fahrenheit) above the 20th century average—melting glaciers; increasing sea levels; drowning coastlines and low-lying regions that are home to most of the world's population; disrupting farming and food supplies; throwing tens of millions of people into poverty; and turning 200 million people into environmental refugees.

Environmental science findings require action. It only took a modest 3-to-8 degree warming to bring the world from the frigid depths of the last Ice

Age into an era of warmth, says Sivan Krtha, a senior scientist at the Stockholm Environmental Institute, specializing in climate risk. Warming on that level, he judges, raises "The specter of a future where the surface of the earth is again radically transformed."

Environmental Apocalypse

This warning is echoed by Kate Sheppard in the title of her article, "Scientists Warn We're Ever-Closer to the Apocalypse" (*Huffington Post,* January 23, 2015). Exactly how that might eventually occur remains unknown, but surely will involve the overlapping forces we have elaborated in the foregoing chapters, including:

- Human reduction of habitat areas by deforestation, conversion of grasslands to other uses, and climate change.
- Alien bird, animal, viral and mycennial species invasions of continents and islands, causing damage to humans and environments, creating warlike struggles between species for survival, with the annihilation of some, the temporary domination of others, and a fall in species diversities.
- Pollution caused by human and natural conditions, inflicting heavy damage to both fresh water and seascapes, destroying fish and other aquatic creatures, as well as setting off flows of contaminated upstream agricultural wastes, deadly effusions from mining and oil exploration, and human plastics and toxins, creating anoxic 'dead zones' in marine waters.
- Worldwide exponential population growth, leading to displacement of forests and frontier territories to grow foods; impacts on both inland seas and ocean basins; over-fishing and undermining crustacean life; creation of fish farms spreading Polychlorinated Biphenyl (PCBs) and other poisons; and, the latest danger, sea-breeding of escaping bioengineered salmon from inland tanks in facilities in Canada and Panama—risking the decline in diversity by combining injected genes of Pacific Chinook with wild Atlantic varieties, making them grow faster.

As these forces merge or overlap on a global scale, they become lethal to most living species.

Earth Burns

Your author's study of Gaia and the Fate of Midas turned on the ways in which the technological sphere impinges on and destroys the biosphere. Dozens of biologists, geologists and anthropologists also have elaborated on the links between various technologies and environmental peril now and in the future.

These technologies have accompanied the degradation of the biosphere. Humankind has developed and expanded the technological sphere, that in turn has shaped and taken charge of the biosphere's atmosphere, lands and seas, yet lost control over heat and waste on a global scale. Thus in the post Holocene era that had lasted some 11,700 years, the emergent Anthropocene has been beset by:

- Modern fossil fuels and fertilizers that have multiplied concentrations of microbes creating nitrous oxide above natural levels in Earth's soils and oceans
- Agricultural expansion, natural gas production emissions, and permafrost thaw that together have enlarged methane expelled into the atmosphere
- Post World War II continuing the industrial revolution's concentration and escalating release of carbon dioxide to the atmosphere
- Atomic and nuclear bomb releases on Japan and tests begun in the 1940s, ejecting the isotopes plutonium 239 and 240 into the air that settled in the ground, so that globally in 100,00 years they will decay to a layer of uranium 235. To these are added the fissionable explosions of nuclear power plants at Three Mile Island, Chernobyl, and Fukishima
- Burning black coal since the 1800s, causing the airborne by-product black carbon, increasingly blanketing Earth's rocks and ice formations since the 1960s
- Concrete production overwhelming groundcover since World War II, leaving a lethal cover of the Anthropocentric Age
- Plastic polymers found in rock sediments as well as slowly decomposing in rivers and oceans, scattering their pieces among and within living creatures.

As humans have become a new geologic and anthropogenic force, and since the 1950s have decidedly changed the range of planetary technological residues transformed into rock-fossils, we have permanently destabilized the cycles of planetary life and jeopardized Earth's energy balance—whether or not we name it the Anthropocene or simply the technological sphere overwhelming and endangering the remains of the biosphere.

Earth's Loss of Energy Balance

Earth is out of energy balance—meaning more energy and heat enters the planetary sphere than goes out—as a result of increased atmospheric greenhouse gases, especially carbon dioxide, absorbing Sun radiated to the planet and re-radiated to the atmosphere.

This is exacerbated by the long-term warming of the Eastern Pacific Ocean, increasing the chances of a torrid *El Niño* comparable to the two strongest *El Niños* of the past century in 1998 and 1983. Combined, these forces producing torrid heat and rapid evaporation of moisture create atmospheric rivers 300-or-more miles long, capable of plunging down as "water bombs" on terrain where winds, surf and weather patterns carry them. These forces also produce storm surge at the height of coastal tides, further endangering both human and species survival already suffering from toxic ocean wastes.

Further evidence of Earth's resource imbalances:

- In the year 2014-15, cities and industries worldwide are already receiving barely 25 percent of their full water needs for human health and safety.
- And besides such water shortages, as no society that burns fossil fuels can escape heat and and waste, there remains the problem of toxins and affluents dumped into air and water supplies by oil companies' drilling, fracking, refining and delivery operations, as well as by petroleum-base fertilizer run-off to seas, rivers and lakes; and by burning tires ejecting lethal substances into the air; and other forms of garbage disposal onto the land and into the sea.

Human Costs

We can expect that the human cost will worsen as Anthropocentric forces continue to invade air, soil and water.

- Severe drought during the last half of the 21st century will plague the U.S. Southwest and Central Plains with extremes of water loss greater than any dry period known within the past 1,000 years.
- Too many people concentrated in certain geographical areas, like California; the failure of crops to sustain expanding populations, and their needs for more fresh water in a time of extreme global heat, evaporation exceeding snow-melt and sources of rain, fog, and other precipitation.
- In addition, regional impending disasters, as from California's decaying levees which protect fresh water from salt water during storm surge, earthquakes and the overdraw of the San Francisco Bay Delta watershed.
- Add to that the unfunded Government plans to build $14.5 billion tunnels to carry fresh water to sustain 25 million Californians from Northern California's Cascade Mountains to Southern California's Tehachapis, depriving fish and other wildlife of their life source.

California is at particular risk. At a news conference, March 19, 2015, California Senate President Pro Tem, Kevin de Leon, warned: "There is no greater crisis facing our state today then our lack of water." And in *The Los Angeles Times*, March 12, 2015, Jay Famigletti, a scientist with NASA's Jet

Propulsion Laboratory in Cañada Flintridge, California, wrote: "Right now the state has only about one year of water supply left in the reservoirs, and our strategic backup supply, groundwater, is rapidly disappearing. California has no contingency plan for a persistent drought like this one (let alone a 20-plus-year mega-drought), except, apparently, staying in emergency mode and praying for rain."

Though in 2016, Northern California was temporarily relieved by *La Niña* downpours of cooling rains, state maps indicate that the areas of California hardest hit by the *El Niño* mega-drought were those that grow a large percentage of U.S. food. California supplies 50 percent of the nation's food and more organic food than any other state. Western growers estimate that in 2014, 500,000 acres of farmland were left unplanted for lack of water, possibly increased by 40 percent in 2015. Farm job losses in 2014 were 17,000 and even more in 2015-16.

The future for water shortages is predicted to be worse yet, as farmers with contracts from the Central Valley Project, a large federal irrigation system, received no water for the second half of 2015, at a loss of $2.2 billion. Farms went broke; wells went dry; food production fell; nut trees died; farm workers lost jobs; and the great internal watering hole that once sustained Imperial Valley crops and farm families became the haunt of voracious flocks of birds and wild animals searching for a drop of salty water.

Cause-Effect: Evolution Itself at Risk

How can we understand the cause-effect of the Anthropocentric global water crisis and the effects of the expropriation of natural resources? From atmospheric precipitation to terrain and subterrain, natural flows of pure water to needy populations have been cut off, cinched with spigots at their sources by new robber barons, avaricious individuals and corporate groups whose power derives from their wealth—allegedly including Goldman Sachs, J.P. Morgan, Chase Manhattan, Monsanto, and some members of the Bush family—purchasing water rights all over the world, viewing fresh water as their own legal commodities to be bought and sold like rice, cotton or grains.

Populations unable to purchase fresh water must go without and turn to polluted sources. And lacking sufficient water in their various environments, other species are being eliminated without offset to balance the loss by evolution of new species. For example, *The World Wildlife Fund* reports that populations of more than 100,000 vertebrate species declined by 52 percent on average between 1970 and 2010. In South America, vertebrate species declined by 83 percent.

Once upon a time, mainland and island species were protected from invasive ones. But no more, as our Chapter 13 proves. During the 16-year period, 1993 to 2009, the human footprint on the global environment increased by 9 percent, writes Oscar Venter, lead author in *Nature Communications*, August 23, 2016.[7] Another study shows that today three-quarters of the Earth's land surface faces measurable human pressures, while only 3 percent of the world's biodiversity hotspots remain unaltered. "There's little wonder why there's a biodiversity crisis," said James Watson, co-author of this study and director of science and research initiatives at the *Wildlife Conservation Society*. The human invasions of the Galápagos, for instance, came with their goats, donkeys, cats and rats that do more destruction to native species than are saved by such projects as the introduction of ladybugs driving back cottony scale bugs that feed on indigenous trees and plants. Though 95 percent of the species that existed on the islands during Darwin's time are still present, here and elsewhere the catastrophe of invasive human and other species is still unfolding on Earth.[8] As head of the Universal Catholic Church, Pope Francis says, evolution is at risk from invasive species infringing on the biosphere.

Other forces are blocking evolution as well. The powerful plutocrats of oil, coal and gas wealth, for example, are causing the evaporation of greenhouse gases, turning the skies into a heat blanket absorbing the Sun's radiation re-radiated from Earth's face. By operation of the second law of thermodynamics, this has spread heat to the cooler portions of Earth, leading to warming and further evaporation of greenhouse gases, and more melting glaciers, sea ice and tundra.

Heat also slows down the circulation of the global ocean conveyor belt, which is a constantly moving system of deep-ocean motion driven by rising temperatures and salinity. More and more, the heat will lead to a sharp sea level rise as part of the larger global increase in ocean thermohaline temperatures and salinity, driving the U.S. East Coast Gulf Stream—producing coastline overflows in cities like New York and Boston.

Coastlines Overflowing

No surprise, the sea is rising. And once the salty brine invades the land, its soils may never recover to produce sufficient plants and foods to sustain animal life on a scale sustaining multiplying populations.

Continental coastlines overflowing and islands submerged below the sea may signal the human culmination of the Anthropocene, recalling victorious Rome plowing salt into the soils of Phoenician strongholds to prevent farming in what has become today's desert reach of Tunisia. At that time, Romans poisoning the soil with salt defeated a nation: Without food to sustain its galley slaves, chained to oars, moving its war ships, Phoenician power was no more. No food, no life, no empire. Today bioengineering is poisoning the soil, which also degrades it and disintegrates its essence. Salt can destroy.

But in our time, it will be brought to land by ocean overflow—again caused by humans creating the forces we have described.

Disappearing Landscapes

We already know the cyclical aeration of greenhouse gases from burning fossil fuels has a direct impact on the planet's accumulation of heat, as these gases capture solar rays that have been re-radiated from Earth—at once melting ice at the poles, accelerating evaporation and dehydrating soils that can no longer sustain plant, animal or mycelial life as the vegetable part of a fungus.

As we have seen, based on scientific advances, the industrial and technical sphere are moving fast-forward with a grim finality of intense heat, destroying the biosphere in a time-frame that the natural world and species evolution cannot sustain, or rehabilitate themselves within.

What Destiny Awaits the Tree of Life? Examples to Ponder

A modern tree-cutting saw run by gasoline or electric-current cables can destroy in a moment a 2,500 year old redwood, oak, or cedar. But a new giant sentinel equal to the marvel just destroyed will take from today to an unknown tomorrow to grow—or perhaps never grow.

Where oil drill bits rip apart the forest floors of roots and fungal life to extract oil—black gold—this destruction can also lead to crude petroleum spills, or to salt water invasions from nearby seas, or to land subsidence. The oil pollutes land that may never to be rehabilitated.

Consider the proposed erection of dams in the Amazon, fitted with sluices to generate electricity from overflowing river waters, flooding over land that has been farmed for over 10,000 years by indigenous communities, never to be resurrected—despite bitter strife and pending guerilla warfare in defense of land rights.

There is more to this. Ten U.S. timber, financial and industrial corporations are planning to build a dozen gigantic dams along the Amazon river basin, operating with approval of both the U.S. and the Brazilian and Peruvian states, using military power against indigenous people—evicting, criminalizing, incarcerating them, threatening their lives and families—as their leaders have described this lawless order to Amy Goodman on "Democracy Now" at the U.N. Climate Conference, (COP), in Lima, Peru, December 8-12, 2014.[9]

As with the British oil company fighting to drill for oil in Verunga, in Congo, these corporations have their own armies to arrest, torture and murder those opposing the cutting and destruction of age-old trees and the mining of resources, to say nothing of the wildlife that is also devastated.

The environmental impacts on soil, water and life are unending. Mine tailings alone despoil and poison river waters; lead to sickness and death of animals, plants and people; block river flows, causing drought, so nothing grows and entire communities face lack of potable water, dehydration and hunger.

Protest, bitter strife, and pending guerilla warfare are on the future calendar. While representatives of 500 transnational corporations were meeting in Lima, Peru, to set their own invasive agenda, at the same time, the United Nation's COP15 was officially in session, the streets were filled with citizens and community protestors from over 300 nations. They were penned within five corridors of heavily-armed, Peruvian crowd-control police, keeping the protestors away from the corporate meetings nearby, as well as from the official COP sessions held at the 1975 army detention center—the very place where the government had tortured and killed hundreds of indigenes opposition forces trying to protect the forests.

By the Amazon River live 14 to 15 of Peru's native communities with no contact with the central Peruvian government. Another 1,135 tribes are being threatened by tree-cutting and mining companies in their own territories.The government is unable or unwilling to protect them against illegal expropriation of land, deforestation, and the mining of gold and other metals. And aligned with the invaders, the Peruvian government has passed immunity laws to exonerate their own armies from liabilities for torture and murder of Earth's defenders on their own land.

Though a new pattern of self-defense is emerging among indigenous peoples, set on securing Nature's realm, will they become Earth's only defenders of territories that have been legally set aside as their preserves? Can indigenous peoples alone stop the heating of the planet's atmosphere or the degradation of soils needed to plant or to hunt in forests?

Up till now, they have had their own defending weapons—casting spells on evil invaders, shooting poison darts and arrows—against workers representing invading logging and mining companies, while U.S. and Canadian transnationals are trying to cover up their illegal forest-cutting and mining tracks—removing their responsibility for the devastation by nominally passing their vested interests and lethal practices to Chinese companies invading Brazil, Peru, and a half-dozen other Latin American countries.

Pope Francis blames many of such injustices by the global economic system on invasive international "corporations, loan agencies, certain 'free trade' treaties, and the imposition of measures of 'austerity', which always tighten the belt of workers and the poor."[10] Again, what the future holds is unknown. But the aligning puzzle pieces make securing the biosphere unlikely in the near future.

3. CHANGING THE RHYTHMS OF THE NATURAL ORDER: TECHNOLOGICAL INVASION VS. CLIMATE INTEGRITY

The world's 500 major fossil fuel companies which explore, drill, mine, and burn forests, have changed the rhythms of the natural order. They have effectively altered the seasonal sequences of agricultural production. Peasants and farmers now plant seeds that do not germinate into crops, as the skies are befouled by debris and smoke blocking out sunlight.

Rains do not come when sunlight can't reach groundcover to evaporate and cause plant transpiration sufficient for the accumulation of aerated moisture, as during the once-traditional rainy season. The technological sphere drilling oil, fracking gas, mining coal, burning fossil fuels, overwhelms the skies, drying crops and emaciating barnyard animals, causing insufficient food supplies to feed local communities and trade at crossroads to earn enough for their other needs.

These fossil fuel producers also release atmospheric particle matter that blocks and poisons animal and human respiration, especially impacting the health of children, asthma patients, pregnant women, and urban populations trapped between mountain-like skyscrapers amidst the descending toxic web of wastes.

Dangers, Sickness, Death

Then there is the deadly trail of excavating and refining fossil fuels, as well as the unsafe delivery of energy to ultimate users. One example of this is the procedure of extracting natural gas under explosive friction pressure. Such hydraulic fracturing (fracking) extracts this gas as a source of energy by drilling or injecting 478 different chemical fluids deep into terrain at high pressure, breaking shale rocks and releasing natural gas. (In contrast, using conventional technologies, the shale layers of organic-rich, fine-grained sedimentary rock of low permeability cannot otherwise be reached.)

Sand, water, methane and still-unpublicized chemical derivatives (like salt, benzene, lead and other acids) are pumped under high pressure into drilling wells based on scientific advances. A point is reached when the pressure exceeds the rock strength, the shale is desiccated, and the fluids released open or enlarge fractures, allowing the natural gas to flow relatively freely from the rock pores to production wells that are often several hundred feet away, bringing the volatile gas and oil to the surface to be contained in isolation or storage tanks.

These releases impact nature, disrupting Mother Earth, laying the conditions for a potential future disclimax of the essential conditions supporting fresh water and air purity for animals and community life, as well as for the surrounding environment. Fracking has already inflicted sickness and death on miners, transportation workers, and those needing pure water and a minimum of 17% oxygen-laden air. Though shale gas provided only 1% of total U.S. gas along the path of pipelines and well resources in 2000, by 2050 the output might reach 20%, and promises to reach 50% by the end of the century. The U.S. thereby hopes to become the Saudi Arabia of the West.

Industrial Society and Sea Level Rise

The obvious question is this: To offset any dangers caused by burning gas as well as coal and oil which heat Earth's atmosphere, must not communities and nations financially support one another to escape the forces that have already built a greenhouse atmospheric cloud surrounding and heating the planet? This heat leads to melting polar regions; and sea level rise will leave less and less land for Earth's increasing population to survive; more hunger, malnutrition and a Darwinian war of the fittest.

There is much speculation about future ocean levels. During the course of two-and-a-half centuries of the continuing Industrial Revolution, the atmosphere has continued to accumulate greenhouse gases and absorb re-radiated solar heat from land and sea. This blanket of atmospheric heat seems certain to melt ever more glacial ice and permafrost, eventually raising sea levels by at least 30-90 meters (90-270 feet) in the 21st century. And as the sea itself continues to absorb most solar radiant heat, the heat also causes the water to expand and thus contributes to sea rise.

Today the central question remains, how to offset global radiation from the Sun's heat causing warming and rising seas, says Nnimmo Bassey, the head of the Mother Earth Foundation. A historic check reveals the ocean has been rising slowly and relentlessly since the late 19th century, one of the hallmark causes being the changing temperatures and climate of Earth. The average global sea rise has been about eight inches since 1880, but the local rise has appeared higher in some places where the land is also sinking, as in Louisiana and the Chesapeake Bay regions of the United States.

In 2011-12, sea rise has appeared to have accelerated, raising the total to about a foot per century; and many scientists expect a further elevation of sea levels as the warming of the planet continues. The oceans conceivably could rise another foot over the next 40 years.

Sea level rise has already cost governments and private landowners billions of dollars as they have pumped sand and stone onto eroding beaches and repaired the damages from storm surges. Experts confirm that a few inches of sea level rise can translate into large incursions by the ocean onto shallow coastlines.

Other experts now expect sea levels to rise by as much as 1 to 2 meters (3.3 to 6.6 feet), depending on how much additional carbon dioxide humans pump into the atmosphere from burning fossil fuels and from land-use changes. The figure could be higher still, with some researchers and *Christian Science Monitor* experts arguing (2015) that ice sheets atop Greenland and West Antarctica may lose ice faster than currently anticipated.

By any scenario, continental coastline seas will either gradually or quickly rise—as they have already by some 3.9 inches on the U.S. Northeast Atlantic coast in the two years 2009-10; and 5 inches on the U.S. Western coast near Portland, as Jianjun Yin wrote in *Nature Communications*, February 2015.

Low-lying islands and atolls are being slowly submerged. The Pacific island Kirbati is already suffering storm surge, tropical cyclones, increasing erosion, and saltwater invading its wells and fresh water supplies. King tides twice each year arise when the moon is closest to Earth; and as the surrounding sea rises, Kirbati's 11,000 people have suffered higher temperatures disrupting food supplies. Without pure water, dengue and other dangerous fevers have appeared. Planning to migrate within decades, they have even purchased lands in Fiji.[11]

Imbalance of Food Supplies

Food supplies remain critical. A stark reminder, February 11, 2015: land-produced food will not survive climate change droughts that already are more severe than those over the last 1,200 years. In his classic study of *Climate Cycles and Tree Growth*, Vols. I-III, (Washington, D.C: Carnegie Institution, 1936), A. E. Douglas documented tree-ring records of eleven-year cycles of lack of rain, rising local temperatures, the shriveling of vegetative life and food supplies leading to the exodus of the Native American Anasazi in the Southwest.

Will our present balance of food supplies and species' traditional geographic settlements again be upended? Following Douglas's comparison of tree rings past for signs of a drying era, which drove away the southwestern Indian tribes, with the present consecutive eleven-year drying period in the same Southwest geography, it may be possible to predict the future of this desert. No water for cattle, agricultural crops or human populations would bring on irremediable drought conditions in an already very dry desert climate.

The logic is that as deserts are enlarging and populations are growing in the U.S.' Southwest and Great Plains, there will be a sharply elevated demand for water, creating droughts for decades over consecutive years. Such projections also are impacted by ground water loss, failing precipitation, and higher temperatures.

Clearly, drought caused by searing temperatures again would make it impossible to maintain present tiered interlocks of feeding and breeding chains, and nourishment patterns of species and communities. Deadly cascades of plants, animals and fungi would be quickened, unhinging the survival of unnumbered symbiotic species.

This oncoming, so-called Drought Age takes on Biblical proportions.

"The 21st century projections make the previous mega-droughts seem like quaint walks through the Garden of Eden," says Jason Smerdon, climate scientist at Columbia University's Lament-Dohert Observatory.[12]

Will This Lead to the Extinction of Species?

Those who calculated past causes of mass extinction once believed the ecosphere was supported by a momentary static and balanced ratio of about 100 plant species for every ten invertebrates and one vertebrate—the 100-to-ten-to-one supply ratio. If the proportions temporarily changed, as plants suffered a drought, for example, there would be too little food to support dependent invertebrates and vertebrates, and their numbers then would fall until a new equilibrium level was established.

In contrast, if adequate Sun radiation, water and soil nutrients enlarged plant growth as a food supply for invertebrates, these conditions would also enlarge their populations as a nutrient supply for carnivores—guiding a new balance.

In our time, these ratios have been challenged and undercut by human pollution of water, land and atmosphere; the massive invasion of alien species as well as an over-population of some species drawn to the surplus food resources, seizing territories and causing the habitat loss of indigenous species. The biological ratio of species varieties, types and numbers are constantly being altered by the availability of food supplies for predators and prey.

Charles Darwin, Charles S. Elton and Elizabeth Kolbert have documented the effects of invasive species. Almost all of them came in the wake of extended colonization, including plants, fungi, animals and humans that are not native to a specific location, and which damage the environment and indigenous species. My study, *Gaia and the Fate of Midas*, traces the details of such colonization, particularly how it transpired in Hawaii. And Ernst Mayer also has explained its historic process: a predator species with an enlarged food supply of prey soon becomes the surplus prey of still other predators.

Disease organisms can also establish a foothold. Even a single pathogen of uncertain origin can invade new hosts, causing their destruction. For example, Chytrid fungus, possibly of African origin, has invaded Australia, New Zealand, Europe, North America, the American Tropics and Africa, resulting in the reduction and loss of scores of species of frogs.

Think of humans as an invasive species, especially as seen in the last century, when human domination of land and sea has reduced Earth's biodiversity through the extinction of at least 10 percent of its estimated total plant and animal species.

The present ratio of all known plant species (307,700) plus known fungi (48,500) compares with known animals (1,371,500). Thus such plants and fungi are only 21% of the animals' potential food supply, indicating a shortage of food for any enlarging animal population. Both invertebrate and vertebrate (including humans) make up the shortage in part by also consuming or destroying one another.

Obviously in our moment of the Anthropocene, the swings of climate (escalating thermal temperatures and irregular rising evaporation reimbursing precipitation) affect plant and animal transpiration and survival, and force the movement or displacement of many species.

As their food needs exceed the supposed earlier 100-to-ten-to-one supply ratio, the natural balance of food supplies among predator and prey are constantly being altered, affecting the frequency of death and reproduction ratios of all species. As well, anthropogenic heat has overcome plant and animal maximal thermal limits, setting the scale of either their demise or movement along isolines to higher elevations, or along longitudes reaching towards the remaining polar ice caps.

In terms of today's estimated global population of 450,000 plant species plus 10 million fungi species, compared to 11 million animals of all varieties, there also has been a constant rebalancing of food supplies with the needs of consuming invertebrates and vertebrates; so much so that the sharp decline in animal species has also readjusted Nature's plant and fungi species populations, causing an overall decline in diversity.

Extinction of Species

How many and which species are being extinguished? Of the estimated unknown number of living species, there are 2-to-11 million animals, some 1,371,500 described in anthropological texts; 600,000 to 10 million varieties of fungi, only 8,500 partially described; and 307,700 to 450,000 plants. *Nature*'s December 15, 2014 status report on biodiversity places the overall present death-rates of amphibians at 41%, mammals at 26%, and birds at 13%, caused by intentional and often blinded human activity.

Each year, millions of hectares of wild habitat are replaced by agriculture. Native plant populations are also displaced by invasive species, often fueled by human ignorance. Over-fishing and the dumping of pollutants are destroying marine ecosystems, too.

Increasingly carbon dioxide emissions from power plants and factory operations, issuing carbon, moisture and debris, are evaporated into the air, and often descend into the seas, making them warmer and acidic—so far destroying one-third of coral reefs, which support more life forms than any other ecosystem on Earth. The process, called "bleaching," occurs when the warmer, acidic water weakens coral that then lose their algae, which provides oxygen and nutrients.

Coping with Rising Temperatures

Rising global temperatures are also destroying habitats where fungi, plants and creatures cannot adjust. The *Planet Index* issued by the World Wildlife Foundation's *Living Planet Report of 2014*, reveals that hunting, fishing and other forms of resource exploitation are the major forces causing the 37% overall decline in animal populations. With habitat degradation and climate change, 31% of species have gone the way of the Dodo. And with habitat loss due to agriculture and extended human living space, 13% of species have expired. Among the other forces causing species declines are ensuing climate change (17%), invasive species (5%), pollution (4%) and disease (2%). All these changes are challenging species variability and causing decline.

When species reach their thermal limits and expire, most forever after will remain Earth's unknown bioforms. And from unbearable heat, our species also is destined to decline in number, face known and unknown forces, and go extinct in the far distant future.

Humans are particularly vulnerable: Dependent on minerals and iron-enhanced plasma for placental reproduction; born unclothed to suffer worsening environmental conditions; requiring oxygen between 17 to 21 percent of the air that we have diluted with greenhouse gas poisoning; and, dependent on food from agriculture, fishing, and hunting which we are destroying—all these factors give our species a million years longevity at best, some anthropologists declare.

Which geographies first will be awash in rising seas? How will species be lost or somehow escape, adapt or develop a way to cope? Humans may prove the least adaptable, although a few warning signs may jar some awake to save themselves, if they can.

Dangers at Sea

Returning to the past of great empires that explored the ocean's farthest reaches and colonized most of then unknown worlds, today the global economic successes among the seaways turn on a fleet of more than four million fishing and small cargo carriers; 100,000 large merchant ships hauling 90 percent of the world's products; trade in capital investment equipment; smog, fog and carbon atmospheres over harbors and sea lanes—and exploited seamen, enslaved children, unnumbered prostitutes, and desperate stowaways seeking freedom and work.

From his dismal sea log, "Men and Laws, Thrown Overboard," Ian Ubina writes in the *New York Times*:

- From the sinking of an average of a large ship every four days, between 2,000 and 6,000 sailors are lost annually.
- Intentional dumping of engine oil and sludge exceeds the spills of Deep Water Horizon in the Gulf of Mexico and the Exxon in Alaska.
- More air pollutants are released by ships than all the world's cars.
- Illegal commercial fishing has plundered world marine stocks, causing the decline of two-thirds of predatory fish.
- Protection of seamen is practically non-existent despite the promise of rules in 2017 mandating that ship-owners carry insurance or show other proof that they can cover the cost of sailors marooned in port, as well as seafarers' death and long term disability entitlements.
- A new ship-owner's law of the sea and new flag of country identifications make ships immune to a ship's past misdeeds.

Other Known Dangers

Other known dangers already threaten by seas, inundating islands and continental fringes.

In the U.S. lower 48 states alone, about 3.7 million people live within the 1-meter level (3.3 feet) of high tide, at risk of being hit by more frequent coastal flooding and storm surge in coming decades.

The worst flooding from storm surges can occur when a storm strikes at high tide. So, using an area's average high-tide mark as the waterline offers a better risk-calculation than a uniform 1-meter rise above the average sea level all along the coastlines. And if the pace of the sea rise accelerates as much as expected, coastal flooding at levels that were once exceedingly rare could become an every-few-years' occurrence by the middle of this century.

Florida, by far the most vulnerable state in the U.S., is populated by roughly half of the nation's at-risk people living near the coast on the porous, low-lying limestone shelf that constitutes much of that state.[13] But Louisiana, California, New York and New Jersey are also particularly vulnerable; and virtually the entire American coastline is at some degree of risk. The at-risk size or percent of the affected nation's population is largely guesswork. But the land below the horizontal 3.3-foot line from the sea is expected to be permanently inundated one day, possibly as early as 2100.

Long before inundation occurs, however, life will become more difficult in low-lying zones because the rising sea will make big storm surges more likely. At heavy risk are those populations living within one meter (3.3 feet) of the mean high tide level, as estimated in 2012 tidal data set from the National Oceanic and Atmospheric Administration. And in a worst case scenario, a Last Wave might well wash away or submerge great swaths of land as the remains of the present Ice Age melt come to a millennial close

3. CHAOS AND DETERIORATION

Steps to prepare for a future disclimax are inadequate compared to the dangers of destruction. Deepened seas and storm surge have run over onto the land. Coastal cities and beaches are regularly battered by 50-foot or higher waves. Engineers are nonplussed as they construct ever-higher seawalls to defend against fierce winter storms and rising sea levels. Beaches are being washed away, caused by human alterations of the shoreline, as already seen in Europe, North America and the Philippines in 2014. By 2016, the beaches and coastlines of the U.S. Atlantic Coast and Gulf of Mexico were eroding and being swamped under the pressures of filled-in marshes, storm surges, and rising seas, wrote Justin Gillis of the *International New York Times* on September 3, 2014.

So say two leading marine geologists, Andrew Cooper, professor of coastal studies at the University of Ulster, and Orrin Pilkey, professor of earth and ocean sciences at Duke University in Durham, North Carolina.

Their view, in the study called *The Last Beach,* sees beaches as nearly indestructible; but seawall construction that disrupts the natural movement of sand and waves actually hinders the process of sand movement along the shoreline. The walls, unable to absorb the sea's energy, crumble, causing the existing beach to disappear, especially in trade and tourist areas, as a growing proportion of the world's populations pour into places like Florida's densely "developed shorelines," Spain's Costa del Sol, Australia's Gold Coast and Brazil's Rio de Janeiro.[14]

As we will see, higher sea levels, storm surge, rebuilt seawalls and defense of property rights next to or on the beach actually multiply beach destruction. They tear apart the near-shore food chain that originates with tiny organisms living between grains of sand and surviving on periodic seaweed influx; they also destroy turtle and bird shoreline nesting, and shorten the life of artificial beaches that become ever less stable.[15]

Inadequate Preparations for Storms

Is it possible to save coastal cities by preparations for storms without retreating inland? Only in a handful of places have modest steps been taken to prepare for a devastating storm surge. New York City is one: Pumps at some sewage stations have been raised to higher elevations, and the city government has undertaken extensive planning. But the city—including substantial sections of Brooklyn, Queens and Staten Island—remains vulnerable, as do large parts of Long Island, Connecticut and New Jersey.

"Sea level rise is like an invisible tsunami, building force while we do almost nothing," observes researcher Benjamin H. Strauss, with other scientists. "We have a closing window of time to prevent the worst by preparing for higher seas."

Together with engineering and existing ecosystem protective devices, New York City is preparing for a six-foot rise in sea levels by 2100. With this expected rise, nearly 20 percent of the City's 469-square mile land area has a one percent chance of flooding in any given year.

In 2013, a little less than 11 percent of the City was covered in the flood plain; but in the 100-year expected flood plain, twice as much of the City would lie in the expanded flood plain zone—fostered by seas coming in the wake of projected heat waves of 5.3-to-8.8 degrees Fahrenheit, as well by extreme precipitation.[16]

These calibrations are based on the 2010 U.S. census and on improved 2014 estimates, compiled by federal agencies of land elevation near coastlines and of all levels throughout the country.

Still, "Everything you do has to be a calculation of the risks and benefits and costs you face—you can't make a climate-proof city," says Adam Freed, former deputy director of New York City's Office of Long-Term Planning and Sustainability. Unlike New Orleans, New York City is above sea level. Yet the city is second only to New Orleans in the number of people living less than four feet above high tide—nearly 200,000 New Yorkers, according to the research group Climate Central.[17]

The waters on the city's doorstep have been rising roughly an inch a decade over the last century, as oceans have warmed and expanded. But according to scientists advising the city, that rate is accelerating because of environmental factors; and levels could rise two feet higher than today's by mid-century. More frequent flooding is expected to become an uncomfortable, if not dangerous, reality.

With higher seas, a common storm could prove as damaging as the rare big storm or hurricane is today, scientists say. Were sea levels to rise four feet by the 2080s, for example, 34 percent of the city's streets could lie in the flood-risk zone, compared with just 11 percent now, according to a 2011 study commissioned by the state.[18]

In 2009, researchers with the U.S. Environmental Protection Agency, the Woods Hole Oceanographic Institution, and several regional planning agencies reviewed development plans for communities up and down the East Coast. Nearly 60 percent of coastal lands lower than 1 meter above sea level have been slated for some form of development. Less than 1 percent of the land has been earmarked for conservation, indicating private development is politically favored over securing the public domain.

Such development projects ignore sea level rise. In addition to building on vulnerable lands, they block wetlands from moving inland. Wetlands not only serve as nurseries for marine life—an environmental asset—but they also serve as buffers against storm surge.[19] If wetland barriers are weakened, there will be little way to deal with rising sea levels, storm surge, or to secure life diversities—as Hurricane Sandy revealed in 2012.

Over the past several years, local governments in southeastern Florida have begun to reassess their options for coping with sea level rise—another project likely too late coming.

Survival: Soil, Sea and Human Tenacity

Hope is strong for a future agreement on measures to stop global heat and climate change. Climate change is experienced by people around the world, said journalist James Fahn, the executive director of Internews's *Earth Journalism Network*, December 10, 2014, and a lecturer at the University of California, Berkeley, Graduate School of Journalism.

But interrpretations of environmental change sharply differ. Controversy rages over the causes of the worst California drought in 1,200 years, for example. Was it the record high temperature and the three years (2012-15) of low rainfall causing high temperatures and accumulated deficits in soil moisture in Central and Southern California, as claimed by a paleoclimatogy and dendrochronology study?

"Our farmers plant at a different time of year than their grandparents did," said Tracy Stone Manning, the chief environmental official from Montana, at the United Nation's Lima conference in 2015. "Our fire seasons are gigantic now, and they're followed by Biblical floods."

Summits like those in Lima and Paris in 2015 offer a chance to let the 200 state representative diplomats catch up with what's happening on the ground—in cities, nations and businesses. The underlying questions rely on the relationship of the conditions of the soil to plants and the conditions of the seas to sustain life.

Every scrap of arable land must be preserved from inundation by the sea if our expanding human population is to be fed and survive as a species. The exodus inland from coastal regions is now the only viable way to feed seven—soon-to-become eleven—billion people.

Need to Protect the Earth, Ourselves, Other Species

Humankind must now care for Nature's soils as if our very lives depend on it, for they do. Our seemingly singular intelligence and selfish genes have driven our species to invent ways to secure ourselves, but not necessarily other species. Humans make tools and machines and use fossil fuels to improve our condition, but today only the foolish and ignorant will continue along this path in the face of clear evidence that we are today dissolving and destroying Nature, species life, and ourselves.

A species so blind to its own destruction lacks self-awareness bordering on irrationality. But evolution and development have produced among our species the means for survival, if only we will logically and sensibly use them.

Such a paradigm shift requires that we humans reach an essential understanding of energy cycles sustaining the natural world. For Western thinkers, these concepts can be encapsulated as Mother Earth's renewal of life diversities in a masterstroke of interlocked cycles of Earth's terrain, seas and atmosphere—in which plants and animals transpire water and carbons that are evaporated to the air; the atmosphere cycles moisture that returns as precipitation to land and sea; and species are reimbursed with oxygen and sunlight to photosynthesize food distributed among an unending chain of land and sea varieties, or by chemosynthesis of ocean chemicals, to evolve on a branching tree of life.

Nowhere is this interconnection clearer than in the U.S. Midwest where the temperature swings during summer can range from 60-to-114 degrees Fahrenheit, rain can come in torrents or not at all, and winters like 2013-14 can bring ice, sleet and snow, thermometers going below subzero 40 degrees. Spring and summer corn, soybeans, sunflower and other crops may return to life or be totally turned to Sun-baked crisp.[20]

But such processes and calamities are not caused by Nature alone. As we have seen, soils destroyed by farms using chemical fertilizers and bioengineered seeds, without proper crop rotation and fallowed fields, will reduce food supplies in decades coming. Oil and gas effluents dumped on decimated forests to clear and burn them for mining and petroleum extraction also will destroy soils for future agriculture—the very purpose of the deforestation. Species transferred from islands and continents, invading other regions with their own species chains, set loose competititve wars for survival, forever upsetting the previous evolutionary balance.

The quickened pace of such technical advances and species transfers far exceed the pace of renewal of evolution and the biosphere—with the tide of conditions supporting life ebbing out and not being replenished, destroying many forms of life. Furthermore, a blind spot in using lethal technologies to sustain food production could be our species' undoing.

Farm Technologies vs. Mother Nature

Most U.S. farmers limit themselves to a few cover crops as they leave some of the land fallow for one or two intensive planting seasons. They may plow and turn over the soil and last season's remaining shaft, but leave the land unseeded to restore its moisture and fertility.

There is also another way. Breaking with tradition, a relatively few farmers mix as many as 14 different plant species for ground cover, mimicking Mother Nature without using synthetic chemical fertilizers and herbicides. But they may also till the land, which rips the soil up with steel blades that bring in weeds, disturb soil microbiota, and leave loam vulnerable to weather erosion from drought, floods and winds.

Both fallowing the land and using cover crops absorb atmospheric greenhouse gases, actually improving the health of the soil. Biological systems of cover crops also marginalize root-destructive insects and disease-causing molds, preserving the next season's domesticated crops. These methods have been successfully practiced by Middle Eastern and North African farmers for centuries, yet have been lost to Western farmers in the rush to use oil-based chemical fertilizers and plow-tilling in the mega-farming rush for profits.

An alternative agricultural method is this:

1. After the harvest of, say corn and soybeans, the farmer plants a mixed-species cover crop, which sucks carbon dioxide out of the atmosphere and, for some legumes, also enriches the soil with life-giving nitrogen trapped from the air and stored in root nodules.
2. As the enmeshed cover crops die, they leave a matrix of organic matter that improves soil fertility, prevents weeds from growing, and acts as an insulating thermostat moderating soil temperature.
3. Rather than disturb soil-enriching microbes, worms and other organisms that consume biomass and generate and improve soil mass and fertility, the soil is not tilled by plowing before planting cash crops like corn or soybeans, thereby preventing buried organic matter from oxidizing into aerated carbon dioxide.

This alternative takes time, but the cover crops eventually create a layer of carbon-rich humus topsoil that is resistant to both drought and erosion in the face of the present Anthropogenic Era of rising temperatures and drought.[21] Call it human understanding aiding natural ways, or boosting Gaian energy flows—either way, this alternative presents a temporary path to sidestep the present period of a very dry, torrid Earth.

More Ideas to Protect the Earth

For believers in the Gandhian capacity for reflection, securing Nature's realm also is grounded on four principles: self-rule, homespun local knowledge, the force of truth, and the uplifting of all. Consider the work inspired by Gandhian principles as promoted by renowned food and anti-globalization activist Vandana Shiva's University of the Seed, co-founded with Satish Kumar in 2001. This university has concentrated on a farm-preserving, wild diversity of indigenous seeds in cooperation with thousands of farmers across India and the world. They are committed to the organic principles of working with Mother Earth, rather than waging war on her with chemicals. They say she should not be tampered with, neither by scientists nor by corporate bioengineers.

Indeed, many countries ban the cultivation of all genetically modified crops, following the golden rice's media debacle early in 2000. Vandana Shiva called the argument a "Trojan Horse," whose purpose was to gain public support for all manner of genetically-modified crops that would benefit multinational corporations at the expense of poor farmers and consumers.[22]

"Gandhi and Globalization" is a course co-taught annually at the University of the Seed for ten short, intense days in November and December. Vandana Shiva, Satish Kumar, and Samdhong Rinpoche (the first Prime Minister of Independent Tibet) designed this course for students coming from all continents, speaking in multiple tongues and joined by a shared passion for both Gandhi's ways and to bring about the end of the era of globalization and neo-colonialism.

Hundreds of other religious and metalogical systems affirm the same principles in relation to the cycles of Earth's flows of energy that sustain the planet's food production and species evolution, development and diversities. But in the rush for maximizing production regardless of environmental risks, these systems and principles have been forgotten.

Is Entropy Possible on Earth?

Can we humans reverse our present course and instead prevent regional disorders, eventual global chaos, and devastation on Earth? Possibilities to save Earth depend on whether or not it is a closed or an open system to the universe.

In his classic study, *What Evolution Is*, Ernst Mayr, writes:

> It is sometimes claimed that evolution, by producing order, is in conflict with the "law of entropy" of physics, according to which evolutionary change should produce an increase of disorder. Actually there is no conflict, because the law of entropy is valid only for closed systems, whereas the evolution of a species of organisms takes place in an open system in which organisms can reduce entropy at the expense of the environment and the Sun supplies a continuing input of energy.[23]

But what follows if humankind closes the global system by pouring carbon dioxide and other greenhouse gases and waste debris into the atmosphere—hermetically sealing Earth into a torrid zone on which the Sun radiates, elevating temperatures above the maximum thermal limits of most species on the planet?

Is there no stopping the advance of entropy and ecocide? In trying to prevent recurrent entropies in various geographies, what nations and which peoples will bear the heaviest burdens for cutting greenhouse gas emissions to limit warming to 2 degrees Centigrade above pre-industrial levels?

The official December, 2014, Lima call at the Climate Action Conference concluded that wealthy countries should carry the burden of cutting carbon dioxide emissions; that loss and damage programs for financial aid should be extended to countries threatened by climate change; and that the United Nations should weigh these pledges and determine whether the collective action was sufficient to limit global warming to 2 degrees Centigrade.[24] But, these commitments are yet to be followed as a heavy political thumb alters Nature's scale.

Evolution Will Continue

In the vein of Aristotle, von Humboldt and Charles Darwin, the most acute observer describing Earth's template was Ernest Mayr, who died in 2005, bequeathing to us his unequaled knowledge. He tells us that Darwin saw clearly that there were two aspects of phyletic evolution.

Anagenesis is the "upper" movement of phyletic lineage, revealing gradual change from an ancestral to a derived condition. This is species formation before the branching of its evolutionary line of descent.

The other aspect of evolution, called cladogenesis, consists of the splitting of evolutionary lineages, or, more broadly, creating new branches (clades) of the phylogenic tree. Darwin himself explained that in the struggle for existence, the "habits" of plants and animals would lead to favorable variations tending to be preserved, unfavorable ones to be destroyed. He simply said that the result of this would be the formation of new species. This origin of new species expanding biodiversity became Darwin's basis of his theory.

Cladogenesis always begins with an event of speciation; and the new clade may become, in time, an important branch of the phylogenic tree by diverging increasingly from the ancestral type. Mayr says anagenesis and cladogenesis are largely independent processes—meaning anagenesis preceded cladogenesis—so a new evolutionary body form first appears, and only then for the first time, reproduces the clad.

Darwin uses different words to show that the two steps operate sequentially during natural selection, as well as during observable human breeding of animals. He saw that natural selection thus takes place through species isolation and inbreeding over time. Human selection of animals takes place the same way, by selecting the "best" body forms and then inbreeding them to produce the same quality and features in a relatively short time. Thus one kind of animal, including humankind, through isolation and inbreeding, can become many types of animals as it descends through time and under human direction and the pressures of specialized breeding.[25] (Research experiments on this have already begun.)

Human experimenters can weed out the plants they do not want to reproduce; and also prevent what they consider to be inferior animals from copulating. Humans can control both species origin and speciation—thereafter branching out through natural selection—securing biodiversity. But mixing the genomes of different plants and animals may prove to be a catastrophe for life forms that lose their vigor and reproductive capacities. With humankind's technological sphere impinging on Nature's biosphere and interfering with the sequential and overlapping aspects of natural selection, the dual process of securing species' evolutionary diversity is placed in jeopardy.

Human Choice, Nature's Eternity

The question of how to sidestep the disaster of a nearly lifeless Earth faces our time. The evolutionary natural selection of each species requires essential conditions for acquiring sustenance, disposing of wastes, and breeding. Deprived of natural corridors connecting the geographic areas and essential elements for their survival, species face an uncertain future.

As Aristotle, Darwin, Mayr and the biologist Edward O. Wilson have said, since the origin of life forms on Earth, extinction has overcome almost all species. During the passage of about 3.5 billion years, some 99 percent of all species that have ever existed have come to an end, replaced by others, leaving surviving arrays of multiplying species that also came to an end.[26]

E. O. Wilson, arguably the best known biologist on the planet, views the remains of diverse species and the natural world in tandem, saying that Nature and species will continue to exist and evolve as an eternity, perhaps beyond the grasp of the human mind, securing the longevity of life on Earth.

Homo sapiens face the same destiny, more vulnerable perhaps to extinction than other kinds of animals because of their evolution from pre-human species, passing on their enduring DNA's intellectual structure, leading to their dangerous technological advances and epigenetically acquired culture, signifying the collective remembrances of all humanity. Humans' unprecedented cultural heritage and rational intellectual power could well shorten their own realm and their remaining genetic time-line on Earth to as little as a half million more years.

Natural and Human Barriers

Other species may survive though, by passing through barriers from one geography, large or small, to another. This may be the avenue that plants, fungi and creatures will again take, sustaining themselves and their dependence on one another, securing diversity.

Some 23 centuries ago, Aristotle studied Nature at the seaside town of Phyyha, also known as Kalloni, handing down to us his treatise, *Historia animalium*, tracing the conditions essential for species survival and their diversity—a study of comparative zoology in a very tiny place.

With the natural shift of seasons and drought, he discovered that the strait between the local lagoon and the sea was a passageway for water creatures. In winter the lagoon was colder than the sea, and most of the fish, but not the giant goby, swam out of the lagoon, to return in the summer.

The absence of winter fish in the lagoon meant the fishermen turned to the strait's small, edible urchins which were spawning and rich in eggs. But other natural events also may have altered the fish available through the strait's corridor, due to a third century (B.C.) earthquake on the eastern shore of Phyyha.

Yet, the human touch altered the strait and its species' destiny, too. Fishermen dredging its bottom allowed certain species like oysters to multiply in the lagoon. And people from nearby Chies island also sought to transplant the strait's oysters.

Once there were many scallops, but dredging and drought exterminated them in Aristotle's time. And looking back, evolutionary developmental biologist Armand Marie Leroi says that "the scallop population of Kollani has been waxing and waning for at last twenty-three centuries, and . . . the locals have been complaining about it all the while." As for the common octopus not found in the lagoon in Aristotle's day, the strait has let them in, enlarging diversity—Armand Leroi in our time having eaten several of them himself.[27]

Left Out of Science

Armand Leroi also says Aristotle largely has been left out of the scientific pantheon of Linnaeus, Darwin and Pasteur, and disastrously overlooked by the Scientific Revolution that was built on past discoveries. Only Aristotle had tried to explain living things in all their diversity; working with zoology and botany, observing that each creature reveals to us "something natural and something beautiful," as no other but Darwin would do for more than two thousand years. Whatever the imperfections and inconsistencies of Aristotle's *Historia animalium*, it offers a system of awesome biological scope, identifying living things as most worthy of his attention as both philosopher and scientist.

Evolutionary Passage to Life Eternal

Today biologists are also concerned with the destiny of each living species they are able to identify, tracing evolutionary origins. For some, it seems life on Earth may last an eternity if the conditions essential for various species remain intact.[28]

"The mathematical reality behind evolution is that there has to be a mechanism to keep you the same," says Prof. David Polly of Indiana University, Bloomington. "Evolution does not move inevitably forward towards new morphology and new designs."

If this is so, Darwin's so-called "living fossils" of star fish, redwood trees and sharks will roughly change very little if the habitats in which they have evolved continue to survive mass extinctions; in contrast, even environmental discontinuities can carry along living organisms, freeing them to further evolve after planetary upheavals and barriers have passed.

Continuity in Species' Environments

Explaining the environmental continuities essential for the life of some species, the Fibonacci Sequence, or Mathematical Golden Ratio, expresses the possibilities for their long-term survival as well as organic transformation over eons.

First developed by Leonardo Fibonacci in 1202, the sequence also is found throughout Nature, approximating the precise patterns of evolutionary growth and modification of different species and their phyla. These can be expressed mathematically in Table 14.1.

You need to use your numerical imagination to understand the table in terms of the billions of years of evolution of species. Over time each species

Table 14.1. The Fibonacci Sequence

Species' Time of Evolutionary Existence Start	Cumulative Organic Complexity	Possible Points of Organic Transformation
0 +	1	=1
1 +	2	=3
2 +	2	=4
3 +	4	=7
4 +	7	=11
5 +	11	=16
6 +	16	=22

Source: Richard Krooth, Table Constructed 2016.

becomes increasingly complex, either developing favorable variations that tend to be preserved or unfavorable ones to be destroyed—leading to the formation of new species.

The life sequence starts with numbers 0 . . . and 1 as the initial building blocks on which each species comes into existence. Each species then develops organically and physically over time, represented by a sequential number equal to the sum of the previous two, and so continuing. Thus 0 +1 = 1; 1+1 = 2; 2 +2 = 4; 3 +4 = 7; 4+7= 11, seemingly ad infinitum. A moment is reached where a species' cumulative genetic complexity and form develops slowly or bursts forth organically or structurally.

This approximation of species evolution and developing complexity makes absurd the mathematically assumed constant ratio 1: 1: 1 representing the supposed prehistoric proportions between species of plants, invertebrate animals, vertebrate animals.

Part of the confusion is that early taxonomic explanations and approximations divided species 1: 1: 1 among plants, invertebrates, and vertebrates, whereas in fact plant species are roughly 10 times, and invertebrates 100 times more numerous than vertebrates. Why this mysterious ratio turned on evolution itself is a story of species modification, development and complexity.

Species Design or Body Structure

Many species designs have observably barely changed over millions of years, as their niche environments or passages to safe survivable conditions have remained intact or nearly so.

Oceans, for example, cradle bacterial life and many larger creatures.

Count among them the goblin shark of the Mitsukurinidae family, with its pink flesh and a dagger-like snout protruding from its head, unchanged in 125 million years; or the Australian duck-billled platypus handed down through millenniums; and the 390 million-year-lineage coelancanth lungfish varieties in Africa, Australia, and Indonesia, that emerged from the sea, fitted out with fleshy, limb-like fins, that around 400 million years ago enabled them to walk on land with these appendages as legs, giving rise to all four-limbed land animals on Earth.

Significantly, their skeletons reflect sharp changes over millennia. A second dorsal fin was transformed from spiny to lobed; and they have lost bones around the rims of their mouths and around their scales.

But slowly disappearing habitats, widespread and common in the past, have been transformed in the human age by natural events and our species' miscalculations, which more quickly force species to either change or expire.

Life Depends on Oxygen, Water and Salt

The seas were drained of oxygen by the atmosphere during the Permian Extinction about 245 million years ago, when about 95% of marine species met their doom. Later, drying climates, expanding deserts, and lack of water took down comparable land species. But in the inter-tidal zones, the cycle of water kept hydrogen bound to oxygen (H_2O), available to sustain some species like branchiopod ancestors.

Tadpole shrimp also have survived for 300 million years, bypassing the Permian extinction, reproducing as hermaphrodites fertilizing their own eggs with sperm-producing lobes in their ovaries. They also have diversified. A single egg can start a colony, as likely happened during the post Ice Age 20,000 years ago with melting in Northern Europe; or it could reproduce in the momentarily rain-drenched Sahara, faster than in Europe, or among Australian species able to endure higher salinity in the sea. And, as the ice shelves melted, reducing sea salinity, these types of eggs were able to continue to evolve as a dominant species.

My own observations of living, ancient species has been brief, but important. In 2008, in New Zealand, I met Henry. He is the world's largest captive tuatara, a spiny reptile, then estimated to be over 110 years old, coming from a lineage dating back 225 million years. Evolution has kept Henry living with a slow heart rate minimizing cellular oxidation. I saw him able to stand erect without moving a single muscle for hours, awaiting live insects for lunch.[29] Miraculously, he has recently fathered a child.

I've also seen ancient coelacanths lung-fish pulled from the ocean bottom off east Africa; and as a child I played on the beach with nearly billion-year lineage horseshoe crabs like *Limulus polypherm.*

While humans are fascinated by extraterrestrial incursions and worldly natural terrors such as the Ice Ages, disasters have not wiped out all life on earth. Many ancient species have evolved mechanisms for survival and reproduction. Humans might not live as long as they.[30] Only 40,000 years ago, hominin forebears of our population exploded and evolution sped up, most recently by a hundred-fold, but given our drive for technology to create supermen, we might cause our own undoing.

So let us look at some sea turtles that can live over 1,000 years. At age 61, I rode on Saddle Back turtles on *Isla de la Mujeras* located off Cancun, Mexico, our age difference obvious but their vigor far greater than mine. In California, Redwood giants over 1,300 feet look down on me. Yet humans are not yet known to have lived more than 123 years—a puny genomic achievement compared to the lizard, the turtle and the tree.

Human Carriers and Open Corridors

In modern times, humans carry exotic species into geographic surroundings where they compete with long-established plants and creatures—upsetting species' feeding chains and life cycles, sometimes annihilating them, creating a downward cascade of life. Some people purposefully bring foreign predators into the Florida Everglades, collectors dumping their oversized pythons from the Amazon, Nile and China's great rivers, disrupting the balance of fishes, bivalves, birds and land animals near the water's edge.

At the other reach away from destruction, open corridors for pent-up species well may be their saving grace. As a practical (though seemingly Utopian) reform, E. O. Wilson hopes to build corridors between protected wildlife sanctuaries, possibly to reserve half the world for diverse species to thrive and reproduce without the human stamp of interference from cities, farms, mines and oil well production; somehow, thereby, to preserve our own deep, inherited need to preserve biodiversity.

By maintaining two parallel worlds on the planet—one for humankind, the other for wildlife—Wilson believes humanity could thereby ensure the survival and continued advance of the rest of life and perhaps ourselves. Otherwise, he and others believe, the remains of biodiversity will slide into extinction, turning the present Anthropocene into the human isolation of the Eremocene, which he defines as the Age of Lonelines. It's a grim reminder that the balance of species includes all of us, and that we have been the cause of the elimination of so many others.

Time Lines and the Balance of Species

Look closely. The balance of Nature is both general and unique within the geographic areas that species ocupy. Parallel to continental drift that consolidated Pangaea and its breakup, later sending species across the globe, modern day species' invasions have arrived on the heels of mass transportation and planned, or inadvertent, redistribution across Earth. Both predators and prey are often thrown into battle and live on the edge of existence, which limits their growth and reproductive success in a chain of extinctions.

Examples are many. Butterflies generally live on plant pollens and are prey of small carnivores and spiders that hunt them. Aphids also live on plants and are prey of lady bugs, that are prey of small birds, that in turn are prey of raptors, that are prey of human hunters, who are the prey of invading imperial armies, that are sometimes prey of the bacillus of enzootic tropical viruses—which have led step-by-step to state funding of medical institutes of tropical medicine—first in Paris, London, Spain and Russia—securing the lives of European colonial administrators and soldiers, aiding their reproductive fitness above those of their subject populations and other species that they have destroyed, reducing biodiversity.

Other species that produce food create their own feeding chains, maintaining autonomy from domination by others. In Belize and Costa Rica, I have seen vast colonies of leaf-cutting ants that produce their own fungus for food by fertilizing underground gardens with chewed-off pices of vegetation. Millions of these ants work in cooperative, industrious brigades; yet each ant is put in motion, not by a directing queen, but rather by reacting to chemical symbols from other ants, larvae, intruders, food supplies, waste disposal, and leaves as raw materials. Each ant in fact is an autonomous unit, its reactions dependent on the local environment and its inherited, genetically-encoded rules for its particular variety of ant. Each is seemingly born free, yet genetically predetermined as a slave to work in group production for survival of the ant colony as a living agricultural helotry.

By contrast, humans treat production on land as both their domain and a vested interest to be secured from others—the source of conflicts and wars time beyond memory. Our inherited codex infuses a struggle for survival of our species against one another and other species, Darwin says. The human quest for land and resources as wealth is but an indication of this struggle.

Cattlemen's Quest for More Grazing Range Reduces Biodiversity

Feeding chains interrupted by our species often turn on the human conflict of social classes and their ideological beliefs about alternative ecosystems. As an example, in the United States, Federal support for rangers grazing cattle and sheep remains the chief cause of desertification, with cattlemen and large corporations raising beef and mutton and supporting state legislation to take over the Federal governments' regulation of the prairie for private grazing and coal and oil extraction—the costs to be paid by the taxpaying public.

Preparations for cattle-grazing have eradicated native plants; denuded cover for birds and wild mammals; led to cutting hardwood forests; and opened the way for the mono-culturing of grasslands. In addition, the cattle have polluted springs and streams with rumen and urine.

By contrast, the Great Basin steppe of southern Idaho and northern Nevada, covered with shrubby sage bush growing close to the soil, has evolved to preserve its moisture against heat and wind, and to survive cold, snowy winters. The sage holds the soil in place, creating a habitat for native flowers and bunchgrasses; in turn, those flowers and grasses feed native elk, pronghorn antelope, and mule deer. And those animals, nurturing the grass, feed the great predators of the West, the wolf, the cougar and grizzly, creating "a natural system in perfect balance," as Rachael Carson wrote in *Silent Spring*.

Now that balance has largely been ruined, as cattle herding and fencing have taken the Western sea of natural grass from Federal land-management oversight, replacing it with:

- intensified grazing;
- ruining and degrading soils;
- polluting water sources;
- introducing invasive species of crested-wheat-grass that cattle ranchers prefer as forage;
- letting Asian Cheat-Grass overtake native grasses by more efficiently sucking up water resources and soil nutrients, as well as more prolifically producing seeds. (The Asian Cheat-Grass dries up in summer earlier than native grasses, forming contiguous flammable mats, especially in the wake of fire storms.)

In this way, human profit-driven cattle and sheep grazing has built a landscape and a monoculture that opens the door for invasive species, bringing on ecological collapse.

Predators and the Balance Among Species

In the linkage of plants and animals, humans are the most invasive species on the planet. Our species is the force behind the downward cascade of species diversities.

Consider the North American Monarch Butterflies, traveling two different flight patterns along the East and West Coast routes towards the Mexican highlands, no longer able to find sufficient milkweed plants on which to lay their eggs. This is due to conversion of prairie into cropland and the increasing use of agricultural weed-killer-resistant crops that have reduced the extent of milkweed.

Facing a 90 percent decline in population, with the lowest ever recorded in 2013-14, the Monarch butterflies also face loggers. They are among the Monarchs' most lethal predators in their reproductive bacchanal, cutting trees in their Mexican wintering ground and helping to bring on regional climate change that restricts their life cycle.[31]

There is also a biological law to be considered. If a species anywhere in a food chain acquires a large supply of food, individuals of that species will reproduce more, increasing their population as well as enlarging the population of their predators eating their way through them as prey, bringing the two populations back to their previous balance.

The entire biological domain of flower-pollen-producing plants, the butterflies, and their predators forms a food chain that is part of an even larger ecosystem of these and other kinds of small plants and creatures. Within the larger feeding chain over time, the uneven equilibrium of individual species' numbers tends to be rebalanced—reproducing the overall Balance of Nature.

But what follows when the human species enters the loop at critical points as a profit-driven force? They expand cropland; remove milkweed plants on which Monarchs can no longer lay eggs to reproduce; use agricultural pesticides to protect crops while poisoning the winged; and cut down trees in the Monarchs' warm, mating season.

The Questions Remain

As a dominating species, what unperceived forces could bring us into balance with the Monarchs, preventing their near-total destruction? Without natural predators, other than ourselves and microbial diseases, we humans increase our own food supply and population which, in turn, increasingly becomes dependent on burning fossil fuel, which brings on torrid heat that reaches the maximum thermal limits, and causes the extinction of untold numbers of individuals among all species.

Add sea level rise, enveloping islands and continental coastlines, cutting away their agricultural areas. Consider that these often overlap with the very moment deforestation and burning fossil fuels carelessly and ceaselessly increase global warming.

With no immediate way back to the past in the human struggle for existence among other species, the downward cascade of these others is almost certain. Fewer sea-covered acres to plant, more people to feed, together might ignite conflicts and wars for remaining territories, resources and food supplies, ending in battlefields of death, and pandemics and starvation—Monarchs long forgotten.

Hope for Species Survival

Picture a worst case scenario. The techno-sphere on land, impacting the atmosphere and the sea, is jeopardizing the ecosphere, tainting the biosphere, and undermining the symbiotic relationship between species. And this process continues to overwhelm species diversity and our human sphere.

How will humankind and species diversity survive the onslaught? How can we hold accountable those who are most responsible, and force them to reverse the way they mistreat planetary life?

The largest 500 fossil fuel and cement companies along with trans-national enterprises, owned in part by 556 U.S. billionaires and their 62 counterparts elsewhere—with as much wealth as the bottom half of the world population of 3.6 billion—are jeopardizing the breadth of life of almost every known world species, as well as undermining the bio-geological structures of the planet, seemingly caring not for any forms of life.[32]

In jeopardy meanwhile are some 3% slivers of global geography still untouched by our species, seemingly the only hope for species survival and diversity—providing small windows for emergent life. The saving grace for evolution and diversity may be their geographic insularity, shielding Nature from our species' intrusions.

Though there are ever-fewer territories and seas on Earth where species can continue to evolve and evade the lethal hand of humankind during the present Anthropocene, some among an estimated unknown 340,000 plants, 9.5 million fungi, and another 9.5 million animals may be able to adapt and slip through the zone of poisoned seas, torrid skies and searing terrain. And although humans are still unable to fully comprehend the origin, development and reproductive fitness of millions of individual species, biologist Edward O. Wilson warns us that the precautionary principle should be for humans to move as slowly and carefully as possible, if at all, into Earth's surviving wild domain.[33]

Human Longevity and Myopia

What if our species survives for another million years or more, retaining our blind spot on the survival needs of other creatures, plants and fungi? Or might humans first expire on the planet, leaving the surviving realm of diversity to other species?

Nature, somehow, is often seen by us as background-staging. If fish harvests rise, for example, gross national product (GNP) appears to increase, even though the supply stock of fish declines. If logging intensifies, gross domestic product is increased even if forests are denuded. GNP is simply a blind economic measurement that is numerically impervious to Nature's destruction and constraints.

Over the last 250 years or so, allied to economic growth and science, there have been improvements in the quality of some environments and the quantity of resources, potable water, food supplies, and medical security in industrial nations from water-and air-borne diseases. But there are limits to sustainable patterns of economic growth and social development, as the rates of human exploitation of fossil fuels and other resources may suddenly and dramatically change the character of the climate, as seen today in torrid skies, melting glaciers, and rising seas. Together they breed degradation of Nature's realm, on which other species rely to survive and reproduce.

Our Species Myopia

We collectively seem unable to accurately assess the effects of our impact on Nature and ourselves. Alone during the 20th century:

- world population grew by a factor of four, to more than 6 billion
- world output enlarged 14 times
- industrial output increased by a multiple of 40
- the use of energy grew 16 times
- methane-producing cattle kept pace with the human population
- fish-catch increased by a multiple of 35
- Growth in the industrial and economic realm led to an increase in carbon and sulfur dioxide emissions by more than 10.[34]

If humans continue to support themselves by degrading Nature's sustainability, we as the dominant species on Earth will remain so—but just so long as our species can steadily drain Earth's energy. The drain has limits, though. As the number of plant and animal species sustainable in a given habitat decrease with the loss of roughly the fourth-square-root of an area, say by 90 percent, the number of its species is cut in half.

For example, rain forests taken away for agriculture, mining, and the spread of growing human populations by 80 to 90 percent of the area in Madagascar, the Philippines, or the Atlantic coast of Brazil, has meant the number of species limited to the rain forest must plunge from half the original population to nearly zero.[35]

RETURNING TO NATURE

The many causes of the fall in biodiversity today turn on the wheel of financial capital expanding its range, from cutting forests and destroying habitat to laying the foundation for invasive species; to pollution, and continuing human population growth; to accelerated over-consumption of resources without replacement, and relentless hunting and ocean fishing, often removing the last key individuals among each species—all these overlapping with burning fossil fuels causing climate change, interrupting the evolution of species.

As the technological sphere advances, the biosphere continues to yield. Oceans are lapping at the threshold of our civilization. Atmospheric greenhouse gases are concentrating. Global surface temperatures, global ocean heat content, and global sea levels all are exceeding previous record highs.

Great storms are on the horizon. *El Niño* heating of surface temperatures in the equatorial Pacific is spreading eastward, changing weather conditions around the globe. Declining global ice and snow cover are harming Arctic animals and fish species; and powerful cyclones have led to flooding on storm surge scales, as well as severe drought and raging fires. So said the National Oceanic and Atmospheric Administration, *State of Climate,* report, August 2, 2016.

Evolutionary life on Earth is changing. Nature's cycles have catalyzed the evolving generations of species complexity of cellular matter.

Wastes from the techno-sphere and once-living plants and animals are deposited in aging layers of rock-time fossils. These remains act either as poisonous barriers or essential material conditions for the emergence of future species evolution and reproductive success.

Evolving species have either met their end or surmounted these fossilized rocks of ages up to now. Their remains will become either death-stones or living-witnesses long after our species goes extinct. Though there may be few surviving humans to perceive it, future biological diversity likely will survive us among billions of sea and sub-terrain species. There will not be a definitive sixth extinction of all forms of life on Earth. Even a few of us supported by our powerful DNA may live on.

Species Diversity Will Live On

Walking along California's Alameda Island Cove Beach mud flats, you can watch plovers with their evolved, long, curved bills feeding three or four inches deep on micro-organisms. And if you could see 75 meters beneath South Pacific Ocean seabed sediments, you would also discover aerobic, oxygen-dependent microbes reaching downwards to the igneous interior, their gaseous flows rising to the surface.[36]

Many thousands, if not millions, of species are being modified through an evolutionary drama. The discovery of some 30 new insect species in heavily urbanized backyards in Los Angeles could be multiplied thousands of times on a global scale.[37] There is no end to evolution. And evolutionary transmutation of Earth's species will likely go on and on, perhaps eternally until the Sun burns out.[38]

Rock-Strata Represents a Time Interval

During the emergent Anthropocene epoch, the human impact is gradually leaving a clearer mark, fossilized in strata, that in the far future can be recognized tens or hundreds of millions of years from now by some intelligent being. This unit of fossilized geologic strata equals a geologic time in the past, a "rock-time" interval defining a new course of evolutionary and social history.

During the 200,000 year Holocene, the natural conditions of the biosphere empowered humankind with the intelligence to build a robotic technosphere that could almost irreversibly create the Anthropocentric superman able to "jumpstart the transition to electronic (and potentially immortal) entities that transcend our limitations and eventually spread their influence far beyond the Earth,"says Lord Martin Rees, the astronomer royal and former president of the British Royal Society.

Rock-time has already begun for our era. But for the Anthropocene epoch to have deep geologic meaning, it must clearly represent its own, accumulating rock-time evidence. We already have accumulated:

1. Radionuclides released from nuclear bombs dropped on Japan, tests in the 1950's, and more recent power plant effusions of radioactive elements released into the stratosphere before settling down to Earth.
2. Unburned carbons and microscopic debris emitted by power stations, coal mines and oil and gas drilling operations around the world since the mid-20th century.
3. Plastic pollution, aluminum derivitives, and several centuries' of concrete particles distributed worldwide, as well as high levels of nitrogen and phosphate in soils derived from artificial fertilizers.
4. The detritions of undiminished wars, with ever-more terrible machines, weapons and explosives deposited on Earth and aerated, eventually plaguing all species. Today wars are on course to see extinction of 75% of all species in the next few centuries, if current trends continue.

These global rock-time discharges may well eventually interrupt and close down natural processes,. They are likely to block the formation of amino acids as well as complex cellular-molecule-proteins required for the operation and regulation of living body tissues and organs; and disrupt the flow of trace minerals that every sprouting seed and towering tree have helped raise from deep within the soil to Earth's surface for plant growth essential to the life of all plants and animals. Without these elements sustaining the growth of plants that feed and nourish all invertebrates, which in turn feed and sustain all vertebrates, most life on Earth will face extinction.

These dynamics will also impact the ocean's interdependent feeding chain that will be heated, acidified and irradiated to the moment of annihilating phytoplankton, zooplankton, krill, coral and other shell-forming creatures, leaving a dead sea. Only chemosynthetic life will be likely to survive.

Are not these annihilating forces, endangering plant and animal life, comparable to the deposits of the metal iridium in sediments—the very hard and brittle chemical/metal of the silvery-white platinum group—that were dispersed around the world from the meteorite that collided with Earth to end the Cretaceous-Dinosaur age, blocking and taking millions of years for life to evolve again?

Yet, What If . . .

At some critical point during the Anthropocene and the ensuing Eremocene era of loneliness, there may well be mass starvation and a Malthusian death march. Yet what if you and I somehow could sidestep these perils and protect and reimburse Nature's energy? What if today we were brave enough to risk our very lives to stop the masters of war, the burning of fossil fuels, the spew and aeration of chemicals and poisons, and the irradiation of our planet?

We might then be able to revisit Darwin's wonderment, viewing our collective surroundings as majestic, clearly part of Mother Earth's natural cycling of matter and species life. We might also see one eon or epoch folding into another, species evolution and modification moving forward with all their nuances and shortcomings, understanding that the peoples of the world, with a common heritable genome, still might together act in time to preserve Gaia's cycles of energy and species evolution and diversities.

Should we fail to do this, seven-tenths of Earth, its surface-seas, may well up as a last wave over fertile land, deepened, churning waters surely consuming us along with most other species—those that survive evolving into elaborate, still inconceivable varieties in the future.

NOTES

1. Jan Zalasiewicz, "What Mark Will We Leave on the Planet?" ScientificAmerican.Com., September 2016, Vol. 315, No. 3, 32.

2. Nova, "Australia's First Four Billion Years," April 17, 2013, 9-10 PM, with geologist Richard Smith. Reported online in the journal *Nature*, "The African coelacanth genome provides insights into tetrapod evolution."

3. This and the following two sections are based in part on Robert M. Hazen, The Story of Earth: *The First 4.5 Billion Years: From Stardust to Living Planet* (New York: Penguin Books, 2013). Chapter 9, traces the break-down of rock into clay, releasing calcium and other elements that combine with carbon. See also Nova, "Australia's First Four Billion Years," April 17, 2013, 9-10 PM, with geologist Richard Smith. And see Henry Fountain, "Turning Carbon Dioxide into Rock and Burying It," *The New York Times*, Science, February 9, 2015; in turn based on the Smithsonian Institution, National Museum of Natural History; and the Global Volcanism Program, The University of Iceland, The Institute of Earth Sciences, 2015.

4. David Grinspoon, "How Long Will We Last?" *Scientific American*, September 2016, 76-79; Krooth, Gaia, chapters 25, 26.

5. See Kathleen Maclay, *Scientific Reports,* June 9, 2016; citing Solomon Hsang and Alan Sobel, "Climate change could trigger tropical evacuations, researchers advise," *IUVC Berkeley News*, June 9, 2016. See Krooth, herein Chapters 1-14. Examples of overlapping disasters displacing populations are many. In Mid 2016 alone:

- Houston, Texas rivers over-flowed
- Deady floods recurred in the Midwestern U.S.
- Scorching heat of 164 degrees Fahrenheit (73 Centigrade) appeared in parts of Iran and Iraq
- Blazing temperatures upsurged in Tennessee and Indiana
- Australia's New South Wales beaches lashed by giant waves
- Wild fires in California and Saskatchewan
- Clusters of earthquakes upend Alaska

- Threatened eruptions beset Mexico's Volcano of Fire
- Oregon, Washington, and parts of Northern California faced overdue 9.2 Richter scale earthquake to be followed by a possible tsunami leaving 13,000 dead
- Moscow suffered severe thunder, lightening and one-day rainfall over 40 percent of what it typically receives in an entire month
- Spreading El Niño heat led to India and Vietnam drought, failed harvests and a spreading reign of death
- Floods in Paris from the Seine overflow, seeping into the Louvre's treasure trove of some of the world's greatest works of art.

How can we account for such violent acts of Nature? Alexander von Humboldt discovered that the atmospheric realm is linked to Earth's terrain and seas in a way that changes in one sphere affected the others. This connection of forces has existed during the entire span of the planet's existence, creating the conditions for both natural disruptions and the emergence of life. Extraterrestrial meteorite invasions leave traces of a crystaline, carbon mineral called Ionsdaleite. (BBC.com/earth/story/20160706-in-siberria-in1908). Current environmental disasters have also been linked to fossil fuel pollution, making global weather patterns hotter and more extreme; as well to western cultural beliefs in apocalyptic Biblical prophesy (Malach 4:1) as a way of seeing—a burning furnace on Earth, arrogant and evil doers set ablaze, left with neither root nor branch.

What is certain: Should the Sun's heat re-radiated from Earth to the atmosphere raise the planet's temperature by 6.4 to 9.5 Celsius (about 8 to 12 degrees Fahrenheit) above the 20th century averages by the year 2300, all glaciers will melt; seas will rise to 30-90 meters (90-270 feet); islands, coastlines and low-lying regions will be drowned; 200 million people could become environmental refugees; food supplies and agricultural land will be submerged, along with most species, environmentalists, and those who symbolically rely on apocalyptic prophesy. As Alaska's permafrost, that has been solid for tens of millions of years, thaws, roads, buildings and other structures will collapse; and the permafrost will release methane and carbon dioxide, contributing to global warming. See Jake Ellison, "Destruction of Alaska Continues Under Record Heat," *SFGate*, June 9, 2016.

6. The several above paragraphs are based on Axel Timmermann and Tobias Friedrich, "Late Pleistocene climate drivers of early human migration," nature.com, doi:10.1038/nature/19365 (September 21 2016); Wu Liu, et al, "The earliest unequivocally modern humans in southern China," Nature.com 526. 696-699, doi:10.1038/nature15696 (October 14, 2016); Luc Pagani, et al., "Genomic analyses inform on migration events during the peopling of Eurasia," *Nature.com*, doi: 10.1038/nature19792 (September 21, 20); and *Nature Climate Change*, May 18, 2016, Study led by Katarznia Tokarska, a climate scientist at the University of Victoria, British Columbia. https://mail.yahoo.com. See also Hsang and Sobel, 'Climate change could trigger tropical evacuations, researchers advise," *IUVC Berkeley News*, June 9, 2016. And see Oliver Smith, et al.,"Sedimenary DNA from submerged site reveals wheat in the British Isle 8,000 years ago," *Science*, Vol. 347 No. 6225 (February 27, 2015), 998-1001. See also E. O. Wilson, *The Meaning of Human Existence*, 128-130. See also Stephen Rahmstorf, et al., *Nature Climate Change*, March 23, 2015 (Doi:10.1038/nclimate2554). And see Helen Biggs, "US Approves Genetically Modified Salmon for Food," *BBC News*, Nov. 19, 2015. See also Matt Carr. "A 'Showcase of Evolution at Risk," *The New York Times*, December 2, 2014. Opinion.

7. Oscar Venter, *Nature Communications*, August 23, 2016.

8. *Nature Climate Change*, May 18, 2016.

9. Amy Goodman, *Democracy Now*, U.N. Climate conference, COP, in Lima, Peru, December 8-12, 2014.

10. Pope Francis, *Our Common Home, passim.*

11. Mike Ives, "A Remote Pacific Nation Threatened by Sea Rise," *The New York Times, July* 2, 2016.

12. Jayson Smerdon. "Unprecedenterd 21st Century Drought Risk in the American Southwest and Western Plain," *Science Advances*, on line, February 21, 2015.

13. Anthony F. Randazzo and Douglas S. Jones, eds. *The Geology of Florida* (Grainsville, Florida: University Press of Florida, 1977), *passim.*

14. Andrew Cooper, professor of coastal studies at the University of Ulster. and Orrin Pilkey, professor of earth and ocean sciences at Duke University in Durham, North Carolina, *The Last Beach (*Amazon, 2014), *passim.*

15. *The Guardian*, "World's beaches being washed away due to coastal development." December 15, 2014.

16. James Gdrken, "New York City Could See Up to Six Feet of Sea Level Rise by this Century: Report," *Huffington Post*, February 17, 2015. See also Stephen Rahmstorf, et al, Potsdam Institute For Climate Impact Research, Study, March 2015 21, 2015.

17. Mireya Nevarro, "New York is Lagging as Seas and Risks Rise, Critics Warn," *New York Times*, September 10, 2012; Gdrken, "New York City Could See Up to Six Feet of Sea Level Rise by this Century: Report," *Huffington Post*, February 17, 2015.

18. Peter Spotts, "Sea level studies: US coasts even more vulnerable than previously thought," *Christian Science Monitor,* March 4, 2012, http://www.csmonito .com/Environment/ 2012/0314/Sea-level-studies-US-coasts-even-more-vulnerable-than-previously thought?cmpid=addthis_email# .T2Ir3i87V-4.email.

19. *Geophysical Research Letters*, December 6, 2014 9DOI:10.91002/2014GL062433).

20. Richard Krooth, 'Travels in the Midwest," *Interviews*, *Notebooks*, 2014 -15.

21. See Tom Philpott, "Talk Dirt to Me," *Mother Jones*, September/October 2013: 41-45; Timothy Eagan, *The Worst Hard Times* (Boston/New York: Mariner 2006, *passim*; Green Gazette, "The Many Benefits of Grass-Fed Meat," http://googl law 59 dz; Marvin Harris, Cows, . . . Riddles of Culture (New York: Touchstone 1975), *passim*; Krooth, *Gaia*, Chapters 29, 30; Krooth, "Interviews with Abe and Theda Eckstein and Sons, cow-and-corn farmers," Larson, Wisconsin, July 10, 2014.

22. Madhu Suri Prakash, "Vandana Shiva: Teachers for a Living World." Amy Harmon, "Golden Rice: Lifesaver?' *New York Times*, August 25, 2013.

23. Mayr, *What Evolution Is*, 8.

24. Suzanne Goldenberg, "Lima climate change talks reach global warming agreement," *The Guardian*, December 28, 2014.

25. Darwin, *Origin*, Chapter I; Mayr, *What Evolution Is*, 8, *et passim*.

26. Adam Gopnik, "Rewriting Nature: Charles Darwin, Natural Novelist," *The New Yorker*, October 2, 2006; and Edward O. Wilson, "Introduction," *From So Simple a Beginning*.

27. Armand Marie Leroi. *The Lagoon: How Aristotle Invented Science* (New York: Viking, 2014), 16-17. See also Edward O. Wilson, *A Window on Eternity: A Biologist's Walk Through the Gorongosa National Park* (New York: Simon & Schuster, 2014), xviii-xix, 118, 121.

28. See O. E. Wilson, *A Window on Eternity, passim.*

29. Richard Krooth, Seeking Ancient Species, South Island, New Zealand, *Notebooks*, 2012.

30. Ketcham, *"The Ruin of the West: How* Republicans are plundering our public lands," *Harper's Magazine*, February 2015, 23-31. For analogies, see also Wilson, *A Window on Eternity*, 128; and The Pontifical Academy of Sciences, "Sustainable Humanity, Sustainable Nature: Our Responsibility," Proceedings of the Joint Workshop, 2-6 May. 2014, Extra Series 41, Vatican City, 2015.

31. *SFGate*, Associated Press, Washington, "Government Funds Monarch Butterfly Conservation Program," February 9, 2015; Richard Krooth, "Interview with David Thompson Mexican Highland photographer," July, 2015.

32. Oxfam, "Billionaires Who Own the Same Wealth as Half the World," *Press Release* January 18, 2016.

33. Wilson, *A Window on Eternity*, 126-127, *et passim*.

34. The Pontifical Academy of Sciences, "Sustainable Humanity, Sustainable Nature: Our Responsibility," Proceedings of the Joint Workshop, 2-6 May. 2014, Extra Series 41, Vatican City, 2014.

35. Wilson, *A Window on Eternity*, 126-127, *et passim*.

36. Richard Krooth, "Beach Walk," *Notebooks*, March 7, 2014.

37. See Andrew C. Revkin, "Earth's Untallied Biological Bounty, from L.A. Suburbs to Deep Seabed Sediments,"*The New York Times*, March 25, 2015.

38. Darwin's legacy set the foundation, but not the framework, for the future study of evolutionary fitness of species. From his 19th century foundation of knowledge, he did not

envision that mistakes in genetic sequencing might be discovered, driven by evolution to weaken, not strengthen, surviving progeny. On this point, among all his scientific groups, Darwin highly ranked the Scottish surgeon, Hugh Falconer, a paleontologist and Indian tea grower. In 1862 Falconer critisized that, although Darwin had "laid the foundations of a great edifice; . . . he need not be surprised if, in the progress of erection, the superstructure is altered by his successors, like the Duomo of Milan, from the roman to a different style of architecture."

Receiving an advanced copy of Darwin's *Origin* and a request for Falconer's review and criticism, Darwin was stung, replying on October 1, 1862: "far from being surprised, I took it as absolutely certain that very much in the *Origin* will be proved rubbish; but I expect and hope that the framework will stand." (Darwin's letter to Falconer in F. Darwin, 1903, volume 1, page 206. Cited at Stephen Jay Gould, *The Structure of Evolutionary Theory*, 1-2).

It did only in part. For the superstructure was reconstructed in quite a fundamental fashion by DNA analysis that revealed the following many aspects, viz.: (1) DNA sequences of elementary information involves four symbols—the A, C, G and T that stand for DNA's nucleic acids—and the order in which they appear in the genome. (2) Once it became possible to examine that code and specify its sequence, variations could be identified, comparisons made, and changes traced. And these steps following a similar procedure could be used throughout the natural world. (3) Associated traits or diseases—such as hair color, mental illness, taste sensitivities, even sickle cell anemia, etc.—reveal genes controlling them, such as some 4,000 diseases caused by or undefended against by single flaws in human genes. (4) Genomic research does not just identify the source of a disease or trait; it also has helped discover remedies, uncovering defects in the human genome handed down to progeny, making humankind susceptible to weaknesses. (5) And as a supposed offset, human bioengineering already has produced such wondrous Frankenfoods as adding corn DNA to rice DNA, a 2013 bowl of so-called Golden Rice providing an alleged 60 percent of the daily requirement of vitamin A for healthy children.

Largely unanswered in this study is how deeply can epigenetic events and their influence play in changing the human genome, shaping behavior and actions, becoming heritable and having shaped evolutionary prehistory and history among various people in different geographies and cultures.The racial and racist implications of beliefs in genetic inheritance by tribes and groups living in different geographic regions, with alleged genetically-predetermined social inclinations, is neither biologically nor behaviorally verifiable. For another opinion, see Nicholas Wade, *A Troublesome Inheritance: Genes, Race and Human History* (New York: The Penguin Press, 2014).

Post Script

During the Anthropocentric Age of Earth's circular path around the Sun, our species has come to worship its own technological advances, collectively created with intelligence and labor, though this massive sphere has infringed on and jeopardized the biosphere.

We have inherited superior intelligence far above that of most other forms of life on Earth, empowering us to create the means to shape landscapes, invade seas, and use other species for our benefit and survival.

Yet our powers have now turned against us, as we are tested to endure and survive our own blinded quest for ephemerals and accumulated wealth, changing climate into a torrid world cauldron.

We have yet to discover that our invention of tools and technologies have forged perilous environmental conditions we cannot regulate, control or contain, as the global escalation of heat and the overwhelming accumulation of waste together have overwhelmed the capacity of animals and plants to survive on land and in the sea, undermining their biological integrity supporting life.

Today our future is grim. There will be no return to a world unaffected by the massive climate changes we have put in motion. Neither the strongest nor most intelligent living species will evolve with the necessary speed and modification to survive, fulfilling Darwin's evolutionary expectations; and only those among us able to adapt to the onrushing change in climate will likely reproduce, filling in the branches of the Tree of Life.

Our species may construct ever-new technological means to slow, even momentarily stop, the transformation and destruction of the biosphere placing millions of species in jeopardy. But these steps alone will not allow Darwin's Walk to forever fulfill the dramatic evolutionary design of existence, nor stop a Last Wave from enveloping most earthly territory, grinding to bits the planet's existing forms of life into the sands and residues of time.

—Richard Krooth, April 30, 2017

Bibliography

"Accumulation of Micro-plastic on Shorelines Worldwide: Sources and Sinks." ACS Publications, Environmental Science and Technology. http://pubs.acs.org/doi/abs/10.1021/es201811s

Ackerman, Diane. *The Human Age: The World Shaped by Us*. New York/London: W.W. Norton & Company, 2014.

———. *The Zookeeper's Wife*. New York, London: W.W. Norton & Company, 2007.

The Air We Breathe. San Francisco, Conference, n.d. Based in part on explanation of Andes' rarified atmospheric oxygen.

Alaska Geographic. "Arctic National Wildlife Refuge." KQED, June 8, 2013, 9-10 P.M. Holand Hargrave, production editor.

Alfè, Dario. University College London, et al. *Nature*, 2012. The British Library, copy, n.d.

Alvarez, Luis and Walter. "Extraterrestrial Causes for the Cretaceous-Tertiary Extinction." *Science*, June 1980.

Amos, Jonathan. "'The Hoff' Crab is New Ocean End." *BBC*, December 4, 2012.

Angier, Natalie. "Ebola and the Vast Viral Universe." *The New York Times*, *Science*, October 27, 2014.

———. "The Enigma 1,800 Miles Below Us." *The New York Times*, May 28, 2012.

"Are 90 Companies Responsible For Nearly Two-Thirds Of Global Warming?" *HuffPost Green*, Nov. 22, 2013.

Aristotle. *Metaphysics*. The British Library, copy, n.d., 1.1, 980a20-25,

———. *Parts of Animals*. Book I chapter 5. The British Library, copy, n.d.

———. *Politics*. Book. I chapter 2. The British Library, copy, n.d.

Arrhenius, Svante August. *Theories of Chemistry*. Stockholm, 1903, 1905, 1914.

———. *Worlds in the Making: The Evolution of the Universe*. Stockholm, 1903, 1905, 1914. Trans. *H. Borns. New York: Harper & Brothers, 1908.*

Attenborough, Richard. *KQED*, February 2012, 9-10 PM.

Bamshad, Michael J. and Olson, Steven E. "Does Race Exist?" *Scientific American*, December 2003, 80-85. http://www.sciencemag.org/search?author1=B%C3%A4rbel+H%C3%B6nisch&sortspec=date&submit=Submit"Bärbel, Hönisch. Columbia University paleoceanographer. *Scce*, March 2, 2012: Vol. 335 no. 6072: 1058-1063. DOI: 10.1126/science.1208277.

Barker, J. Ella. *Rise and Decline of the Netherlands*. London: Smith Elder, 1906.

Barnhardt Bz. Posted Jan. 10, 2015. Cached. http://www.barnhardt.biz/2015/01/09/.

Barraclough, Timothy G. Professor of evolutionary biology. With Diane Lawrence, graduate student in biology, et al., at Imperial College London. "Microscopic Neighbors, Evolving Together." *PLoS Biology*, May 15, 2012.

Bartholet, Jeffery. "Swept from Africa to the Amazon." *Scientific American.com*, Vol. 306, No. 2 (February 2012): 44-47.

Basak, Sonali and Bloomfield, Doni. "'Once-in-Three-Billon-Year' Jump in Bonds Was a Warning Shot, Dimon Says." *BloombergBusiness*, April 10, 2015.

Bauerlein, Monika and Jeffery, Clara. "The Speedup." *Mother Jones*, July/August 2011, 18-25.

BBC News, Science and Environment. "Ancient American Genome Mapped," February 13, 2014.

BBC. "Extracting 2.5 billion Year Old Oxygen from Ancient Rocks." Broadcast on BBC Two at 21:00 GMT on Tuesdays 7, 14 and 21 February 2013, catch-up via BBC. http://www.bbc.co.uk/programmes/p00ns704.

Beckwith, Martha. *Hawaiian Mythology*. Honolulu: University of Hawaiian Press, 1970.

Belli, Brita. "Defender of the Seas." *E-The Environmental Magazine*, January/February 2012.

Berg, Ed. "Analysis of the Jigsaw Lake Sedimentary Peat Macrofossils, vers. 2.0," April 13, 2011.

———. Destruction of White/Lutz Pine. Cited at Richard Krooth, *Gaia and the Fate of Midas*, xxxi, fn. 18.

Berkeley Resource Group, *Species Survival Rates*, Box 9515, Berkeley, CA., 94709.

"Berlin Golden Hat." Vorderasiatishches Museum zu Berlin: Antkensammlung (Pergamonmuseum); Museum für Vor-und Frühgeschichte (Nues Museum), Berlin Bodestraße, 2000.

Bertnesss, Mark D., Brisson Caitlin P. et al. "Herbivory Drives the Spread of Salt Marsh Die-Off." *PLOS ONE*, March 20, 2014. DOI: 10.1371/journal.pone.0092916.

Bibi Farber, "Above Ground Mining," *Nextworld TV*.

———. NextworldTV, March 16, 2012.

———. "The Story of Bottled Water," Next World TV.

Bierman, Paul and Rood, Dyland. *Science*, April 21-25, 2014.

Biggs, Helen. "US Approves Genetically Modified Salmon for Food." *BBC News*, November 19, 2015.

Black, Richard. "Green Tuna Goes FAD-free." *Earthwatch, BBC*, March 9, 2011.

———. "The Missing Link," *BBC*, January 21, 2011.

———. "Prince Charles optimistic for fisheries' future," *BBC News: Science and Environment*, February 3, 2012. http://www.bbc.co.uk/news/science-environment-16875215.

Bodo, Murry, O.F.M. *Tales of St. Francis*. Cincinnati, Ohio: St. Anthony Messenger Press, 1988.

Borenstein, Seth. "Skeptic Finds He Now Agrees Global Warming is Real." *AP*, October 30, 2011.

Borgerson, Scott. "The Coming Arctic Boom, As the Ice Melts, the Region Heats Up." *Foreign Affairs*, July/August 2003: 58-89.

Boyce, Nell Greeenfield. "Big Data Predicts Harm if Climate Warming Goes Unchecked." *NPR*, November 30, 2015.

Bradsher, Keith. "China Cuts Interest Rates." *The New York Times*, June 28, 2015.

Brannon, Peter. "Scientists Worry that Warming Seas May be Harming the Endangered Right Whale." *Washington Post*, April 30, 2012.

Brass Check TV. Journeyman video. "Attack of the Drones," http://www.brasschecktv.com/videos/ offensive-technology-1/attack-of-the-drones-usa-.html. *http://www.bgs.ac.uk/downloads/start.cfm?id=432 British Geological Survey, 2011-2015.*

Brown, Kyle S., Marean, Curtis W., et al. "An Early and Enduring Advanced Technology Originating 71,000 Years Ago in South Africa." *Nature online*, November 7, 2012.

Bullfinche's Mythology. New York: The Modern Library, Random House, n.d.

Burnet, D. Graham. *The Sounding of the Whale: Science and Cetaceans in the Twentieth Century*. Chicago: The University of Chicago Press, 2011.

Caldeira, Ken. Carnegie Institute for Science, Stanford, California. *Scientific American*, 78-83.

Campbell-Young, Gael J., and Balkwill, Kevin. "Serpentines of the Barberton Greenstone Belt." *Veld & Flora*. Newlands, South Africa: Botanical Society, Vol. 86 (1) (March 2000): 26-28.

Capra, Fritjof. *The Web of Life*. New York: Anchor Books/Doubleday, 1996.

"Carbon emissions show sharp rise." *BBC News*, November 27, 2006. Citing Dr. Mike Rapauch, Co-chair, Global Carbon Project; and Professor Corinne Le Quere of the British Antarctic Survey.

Carey, John. "Global Warming: Faster than Expected? *ScientificAmerican.com*, November 2012: 51-55.

Carroll, Sean B. *Endless Forms Most Beautiful: The New Science of Evo Devo*. New York: W. W. Norton &Co., 2005.

Carson, Rachel. *Silent Spring, 1962, Chapter1*. Reprinted as Rachel Carson, "Fable for Tomorrow." Ed., Ann Krooth, *A Unique World of Ideas*. Boston/New York: Bedford/St.Martin's Custom Publishing, 2013, 127-128.

Chang, Kenneth. "A" Solar Show with Mixed Reviews." *Science Times, The New York Times*, July 1, 2014, D3.

Chappell, Bill. "Vatican Charges 5 Over Leaked Documents, Alleging 'Organized Crime'" *NPR*, November 21, 2015.

Chill, Emanuel. Trans. *Power, Property and History: Joseph Barnave's Introduction to the French Revolution and other Writings*. New York: Harper & Row, 1971.

Choi, Charles G. "Exobiology: Life Not as We Know It. *Scientific American*, Vol. 297 No. 3 (September 2007): 34.

Climate Central. IOP Science. *Environmental Research Letters*, March 14, 2012 Number 1 (January-March 2012): (010201-019502) (Vol. 7, 2012) http://iopscience.iop.org/1748-9326.

Coleridge, Samuel Taylor. "The Rime of the Ancient Mariner." *Immortal Poems of the English Language: British and American Poetry from Chaucer's Time to the Present Day*. Ed. Oscar Williams. New York: Washington Square Press, Inc, December 1966. This poem is in the public domain.

Compton's Pictured Encyclopedia. *War Supplement: An Alphabetic Reference Book of the European War*. Chicago: F.E. Compton & Company, n.d.

Contribution of Working Groups I, II and III to the Fourth Assessment Report of the Intergovernmental Panel on Climate Change Core Writing Team. Eds. Pachauri, R.K. and Reisinger, A. IPCC, Geneva, Switzerland: 104.

Cooper, Andrew, and Pilkey, Orrin. *The Last Beach*. Washington: Amazon Web, 2014.

Crease, Robert P. Review of Micheal Hitzik, Big Science: Ernest Lawrence and the Invention That Launched the Military-Industrial Complex. New York: Simon and Schuster, 2015; and New York Times Book Review, July 19, 2015.

Darwin, Charles Robert. *Descent of Man. London: British Library Manuscript Copy, 1871. Darwin-online.org.uk. Origin.*

———. *The Expressions of Emotions in Plants and Animals*. London: British Library manuscript copy, n.d.

———. *Journal of Researches,* 1839. *Darwin-online.org.uk.*

———. *Journal of the Voyage of the Beagle*. London: British Library manuscript copy, 1836.

———. *Origin of Species*. 2nd, 3rd, 4th and 5th editions. *Darwin-online.org.uk.*

———. *Origin of Species*. Harvard Classics Vol II. New York: P.F. Collier & Son. 1909. First published November 1859.

———. *Variation of Animals and Plants Under Domestication*. London: British Library manuscript copy, 1868.

———. *The Zoology of the Voyage of HMS Beagle*, 1838-43. Darwin-online. org.uk.

Darwin, Erasmus. *The Botanic Garden*. London: British Library manuscript copy, 1789.

Davey, Monica. "In Midwest, Drought Gives Way to Flood." *The New York Times*, April 25, 2013.

Davies, P.C.W. and Brown, J. Eds. *Superstrings: A Theory of Everything?* Cambridge: Cambridge University Press, 1988.

da Vinci, Leonardo. Exhibit of Codicils. San Francisco, 2010.

———. *Tbe Notebooks*, (1508-1515) Vol. II., Chpt. 25.

Davis, Julie Hirschfeld. "Obama to Call for More Icebreakers in Arctic as U.S. Seeks Foohold." *The New York Times*, September 1, 2015.

Davis, Mike. *The Monster at Our Door: The Global Threat of Avian Flu*. New York/London: The New Press, 2005.

Dawkins, Richard. *The Ancestor's Tale: A Pilgrimage to the Dawn of Evolution*. Boston/New York: Houghton Mifflin Company, 2005.

———. *The Greatest Show on Earth: The Evidence for Evolution*. New York: The Free Press, 2009.

DeBUYs, William. *The Last Unicorn: A Search for One of Earth's Rarest Creatures*. New York: Little, Brown and Company, 2015.

Democracy Now. "Secretive Deal Isn't About Trade, But Corporate Control," May 27, 2015.

Desalle, Rob. Editor, *Epidemic: The World of Infectious Diseases*. New York: The New Press/ American Museum of Natural History, 1990.

d'Halloy, M. J. d'Omalius. *Bulletins de l'Acad*. Bruxelles, 1831, tom. xiii no. 581. Reprint.

Diamond, Jared. "Holistic Extinctions: A Rosetta Stone for Understanding Prehistoric Extinctions." In Paul S. Martin and Richard G. Klein, Eds. *Quartenary Extinctions: Prehistoric Evolution*. Tucson: University of Arizona Press, 1984.

DiChristopher, Tom. "Why Fed Won't Raise Rates Now." *Yahoo Finance*, September 17, 2015.

Dickens, Charles. *Household Words*. London: British Library copy, 1854.

DK Illustrated Oxford Dictionary. New York/Oxford: Oxford University Press, 1998: 398; 554.

Disraeli, Benjamin, "The key of India is London." House of Lords, 5 March 1881.

Dobb, Maurice. *Political Economy and Capitalism*, Rev. 2nd ed. London: Routledge & Kegan Paul Ltd., 1940.

"Docent Lecture Tour." Oakland Museum, Oakland, California, May, 2012.

Donne, John. *Holy Sonnets*, no. 9, L. I.

Douglass, A.E. *Climate, Cycles and Tree Growth*, Vols. I-III. Washington, D.C.: Carnegie Institution, 1936.

Draymond, H.G. "The Fascination of Arctic Exploration." *The African Monthly*, February, 1907, 172.

Dubois, W.E.B. "The Negro Farmer." United States Department of Labor. *Bulletin* No. 9: 69.

Dugan, Patrick. General Editor, *Guide to Wetland*. Buffalo, New York: Firefly Books, 2005, 14-15.

Eagan, Timothy. "The Fires This Time." *The New York Times*, July 8, 2012.

———. *The Untold Story of Those Who Survived the Great America Dust Bowl*. New York: Houghton Mifflin Company/A Mariner Book, 2006.

The Economist. "Special Report: Personal Technology. Beyond the PC," Oct. 8th-14th: 2011:1-20.

Editors. "Special Report: The Future of Medicine, Boosting the Body's Healing Powers." *Scientific American.com*, April 2003: 48-67.

E-The Environmental Magazine. "Seas," January/February 2012: 18-25.

———. "Fighting for Air," May/June 3013: 21-22.

Ehrenreich, Barbara. *Dancing in the Streets: A History of Collective Joy*. New York: Metropolitan Books, Henry Holt and Company, 2006.

Ehrlich, P.R. and Holdren, J.P. "Impact of population growth." An exchange. *Science*, 1971.

Ekman, Paul. "Survival of the Kindest." *Shambala Sun*, November 2010: 61-63.

Elkin, A.P. *The Australian Aborigines*. Sydney, Australia: Angus and Robertson, 1938. Fifth Edition, reprinted 1966. National Library of Australia, card number and ISBN 0 207 13009 4.

Elton, Charles S. *The Ecology of Invasions by Animals and Plants*. Chicago and London: University of Chicago Press, 1958.

Engels, Friedrich. "A Fair Day's Wages for a Fair Day's Work." In *The Great Social Struggle*, Vol II, 86-92. Edited by Richard Krooth, Santa Barbara: Harvest Publishers, 1979.

———. "The Wages System." In *The Great Social Struggle*. Vol II, 86-92. Edited by Richard Krooth, Santa Barbara: Harvest Publishers, 1979.

"Evidence linking Arctic amplification to extreme weather in mid-latitudes." Carl Nolte. *Geophysical Research Letters*, Vol. 39, L06801, doi:10.1029/2012GL051000, 2012, 17 March 2012.

"The Evolution of Cooperation." *Scientific American*, July 13, 2013.

Fackler, Martin. Japan's Nuclear Refugees, Still Stuck in Limbo." *The New York Times*, October 2, 2013. http://p.nytimes.com/email/re?location=InCMR7g4BCKC2wiZPkcVUtypgPg94Azy&user_id=4fbf4d51f4030af146ba6279992caaa4&email_type=eta&task_id=1380744558392277&regi_id=0

Farber, Bibi. *NextworldTV*, March 16, 2012

———. "The Story of Bottled Water." *NextWorldTV*,

Farrand, John Jr. *How to Identify Birds*. New York: McGraw-Hill Book Company 1988. "Fresh Water," 36-37.

Fibonacchi. Table. © Richard Krooth 2016.

Fimrite, Peter. "Sonoma County's Cedars a rare geologic wonder." *San Francisco Chronicle*. Jan. 16, 2012.

Flannery, Tim. *Here on Earth: A Natural History of the Planet*. New York: Atlantic Monthly Press, 2010).

———. *The Weathermakers: The History and Future Impact of Climate Change*. Great Britain: Allen Lane/Penguin Books, 2005.

Fleming, Nic. "Plants Have a Hidden Internet." http://www.bbc.com/earth/story/20141111-plants-have-a-hidden-internet.

Fountain, Henry. "Australia Heat Tied to Climate Shift," *The New York Times*, March 3, 2017.

———. "Turning Carbon Dioxide into Rock and Burying It." *Science*, *The New York Times*, February 9, 2015,

———. "When Giants Had Wings and 6 Legs." *Science Times, The New York Times*, February 3, 2004, D1, D4.

Francis, Jenifer and Vavrus, Stephen. The University of Wisconsin-Madison. Rutgers Marine and Coastal Sciences, 2015.

Francis, Pope. "Meeting of Latin American and Asian landless peasants and other social movement activists," October 2014.

———. On Care of Our Common Home, Encyclical Letter Laudato Si' of the Holy Father Francis. Washington DC: United States Conference of Catholic Bishops and Vatican City, June 2015.

French, Marilyn. *Beyond Power: On Women, Men, and Morals*. New York: Summit Books, 1985.

Friedman, Mitchell. *Here's Every Single Animal That became Extinct In the Last 100 Years (Photos),* September 16, 2014.

Friedman, Thomas L. "The Other Arab Spring." *The New York Times*, April 7, 2012.

Fromm, Erich. "Freedom in the Work Situation." Chapter in *The Great Social Struggle*, Vol. II, 92-107. Edited by Richard Krooth. Santa Barbara: Harvest Publishers, 1979.

Gdrken, James. "New York City Could See Up to Six Feet of Sea Level Rise by This Century: Report." *Huffington Post*, February 17, 2015.

Geophysical Research Letters, December 6, 2014. 9DOI:10.91002/2014GL062433.

Gerth, Hans H. "Lectures." Sociology Department, University of Wisconsin, Madison, 1962-1965.

Gillis, Justin. "Arctic Methane: Is Catastrophe Imminent?" *The New York Times*, December 20, 2011.

———. "Scientists Trace Extreme Heat in Australia to Climate Change." *The New York Times*, Sept. 29, 2014.

———."Sea Ice in the Arctic has Fallen to the Lowest Level on Record." *The New York Times*, Aug. 27, 2012.

Glaze, Jeff. "Hot, Dry Weather Takes a Toll on Trees, Grass." *Wisconsin State Journal,* Madison, Wisconsin, July 13, 2012.

Goldstein, Miriam. Research-graduate student at the University of California San Diego, Scripps Institution of Oceanography. Royal Society journal *Biology Letters*, May 9, 2012.

Golles, Justin. "Rising Sea Levels Seen as Threat to Coastal U.S." *The New York Times*, March 13, 2012. http://www.nytimes.com/2012/03/14/science/earth/study-rising-sea-levels-a-risk-to-coastal-states.html.

Gopnik, Adam. "Rewriting Nature: Charles Darwin, Natural Novelist." *The New Yorker*, October 23, 2006.

Gore, Al. *The Future*. New York: Random House, 2014.

Gould, Stephen Jay. *The Structure of Evolutionary Theory.* Cambridge, Mass./London, England: The Belknap Press of Harvard University Press, 2002.

Gray, Louise. "Global warming has slowed because of pollution." *Telegraph Media Group Ltd.,* Nov. 26 2010. http://www.telegraph.co.uk/ earth/ environment/climatechange/ 8159991/ Global-warming-has-slowed-because-of-pollution.html.

Graybeil Ann M. and Smith, Kyle S. "Good Habits, Bad Habits" *Scientific American* Vol. 370, No. 6 (June 2014): 38-43.

Greenberg, Paul. "Tuna's End." *The New York Times Magazine*, June 27, 2010, 33.

Greenblatt, Stephen. *The Swerve*, Loose-leaf Copy, 2011. Elaborated by Lewis H. Lapham. "Ignorance of Things Past." *Harper's Magazine*, May 2012: 33.

Greene, Brian. *The Elegant Universe: Superstrings, Hidden Dimensions and the Quest for the Ultimate Theory*. London: Johnathan Cape, 1998.

Greene, Charles H. "Climate Change: The Winters of Our Discontent." *ScientificAmerican.com*, December 2012: 51-55.

Greene, Jeffrey with Moline, Karen. *The Bird Flu* Epidemic. New York: Thomas Dunne Books/St. Martin: The New Press, 2005.

"Greenhouse Gas Emissions Reached New Highs Last Year. Again." *HuffPost Green*, November 6, 2013. Reporting by Tom Miles; editing by Stephanie, Nebehay and Alistair, Lyon.

Greenwald, Glenn. *No Place to Hide: Edward Snowden, The NSA, and the U.S. Surveillance State*. New York: Metropolitan Books/Henry Holt and Company, 2014.

Gribbin, John and Mary. *He Knew He Was Right: The Irrepressible Life of James Lovelock and Gaia*. London: Allen Lane, 2014.

Grinspoon, David. "How Long Will We Last?" *Scientific American*, September 2016, 76-79.

Gruber, Howard E. *Darwin on Man: A Psychological Study of Scientific Creativity*. Second Edition. Chicago: University of Chicago Press, 1974, 1981.

Guardian. "Climate Change," December 10, 2010, 07.15 EST.

———. "World's beaches being washed away due to coastal development." December 15, 2014.

Hafner, H. Annamalai, Sooraj, Jan K. P. and Pillai, P. "Global Warming Shifts The Monsoon Circulation, Drying South Asia." International Pacific Research Center (IPRC), University of Hawaii. doi: http://dx.doi.org/10.1175/JCLI-D-12-00208.1

Halfar, Jochen, and Fujita, Rodney M. "Danger of Deep-Sea Mining." *Science*, Vol. 316 (18 May 2007): 987.

Hamilton, Edith. *Mythology: Tales of Gods and Heroes* (Boston, Mass.: Mentor Book/Little, Brown & Co., 1940, 1942.

Hanke, Lewis. *Aristotle and the American Indian*. London: Hollis & Carter, 1959.

Hanley, Charles. "Vast African Lake Level Dropping Fast." AP (PST Jinja, Uganda), 9 December 2006.

Hansen, Dr. James. Affiliation: National Aeronautics and Space Administration, NASA Goddard Institute for Space Studies, 2880 Broadway, New York, NY 10025 USA.

———. et al. 1996. "A Pinatubo climate modeling investigation." In *The Mount Pinatubo Eruption: Effects on the Atmosphere and Climate*. G. Fiocco, D. Fua, and G. Visconti, Eds. NATO ASI Series, Vol. I 42, 233-272. Springer-Verlag. Heidelberg, Germany. http://pubs.giss.nasa.gov/cgi-bin/abstract.cgi?id=ha00900o.

———. Hansen, et al. "Perception of Climate Change." NASA Goddard Institute for Space Studies and Columbia University Earth Institute, March 29, 2012.

Harden, Garrett. "Lifeboat Ethics: The Case Against Helping the Poor." *Psychology Today*, September 1974. Update June 10, 2003.

Harden, Garrett and Baden, John. Managing the Commons. San Francisco: Freeman, 1977.

Harden, Garrett. "Political Requirements for Preserving Our Common Heritage." In *Wildlife and America*. Washington, D.C.: Council on Environmental Quality, 1979.

Harmon, Amy. "Golden Rice: Lifesaver?" *The New York Times*, August 25, 2013.

Harris, Richard. "Air Pollution: Bad For Health, But Good For Planet?" *NPR*, November 15, 2011. http://www.npr.org/people/2100631/richard-harris.

Harris, Marvin. *America Now: The Anthropology of a Changing Culture.* New York: Simon and Schuster/Touchstone, 1981.

Harvey, James and Hood, Katherine. *The British* State. London: Lawrence & Wishart, 1958.

Hausmann, Ricardo, et al., "The Future of Manufacturing." *Scientific American.com*, May 2013: 36-51.

Hazen, Robert M. The Story of Earth: The First 4.5 Billion Years: From Stardust to Living Planet. New York: Penguin Books, 2013.

Heller, Peter. *The Whale Warriors: The Battle at the Bottom of the World to Save the Planet's Largest Mammals.* New York: Free Press, 2011.

Herszenhorn, David M. and Gorman, James. "Russian Scientists Bore into Ancient Antarctic Lake," *The New York Times*, February 8, 2012.

Hertsgaard, Mark. "Climate Change: While Washington Slept." *Vanity Fair*, November 8, 2012.

Herzog, Karen. "Livestock Feed Runs Short as Crops Fail in Dry Spell." *Journal Sentinel* (Milwaukee, Wisconsin), July 13, 2012.

Heyerdahl, Thor. *Kon-Tiki: Across the Pacific By Raft.* Trans. F.H. Lyon. Chicago/New York/ San Francisco: Rand McNally & Co., 1950.

Hobson, John A. *Evolution of Modern Capitalism: A Study of Machine Production*, New and rev. ed. New York: Charles Scribner's Sons, 1907.

———. *Imperialism: A Study*. New York: James Pott & Co., 1902.

Hoegh-Guldberg, Ove. Director of Australian Global Change Institute, University of Queensland, co-author *Science* magazine, 2014.

Hogan, John. The End of Science: Facing the Limits of Knowledge in the Twilight of the Scientific Age. New York: Addison-Wesley, 1996, Broadway Paperback, 1997.

Holden, Raymond. "Journey to the Depths of the Sea." *All About Famous Scientific Expeditions.* New York: Random House, 1955, 3-25.

Hönisch, Bärbel. Paleo-oceanographer at Lamont-Doherty Earth Observatory, *Science*, March 1, 2012. http://www.csmonitor.com/tags/topic/Lamont-Doherty+Earth+Observatory

Hsang. Solomon and Alan Sobel. 'Climate change could trigger tropical evacuations, researchers advise," *IUVC Berkeley News*, June 9, 2016.

Huffington Post, "Why is the U.S. Desperate to OK Slavery in Malaysia?" May 26, 2015.

Illich, Ivan. *Medical Nemesis*. Toronto/New York/London: Bantam Books, 1976).

Ingram, J. K. *History of Slavery.* London: British Library copy, 1905.

Inman, Mason. "The True Cost of Fossil Fuels." *Scientific American. com*, April 2013: 58-61.

Innis, Michelle. "Warming Oceans May Threaten Krill, a Cornerstone of the Antarctic Ecosystem,' *The New York Times*, October`19, 2015.

Jacob, Thomas. http://www.csmonitor.com/tags/topic/France. France's Bureau de Recherches Géologiques et Minières. French counterpart to the U.S. Geological Survey. http://www.csmonitor.com/tags/topic/U.S.+Geological+Survey.

John Wahr, a physicist at the University of Colorado at Boulder and a member of the team reported the results in *Nature*, February 9, 2012 http://www.csmonitor.com/tags/topic/University+of+Colorado+at+Boulder.

James, Richard D. "Boom State," *The Wall Street Journal*, July 25, 1974.

Janik, Allan and Toulmin, Stephen. *Wittgenstein's Vienna* London: Weidenfeld and Nicolson, 1973.

Jayawardhana, Ray. "Coming Soon: A Supernova Near You." *Scientific American*, December 2013, 68-73. On subatomic neutrino particles.

Jenkins, Kathy. AP via AEI-CASC "Endangered Species Gallery." blog.sfgsate.com.

Johnston, William R. *The Austrian Mind: An Intellectual and Social History*, 1848-1938. Berkeley, Ca.: University of California Press, 1972.

Joling, Dan/Associated Press. "Alaska Battered by Storm." *The Detroit News*, November 10, 2011 http://detnews.com/article/20111110 /NATION/ 111100398/Alaska-battered-by-storm#ixzz1dQnVYJZk

Jones, Louis Thomas. *Aboriginal American Oratory: The Tradition of Eloquence Among Indians of the United States*. Los Angeles, California, 1965.

Joyce, Christopher. "Alarming Amazon Droughts May Have Global Fallout." *NPR*, February 5, 2011.

Kahn, Brian. November 10, 2011 "http://detnews.com/article/20111110 /NATION/ 111100398/Alaska-battered-by-storm#ixzz1dQnVYJZk" http://detnews.com/article/ 20111110 /NATION/ 111100398/Alaska-battered-by-storm#ixzz1dQnVYJZkis, March 18, 2014, editor in chief, *Climate Central*. See http://www.huffingtonpost.com/2014/03/18/400-ppm-co2-levels.

Kaku, Michio. *Physics of the Future: How Science Will Shape Human Destiny and Our Daily Lives by the Year 2100.* New York: Doubleday, 2011.

Kandel, Eric R. *The Age of Insight: The Quest to Understand the Unconscious in Art, Mind, and Brain, from Vienna 1900 to the Present.* New York: Random House, 2012.

Kaplan, Matt. 'On Glaciers, Balls of Dust and Moss Make a Cozy Home." *Polar Biology,* June 2012 and August 27, 2012.

Keeling Curve. At Al Gore, *The Future*. New York: Random House, 2014.

Ketcham, Christopheer. *"The Ruin of the West: How* Republicans are plundering our public lands," *Harper's Magazine*, February 2015, 23-31.

Kerry, Immanuel. *Bulletin of the American Meteorological Society*, March 2008.

Kiefer, Christian. *The Animals: A Novel.* New York: Liveright Publishing Corporation/W. W. Norton & Co., 2015). 32.

Klein, Naomi. *This Changes Everything: Capitalism vs. The Climate.* New York: Simon and Schuster, 2014.

Knight, Kathleen. Research ecologist. Discussion. U.S. Forest Service, 2014.

Koerth-Baker, Maggie, "After the Trees Disappear." *Science Times, The New York Times,* July 1, 2014, D1, D5.

Kolbert, Elizabeth. *The Sixth Extinction: An Unnatural History*. New York: Henry Holt and Company, 2014.

Krajick, Kevin. "Race to Plumb the Frigid Depths." *Special Section on Polar Science: News, Science*, Vol. 316 (16 March 2007): 1525.

Krooth, Richard. "Across the Pacific." *Notebooks*, December 2012.

———. "Adventures in Florida." December 2012-January 2013.

———. *Arms & Empire: Imperial Patterns Before World War II. Santa Barbara: Harvest Publishers, 1980.*

———. "At Fort Myers, Florida," *Notebooks*, January 2013.

———. "At the Monterey Bay Aquarium," *Notebooks*, May 17, 2012.

———. "*To BBC News* environmental correspondent Richard Black," October 18, 2010.

———. "Fifty Years on the Trail," *Notebooks*, 1965-2015.

———. *Gaia and the Fate of Midas: Wrenching Planet Earth.* Lanham/London: University Press of America, 2009.

———. *The Great Social Struggle*, Vol. I. Santa Barbara: Harvest Publishers, 1980.

———. *The Great Social Struggle*, Vol. II. Editor, Richard Krooth. Santa Barbara: Harvest Publishers, 1979

———. *The Great Social Struggle*, Vol. III. Editor, Richard Krooth. Santa Barbara: Harvest Publishers, 1979.

———. *Empire: A Bicentennial Appraisal.* Santa Barbara: Harvest Publishers, 1976.

———. "In the Australian Cavil," *Notebooks,* 2010-11.

———. "In the Step of Terror." *Unpublished Manuscript* (Berkeley, California, 2012).

———. "Interview with David Thompson Mexican Highland photographer," July, 2015.

———. "Interview with Dr. Edward Berg," Tempe, Arizona, August 2014

———. "Interview with Photographer Lana Husser: Filming the New Orleans Devastation," February, 2013.

———. "Living in the South." *Notebooks*, May 1965.

———. *Gaia and the Fate of Midas*. Lanham/London: University Press of America, 2014.

———. *Mexico, Nafta and the Hardships of Progress* Jefferson, North Carolina: McFarland, 1995. Reissued with Postscript, 2013.

Krooth, Richard, Edelson, Morris and Fukurai, Hiroshi. *Nuclear Tsumami: The Japanese Government and America's Role in the Fukushima Disaster*. Lanham, Boulder, New York, London: Lexington Books, 2015.

Krooth, Richard. "On the Road Again: Traveling the Plains and Texas." *Notebooks*, 2000-2012.

———. "Philosophy and the Wage System." *The Great Social Struggle.* Vol. 1, 54-83. Editor Richard Krooth. Santa Barbara: Harvest Publishers, 1978.

———. "Searching for Micro-species on Australian Beaches," *Notebooks*, 2010-2011.

———. "Talks with the Abe Eckstein Family: Cow and Corn Farmers." Larsen, Wisconsin, July-August 2012-2015.

———. Training in the drafting offices of the City of Chicago, Division of Bridges and Viaducts, 1954-1958.

———. "Travels Along North Africa's Desert Fringe," *Notebooks*, 1962-1965.

———. "Travels Along the U.S. Eastern Seaboard and in the Midwest" *Notebooks*. 2004-14.

———. "Travels in North Africa, Southern Europe, and the Atlantic." *Notebooks* 1962-66.

———. "Travels in Australia." *Notebooks*, 2010-2011.

———. "Travels in the Costa Rica Cloud Forest," *Notebooks*, 1985-1994.

———. "Travels in the Everglades," *Notebooks*, 2008-2011.

———. "Travels in Florida and the Gulf of Mexico; interview with Captain 'Nemo,' Ten Thousand Islands," *Notebooks*, 2005-07, 2011-2012.

———. "Travels in New Zealand," *Notebooks*, 2008-2010.

———. "Travels in North Africa, Southern Europe, and the Atlantic," *Notebooks* 1962-6,

———. "Travels in Western and Northern Europe, *Notebooks*, 1962-2012."

———. "Travels on California Beaches," *Notebooks*, 2012.

———. 'On Sanibell Island, Florida," *Notebooks*, January 2013.

Kump, Lee R. "The Last Great Global Warming," *Scientific American*, Vol. 305 No. 1 (July 2011): 57.

Lakoff, George. "Global Warming Systematically Caused Hurricane Sandy." *Reader Supported News* (rsn) October 30, 2012.

Landau, Elizabeth. "Rare skull sparks human evolution controversy, *CNN*, October 19, 2013, summarizing article in *Science*, 18 October 2013: Vol. 342 No. 6156 pp. 326-331 DOI: 10.1126/science.1238484. A Complete Skull from Dmanisi, Georgia, and the Evolutionary Biology of Early Homo,"

Langlois Nkrosta "West's Superfloods May Wash Over Aging Dams, Scientists Warn," *High Country News*, reproduced at *Yahoo*, March 7, 2017.

Lapham, Lewis H. "Ignorance of Things Past," *Harper's Magazine*, May 2012: 33.

Lasch, Christopher. *The Culture of Narcissism, American Life in An Age of Diminishing Expectations*. New York: W. W. Norton, 1979.

Lawless, Jill. "Scientists discover 800,000-year-old footprints in the UK." *Associated Press*, February 7, 2014. Reporting by Nick Ashton, of the team from the British Museum,

London's Natural History Museum and Queen Mary College at the University of London, *PLOS ONE*.

Lawton, Geoff. Permaculture expert Video (4:37).

League of Nations, *Statistical Yearbook*, 1941-42, 12-22, Table 12. Table reconstructed, Table 12, at Krooth, *Arms and Empire*, 121.

Leaky, Richard and Lewin, Roger. *The Sixth Extinction: Patterns of Life and the Future of Humankind.* New York: Doubleday/Anchor Books, 1995.

Lear, Linda. *Rachel Carson: Witness for Nature*. New York: Henry Holt and Company, 1997.

Lederberg, Joshua. "Infectious Disease as an Example of Evolution." Chapter in *Epidemic: The World of Infectious Disease*. Editor, Rob Desallle. New York: The New Press/American Museum of Natural History, 1990.

Lee, Spike. A Spike Lee Film, W*hen the Levees Broke: A Requiem in Four Acts*. An HBO Documentary Films Event, 2006. DVD Video. Aspect Ratio: 16.9/93973. Four Hours.

Lee, Stephanie M. "New Plan Crafted to Limit Ocean Beach Erosion," *San Francisco Chronicle*, April 6, 2012.

Leonhardt, David. "There's Still Hope for the Planet." *The New York Times*, July 22, 2012.

Leroi, Armand Marie. *The Lagoon: How Aristotle Invented Science.* New York: Viking Penguin, 2014.

Lilley, S. *Men, Machines and History: The Story of Tools and Machines in Relation to Social Progress*. New York: International Publishers, 1965. Rev. enlarged ed., 2000.

Lindberg, Anne Morrow. *Selections from Gifts From the Sea*. Kansas City, Missouri: Hallmark, 1967.

Lineaus. *Philosphia Botanica*, Sec. 77 (1750). British Library copy.

Linley, Dayna. "Fracking Under Pressure." *The Environmental and Social Impacts and Risks of Shale Gas Development,* August 1, 2011, December 11, 2014.

LiveScience, April 19, 2014.

Lonely Planet. "Whitsunday Islands." October 23, 2010. http://lonelyplanet.com/australia/whitsunday-islands/travel-tips-and-articles/76132.

Lovelock, James. *Gaia: A New Look at Life on Earth.* Oxford: Oxford University Press, 1979, 1982, 1995.

———. *The Vanishing Face of Gaia: A Final Warning.* New York: Basic Books/Perseus Book, 2009.

Lyell, Sir Charles. *Principles of Geology*, Vol. l. London: British Library copy, 1830.

Lyons, Charles. "The Dam Boom in the Amazon." *The New York Times*, July 1. 2012. Opinion.

Mabey, Richard. *Weeds: In Defense of Nature's Most Unloved Plants.* New York: Eco/HarperCollins Publishers, 2011.

Macaulay., David. *Building the Book: Cathedral.* Boston: Houghton Mifflin Company, 1999.

———. *Cathedral: The Story of Its Construction.* Boston: Houghton Mifflin Company, 1973.

Macnbab, Ron. "Letter: Establishing Rights Over the Arctic Ocean." *Science*, Vol. 316 (25 May 2007): 1122-1123.

Main, Douglas M. "Safeguarding Mammoth Trees, Champs of the Ecosystem." *The New York Times*, May 5, 2012,

Mandel, Earnest. *Marxist Economic Theory*, Vol. II. Trans. Brian Pearce. New York and London: Monthly Review Press, 1968.

Mann, Charles C. "State of the Species." *Orion*. November/December 2012: 16-27.

Margulis, Lynn. *Symbiotic Planet: A New Look at Evolution.* New York: Basic Books, 1998.

Markel, Howard. *When Germs Travel: Six Major Epidemics that have Invaded America since 1900 and the Fears they have Unleashed.* New York: Pantheon Books, 2004.

Marriott, Edward. *Plague: A Story of Science, Rivalry, and the Scourge That Won't Go Away.* New York: Henry Holt and Company, 2003,

Martin, Ronald and Quigg, Antonietta. "The Plants that Once Ruled the Sea." *Scientific American*, Vol. 308, no. 6 (June 2013): 40-45.

Masters, Jeff. "The Jet Steam is Getting Weird." *Scientific American.com*, Vol. 311 no. 6 (December 2014): 69-75.

Melville, Herman. *Moby Dick or The White Whale*. New York: The New American Library, Signet Classics, 1961. First published in 1851.

McClelland, Mac. "Shelf Life." *Mother Jones*, March/April 2012: 46-55.

———. "The Secret Hell of Online Shopping." *Mother Jones*, March-April 2012.

McGrath, Matt. "Growing parasite threat to finches made famous by Darwin." *BBC*, December 17, 2015.

McKibben, Bill. "Meltdown: A Global Warming Travelogue," *What Matters*. New York: Sterling Publishing, 2008.

———. *Gurdiian UK*, November 1, 2012.

Miler, Kenneth. "Mushroom Manifesto: A crusading mycologist Miller, says fungi are key to human and environmental health and can clean up everything from oil spills to nuclear meltdowns." *Discover*, 07/08. http://www.nextworldtv.com/videos/inspiration-1/ fantastic-fungi.html.

Milgram, Stanley. *Obedience to Authority: An Experimental View*. New York: Perennial Classics, 2004.

Mithen, Steve. *After the Ice: A Global Human History, 20,000 to 50,000 BC.* Cambridge. Mass., Harvard University Press, 2003.

Mohawk, John. "Steps to Sovereignty." *Harvest Quarterly*, No 1 (1976): 36-41.

"Mt. Pinatubo's cloud shades global climate." *The Free Library*, by Farlax, http://www.thefreelibrary.com" http://www.thefreelibrary.com /Science+News-p210.

The Mount Pinatubo Eruption: Effects on the Atmosphere and Climate. Ed., G. Fiocco, D. Fua, and G. Visconti. NATO ASI Series Vol. I 42, 233-272. Springer-Verlag. Heidelberg, Germany.

Mowat, Farley. *Bay of Spirits: A Love Story*. Toronto, Ontario: McClelland & Stewart Ltd, 2006.

Mullaney, Tim. "Apple car: 5 reasons it's intriguing; 5 reasons it's nuts." *CNBC*, February 26, 2015. CNBC.com

Murrray, Brendan. "Summers Says U.S. IMF Inaction Gives China's AIIB Legitimacy." *BloombergBusiness*, April 6, 2015.

Myers, Simon (Oxford University), Hellenthal, Garrett (University College of London) and Falush, Damiel (Max Planck Institute of Evolutionary Anthropology, Leipzig, Germany). *Science*, February 2014.

NASA, Goddard's Scientific Visualization Studio. 2010-11.

National Geographic. Over the course of 125 years. Washington, DC, 2015.

NationofChange. http://www.nationofchange.org/.

Natterson-Horowitz, Barbara and journalist Bowers. Kathryn. "Our Animal Natures." *The New York Times*, Sunday Review, June 10, 2012.

Navarro, Mireya. "New York Is Lagging as Seas and Risks Rise, Critics Warn." *The New York Times*, September 10, 2012.

Ness, Immanuel and Cope, Zac. "Preface." *The Palgrave Encyclopedia of Imperialism and Anti-Imperialism*, vii. Editors Ness and Cope. London; Palgrave Macmillan, 2015.

Nesse, Dr. Randoff. *Evolutionary Medicine*. Evolutionary Medicine.com., 2015.

Newman, Rick. "It's amateur hour in China." *The New York Times,* September 3. 2015.

Nichols, Mike. "Farmers' Prospects Starting to Dry Up." *The Post-Crescent*. Appleton, Wisconsin, July 15, 2012.

Nolte, Carl. "S.F. Bay Area forecast: long-overdue rain, snow." *SF Chronicle*, January 11, 2012.

Norris, Chuck "Horrors of FEMA Disaster Relief." WND commentary, Nov 18. 2012. http://www.wnd.com/2012/11/horrors-of-fema-disaster-relief/.

Nova. "Australia's First Four Billion Years." April 17, 2013, 9-10 PM. With geologist Richard Smith. Reported online in the journal *Nature*.

———. Decoding Neanderthals." production number 4002H, May 15, 2013, 9 PM.

Nuwer, Rachel. "Not All Wetlands Are Created Equal." *The New York Times*, January 24, 2012. Citing *PLoS Biology*, January 24, 2012.

Oakland Museum, Oakland, California. "Lecture Tour." May, 2012.

Oceanic Preservation Society and IMPT Partners. "The Cove." Director Louie Psihoyos, Producer Fisher Stevens, 2009. 96 minute DVD.

Old Farmer's Almanac, 1976.

Optimum Population Trust. *News Release*. December 15, 2009.

Organization for Economic Cooperation and Development. Paris-based, Reports.

Orion. November/December, 2013: 64-71.

Oxfam, "Billionaires Who Own the Same Wealth as Half the World," *Press Release*. January 18, 2016.

Pacheo, Rob. "Nbird's Bone View of Hawaii." *Hawaiian Forests & Trails*, February 2006.

Pagliery, Jose. "Wikileaks claims to reveal how CIA hacks TVs and phones all over the world." *CNN*, March 7, 20017.

Parry, Wynne. "Climate Change Allows Once-Rare British Butterfly to Thrive." *The Christian Science Monitor,* May 25, 2012. "http://ezorigin.csmonitor.com/tags/topic/United+Kingdom" http://ezorigin.csmonitor.com/tags/topic/United+Kingdom.

———. "Ocean Acidification Worst in 300 million years, Study Finds." *Christian Science Monitor*, March, 2012.

PBS. 'The Drones," January 23, 2013, 9-10 P.M.

———. "The Dust Bowl." November 18-19, 2012. 8:00 PM. Produced by Florentine Films and WETA. Directed by Ken Burns; written by Dayton Duncan; Mr. Duncan, Mr. Burns and Julie Dunfey, producers; Craig Mellish and Ryan Gifford, editors; Buddy Squires and Stephen McCarthy, cinematographers; Patricia Clarkson, Kevin Conway, Amy Madigan and Carolyn McCormick, voices; Peter Coyote, narrator.

———. "Interviews with Firefighters." July 7, 2012.

Pearce, Fred. *The Coming Population Crash and Our Planet's Surprising Future*. New York: Beacon Press, 2010.

Peeples, Lynne. "Why Scapegoating Bats Is a Big Mistake to Human Health." 10/31/14. http://www.huffigntonpost.com/green.

Perera, Liz. "Climate Change May Be Hazardous to Your Health." *Catalyst*, Fall 2011: 7-9.

Pernetta, John. *Guide to the Oceans*. Buffalo, New York: Firefly Books, 2004.

Pfeiffer. Eric. "Giant plumes of methane bubbling to surface of Arctic Ocean." *The Sideshow, Yahoo News*, December 14, 2011.

Pimm, Prof. and Jenkins, Clinton. Institute of Ecological Research, Dry Totugas, Brazil, *Science*, May 29, 2014.

Pitman, Nigel and Stotz, Douglas. "Draught in the Amazon, Up Close and Personal." *New York Times*, 12 November 2010.

Plato, *Republic*, VI, 496d-497a. British Library copy.

PLoS ONE, May 7, 2012.

Polar Biology. June 2012.

Pontifical Academy of Sciences, "Sustainable Humanity, Sustainable Nature: Our Responsibility." Proceedings of the Joint Workshop, 2-6 May. 2014, Extra Series 41, Vatican City, 2014.

Poole, John W. "http://www.npr.org/templates/story/story.php?storyId=112535943 "The Amazon Road: Paving Paradise For Progress?"

———. "To Slow Amazon Fires, Scientists Light Their Own." *NPR*, Nov. 6, 2007. http://www.npr.org/templates/story/story.php?storyId=16024346&ps=rs.

"Pope Francis steps up campaign on climate change to conservative's alarm." *The New York Times*, April 28, 2015.

Posner, Gerald. *God's Bankers, A History of Money and Power of the Vatican*. New York: Simon & Schuster, 2015.

Powell, Bill. "In Washington, A Strategic Shift on China—Towards Containment." *Newsweek*, April 29, 2015.

Prud'homme. "California's Next Nightmare." *The New York Times Magazine*, July 3, 2011, 35.

Prakash, Madhu Suri. "Vandana Shiva: Teachers for a Living World."

Pyle, Robert L. *Birds of Hawaii*. Hawaii: Bishop Museum, 2006.

Quammen, David. The Song of the Dodo: Island Biogeography in an Age of Extinctions. New York: Simon & Schuster/Touchstone Book, 1996.

———. Spillover: Animal Infections and the Next Human Pandemic. New York, October 2012.

Rahmstorf, Stephen, et al. *Nature Climate Change*. Potsdam: Potsdam Institute for Climate Impact Research, March 23, 2015. Doi:10.1038/nclimate2554.

Randazzo, Anthony F. and Douglas, S. Jones. Editors. *The Geology of Florida*. Gainsville, Florida: University Press of Florida, 1977.

"Rates a Day After a Market Plunge." *The New York Times*, Sept. 3, 2015.

The Reader's Digest. "Mysteries of the Ancient Americas: The New World Before Columbus." Editors. Pleasantville, New York: The Reader's Digest Association, 1986: 84-91.

Rees, Morgan J. *Trusts in British Industry, 1914-1921*. London: P.S. King and Son Ltd, 1922.

Revelle, R. and Suess, H. "Carbon Dioxide Exchange Between Atmosphere and Ocean and the Question of an Increase of Atmospheric CO_2 During the Past Decades." *Tellus* 9 no.1, 1957.

Revelle, Roger V. "Roger Revelle's Discovery." *The Discovery of Global Warming*. http:/www.aip.org/history/climate/Revelle.htm.

———. Lecture Notes. Mc6 Box 55, Folder "Natural Sciences 118," Scripps Institution of Oceanography Archives.

Revkin, Andrew C. "Earth's Untallied Biological Bounty" "Global Study of Monsoons Finds Ocean Variations Have Driven Recent Shifts." *The New York Times*, March 24, 2013. Opinion page.

———. "The U.S. Icebreaker Gap is about Arctic Needs, Not About Chasing Russia." *The New York Times*, September 1, 2015.

———. *The Empathetic Civilization: The Race to Global Consciousness in a World in Crisis.* New York: Penguin, 2009.

———. "Tracing the Roots of Pope Francis's Climate Plan for 2015." *The New York Times* December 31, 2014.

Robe, H. T*he Rough Guide to Climate Change,* 2009. http://www.guardian.co.uk/profile/rob-ert-henson rt Henson.

Rogers, Alex D. et al. "The Discovery of New Deep-Sea Hydrothermal Vent Communities in the Southern Ocean and Implications for Biogeography." http://www.plosbiology.org.article/info%3Adoi%2F10.1371%2Fjournal.pbio.1001234.

Romero, Simon and Broder, John M. "Progress on the Sidelines as Rio Conference Ends." *The New York Times*, June 22, 2012.

Romm, Joe. "NASA: Human Activity, Not Solar Activity, Drives Global Warming and Returning to 350 ppm is Needed to Stop It." NationofChange. http://www. nationofchange.org/nasa-human-activity-not-solar-activity-drives-global-warming-and-returning-350-ppm-needed-stop-it-13.

Rosenberg, Tina. "In Africa's Vanishing Forests, the Benefits of Bamboo." *The New York Times*, March 13, 2012.

Rosenthal, Elizabeth. "More Guns, More Killing." *The New York Times*, January 6, 2013.

Rutgers Marine and Coastal Sciences, Francis, Jeenifer and the University of Wisconsin-Madison's http://marine.rutgers.edu/~francis/.

Stephen Vavrus. "Evidence linking Arctic amplification to extreme weather in mid-latitudes." *GeophysicaL Research Letters*, Vol. 39, L06801, doi:10.1029/2012GL051000, 2012, 17 March 2012.

Sanders, Bernie. "The Trans-Pacific (TPP) Agreement Must Be Defeated." *Reader Supported News*, January 3, 2015.

Santoli, Michael. "Is this the 'stock picker's market you've been asking for?" *Yahoo Finance,* April 24, 2015.

Savernice, Sabrina. "Middle-Class Areas Shrink as Income Gap Grows, New Report Finds" *The New York Times*, Nov.15, 2011.

Schultz, Rob, "Area's Corn Crop May Be Total Loss," *Wisconsin State Journal.* Madison, Wisconsin, July 13, 2012.

Scripps Institution of Oceanography, San Diego. California.

Secretariat. Convention on Biological Diversity.

Segal, David. "Coal." *The New York Times*, June 21, 2015. Exclusive expose on govrnment inspectors lying.

Semple, Robert B. Jr. "Concealed Weapons Against the Environment." *The New York Times*, July 31, 2011.

SFGate, Associated Press, Washington, "Government Funds Monarch Butterfly Conservation Program." February 9, 2015.

"Seven Billion Person World." *National Geographic Magazine*, November 2011.

Shea, Rachel Hartigan. "Uprooted." *National Geographic*, June 2014.

Sheppard, Kate. "The Ex Files." *Mother Jones*, May/June: 47.

Shiva, Vandana. *The Enclosure and Recovery of the Commons: Biodiversity, Indigenous Knowledge, and Intellectual Property Rights*. New Delhi: Research Foundation for Science, Technology and Ecology, 1997.

Shore, William H. Editor. *The Nature of Nature: New Essays from America's Finest Writers on Nature*. New York: Harcourt Brace & Company 1994.

Shreeve, James. "Human Journey." *National Geographic*, Vol. 209 no 3 (March 2006), 60-69.

Simon, Julian and Ehrlich, Paul. At Paul Sabin. "Betting on the Apocalypse," *The New York Times*, September 7, 2013. http://p.nytimes.com/email/re?location=InCMR7g4BCKC2wi

ZPkcVUgn+eR5Fy6te&user_id=ba268493a1817a25d721081e4beccbad&email_type=eta&task_id=1378605830999521&regi_id=0.

Simpson, Sarah. "Climate: Rising Sun." *Scientific American*, June 2003: 23. Citing Richard C. Wilson, et al., *Geophysical Research Letters*, March 4, 2009.

Singer, Peter. *A Darwinian Left: Politics, Evolution and Cooperation*. New Haven and London: Yale University Press, 1999.

Slavin, Barbara. "U.S. Eyes Stake in Arctic Oil Pact." *USA Today*, 31 July 2007.

Slivka, Kelley "A Sizzling Start to the Year." *The New York Times*, July 9, 2012.

Smerdon, Jason, et al. "Unprecedented 21st Century Drought Risk in the American Southwest and Western Plain." *Science Advances*, on line, February 21, 2015.

Smith, Oliver. "Sedimenary DNA from submerged site reveals wheat in the British Isle 8,000 years ago." *Science*, Vol. 347 no. 6225 (Februay27, 2015), 998-1001.

Smith, Richard. Series on Australia. Nova, *PBS*. April 13, 17, and 24, 2013, 9-10 PM.

Somervell, D.C. and Harvey, Heather. *British Empire and Commonwealth,* Rev. ed. London: Christophers, 1959.

Specter, Michael. "The Climate Fixers." *The New Yorker*, May 14, 2012.

Spengler, Oswald. *The Decline of the West*. One Volume Edition of Two Volumes. Trans. Charles Francis Atkinson. New York: Alfred A. Knopf, 1932.

Spenser, Herbert. *Leader*. March 1852, reprint 'Essays' 1858.

Spotts, Pete. "Ice caps not shrinking as much as once thought, new data show." *The Christian Science Monitor*, February 8, 2012 http://www.csmonitor.com/Environment/2012/0208/Ice-caps-not-shri.

———. "Sea level studies: US coasts even more vulnerable than previously thought." *Christian Science Monitor*, March 4, 2012. "http://www.csmonitor .com/Environment/ 2012/ 0314/Sea-level-studies-US-coasts-even-more-vulnerable-than-previously thought?cmpid=addthis_email# .T2Ir3i87V-4.email" http://www.csmonitor .com/Environment/ 2012/0314/Sea-level-studies-US-coasts-even-more-vulnerable-than-previously thought?cmpid=addthis_email# .T2Ir3i87V-4.email.

Steinmetz, Charles P. *America and the New Epoch*. New York and London: Harper & Brothers, 1916.

Stevenson, Abigail. *CNBC*, April 9, 2015. "Cramer Remix: This Stock may be ready to break-out."

Stevenson, David J. Caltech. *Nature*, 2003.

St. Fleur, Nicolas. "After Earth's Worst Mass Extinction, Life Rebounded Rapidly, Fossils Suggest." *The New York Times*, Science, February 16. 2017.

Storm, Stephanie. "Misgivings About How a Weed Killer Affects the Soil." *The New York Times, Business Day*, September 19, 2013.

Sydney Aquarium. "The Future is Wild at the Sydney Aquarium." Sydney, Australia. my-fun.com.au/futureiswild. Exhibits, 2010-2011.

Szoldza, Paul. "Trump's Cutting $1 Billion from Coast Guard at a Time When It's Doing More than Ever." *Business Insider. SF Gate*, March 3, 2017.

Tauber, Gerald. Ed. *Albert Einstein's Theory of General Relativity: 60 Years of its Influence on Man and the Universe*. New York: Crown Publishers, 1979.

"The Future is Wild," Sydney, Australia, Aquarium, Exhibit, 2011-12.

Thortnon, Ian. Krakatou: *The Destruction and Reassembly of an Island Ecosyste*m. Cambridge: Harvard University Press, 1996).

Tidwell. Mike. *Bayou Farewell: The Rich Life and Tragic Death Louisiana's Cajun Coast*, 2006.

The *Global 2000 Report to the President*. New York: Penguin Books, 1982.

"Two-way traffic." *Kew*, Summer 2003, 15.

Timmermann, Axel and Tobias Friedrich, "Late Pleistocene climate drivers of early human migration," nature.com, doi:10.1038/nature/19365 (September 21 2016).

Tyndall, John. *Heat Considered as a Mode of Motion*. London: British Library copy, 1853.

———. *The Forms of Water*. London: British Library copy, 1873.

———. *The Glaciers of the Alps*. London: British Library copy 1872.

"Ultimate Climate Change FAC." *The Guardian*, www.guardian.co.uk/ environment/series/ the-ultimate-climate-change-faq.
Union of Concerned Scientists. September 14, 2011.
———. "Heat in the Heartland: 60 Years of Warming in the Midwest." *UCS: Climate Change and Your Health, 2012.* UK. http://www.bgs.ac.uk/downloads/start.cfm?id=432.
University of Wisconsin, Geen Bay. Studies, "Biodiversity of Macrofungi in Northern Door County, Wisconsin," 2015.
U.S. Census Bureau, "Historical Estimates of World Population," International Programs, http:/ /www, census, gov,ipc//www/worldhis.html.
U. S. Department of Health and Services, Centers for Disease Control, *Influenza Vaccine Inactivated*, 2013-2014, 42 U. S. C., § 300aa-26., 07/26/2013.
U.S. Geological Survey. "Hypothetical conditions for a Superstorm based on sediment studies and meteorological modeling", 2011. U.S.
Venter, Oscar. *Nature Communications*, August 23, 2016.
Vidale, John. University of Washington. Unpublished Document, 2012.
Voegell, Jeffrey. "Growth in billionaires wealth to taper off." *Bloomberg,* May 27, 2015
von Humboldt, Friedrich Heinrich Alexander, Baron. *Cosmos*: *A Sketch of the Physical Description of the Universe*. Trans. E.C. Otté. Baltimore and London: The Johns Hopkins University Press, 1997, Vol. I.
———. *Journey to the Equinoctial Region of the New Continent; On the Geographical Distribution of Plants according to the Temperatures and Altitudes*, 30 Vols., Paris/Berlin, 1804-1824.
Vostok Research Station. A. M. Yelagin. Released by the director Valery Lukin of the Russian Antarctic Expedition.
Wade, Nicolas. "Tracing Ancestry, Team Produces Genetic Atlas of Human Mixing Events." *New York Times*, Feb.13. 2014. Citing Simon Myers (Oxford University), Garrett Hellenthal (University College of London) and Damiel Anthropology, Leipzig, Germany), *Science*, February 2014.
———. "Fish's DNA May Explain How Fins Turned to Feet." *The New York Times*, April 17, 2013.
———. *A Troublesome Inheritance: Genes, Race and Human History*. New York: The Penguin Press, 2014).
Wallace, John Mitchell. "Weather-and Climate-Related Extremes: Teachable Moments." *American Geophysical Union*, 2012.
Watson, James Dewey. *The Double Helix*. London: Wiedenfield & Nicolson, 1968.
Weart, Spencer. "Roger Revelle's Discovery." *The Discovery of Global Warming*, http:/ www.aip.org/history/climate/Revelle.htm.
White, E. B. *The Second Tree From the Corner*. New York: Harper & Brothers Publishers, 1935.
Whitty, Julia. "The End of a Myth: Exploring a New Vision for Earth's Greatest Wilderness." *Onearth*, Spring 2012: 39-43
———. "Full Green Ahead," *Mother Jones*, Vol. 38, no. 2 (March-April 2013): 16-27.
Wiedenfeld, Kurt. *Cartels and Combines*. League of Nations: Economic and Financial Section. (1926.II.70), (Geneva, 1926).
———. "Overview of Milankovitch cycles." http://geography.about.com/od/learnabout-theearth/a milankovitch.htm.
Williams, Terry Tempest. "One Patriot." Chapter in *Rachel Carson: Legacy and Challenge*. Edited by Lisa H. Sideris and Kathleen Dean Moore. Albany, New York: State University of New York Press, 2008, 31-32.
Wilson, Edward O. "Introduction" *From So Simple a Beginning*. New York/London: Liveright Publishing Corporation, Division of W.W. Norton & Company, 2014, xviii-xviv.
———. *The Meaning of Human Existence*. New York/London: Liveright Publishing, 2014.
———. A Window on Eternity: A Biologist's Walk Through the Gorongosa National Park. New York: Simon & Schuster, 2014.

Wines, Michael. "Algae Blooms in Florida Lead to Record Number of Manitee Deaths." *The New York Times*, April 6, 2012. http://www.nytimes.com/2013/04/07/science/earth/algae-bloom-in-florida-kills-record-number-of-manatees.html

Winterson, Jeanette. "Strange New World." *The New York Times Book Review*, September 20, 2009. A review of Margaret Atwood, *The Year of the Flood*. New York: Nan A. Talese/Doubleday, 2009.

Witke, Glenn S. "Amazon in Peril." *World Rivers Review*. InternationalRivers.org, Vol. 24, no. 2 (June 2009).

"Wilderness." *Onearth*, Spring 2012: 39-4.

Wong, Kate. "First of Our Kind: Sensational fossils." *Scientific American.com*, April 2012: 31-39.

———. "Mystery Human." *Scientific American*. Vol. 314, no. 3 (March 2016): 9-37.

———. "The New Origins of Technology." *Scientific American.com*. Vol. 316, no. 5 (May 2017) : 28-35.

World Wildlife Federation. "Gallery: Discoveries In The Amazon." *Report*, October 27, 2010.

Worster, Donald, ed. *John Mui: A Passion for Nature*. Edinburgh, Scotland, 2011.

Wulf, Andrea. The Invention of Nature: Alexander Von Humboldt's New World. New York; Alfred A Knopf, 2015.

Wyatt, Caroline. 'Pope of the poor makes big impact in Africa." *BBC*, December 1, 20015.

Yahoo Finance, "Gilded Era Ending," May 27, 2015.

———. "Vatican investing," May 26, 2015.

Yardley, Jim and Appelbaum, Binyamin, "In Fiery Speeches, Francis Excoriates Global Capitalism." *The New York Times*, July 12, 2015.

Yergen, Daniel. *The Quest*: Energy, Security, and the Remaking of the Modern World. New York: Random House, 2012.

Youngerman, Barry. *Pandemics and Global Health*. New York: Facts on File, 2008.

Zalasiewicz, Jan. "What Mark Will We Leave on the Planet?" *ScientificAmerican.Com*., September 2016, Vol. 315, No. 3, 32.

Zeebe, Richard et al. *Nature Geoscience*, April 2008.

Zimmer, Carl. "As Humans Change Landscape, Brains of Some Animals Change, Too." *The New York Times, Science*, August 27, 2013.

———. "Christening the Earliest Members of Our Genus. *The New York Times*, October 24, 2013.

Zolli, Andrew. "Learning to Bounce Back," *The New York Times*, November 2, 2012. Op-Ed, opinion.